Geophysical Monograph Series

Including

IUGG Volumes

Maurice Ewing Volumes

Mineral Physics Volumes

GEOPHYSICAL MONOGRAPH SERIES

Geophysical Monograph 74
IUGG Volume 14

Evolution of the Earth and Planets

E. Takahashi
R. Jeanloz
D. Rubie

Editors

 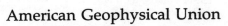American Geophysical Union

 International Union of Geodesy and Geophysics

Published under the aegis of the AGU Books Board

Library of Congress Cataloging-in-Publication Data

Evolution of the Earth and planets / E. Takahashi, R. Jeanloz, D. Rubie, editors.
     p.   cm. — (Geophysical monograph : 74) (IUGG : v. 14)
ISBN 0-87590-465-3
1. Earth—Origin.   2. Earth—Mantle.   3. Planets—Origin.   I. Takahashi, E.
(Eiichi), 1951-   .   II. Jeanloz, R.   III. Rubie, David C.   IV. Series.
V. Series : IUGG (Series) : v. 14.
QB632.E86 1993
552—dc20   93-13344
                                                   CIP

ISSN: 0065-8448
ISBN 087590-465-3

CONTENTS

PREFACE

The past two decades have witnessed a revolution in our understanding of how planets evolve. Space exploration and observations on the Earth's interior, complemented by laboratory investigations and theoretical analyses, have led to entirely new concepts governing our science: giant late-stage impacts, magma oceans, dense primordial atmospheres, continental growth curves, mantle-core interactions, boundary-layer instabilities and the influence of tectonics on climate, among others.

It is the purpose of this book to illustrate the role these new concepts play in defining current thinking about the origin, growth and subsequent evolution of the Earth and planets. The topics range from cosmogony and comparative planetary tectonics to seismic tomography, mineral physics and geodynamics of the Earth's mantle. The articles, based on a Union Symposium at the 1991 IUGG General Assembly in Vienna, Austria, offer a unique blend of geophysics, geochemistry and planetary science. Theoretical, observational and laboratory studies are well represented, and the reader can obtain a broad overview of the state of the art in current research on the internal evolution of planets.

The editors wish to acknowledge contributions from the International Association of Volcanology and Chemistry of the Earth's Interior (IAVCEI), the International Association of Seismology and Physics of the Earth's Interior (IASPEI), the International Union of Geodesy and Geophysics (IUGG), and NASA, which supported the Symposium and the production of this volume. More important, we thank our colleagues who helped prepare and review the articles: their expert advice and unselfish efforts were invaluable.

R. Jeanloz
E. Takahashi
D. Rubie
Editors

FOREWORD

The scientific work of the International Union of Geodesy and Geophysics (IUGG) is primarily carried out through its seven associations: IAG (briefly, Geodesy), IASPEI (Seismology), IAVCEI (Volcanology), IAGA (Geomagnetism), IAMAP (Meteorology), IAPSO (Oceanography), and IAHS (Hydrology). The work of these associations is documented in various ways.

Evolution of the Earth and Planets is one of a group of volumes published jointly by IUGG and AGU that are based on work presented at the Inter-Association Symposia as part of the IUGG General Assembly held in Vienna, Austria, in August 1991. Each symposium was organized by several of IUGG's member associations and comprised topics of interdisciplinary relevance. The subject areas of the symposia were chosen such that they would be of wide interest. Also, the speakers were selected accordingly, and in many cases, invited papers of review character were solicited. The series of symposia were designed to give a picture of contemporary geophysical activity, results, and problems to scientists having a general interest in geodesy and geophysics.

In view of the importance of these interdisciplinary symposia, IUGG is grateful to AGU for having put its unique resources in geophysical publishing expertise and experience at the disposal of IUGG. This ensures accurate editorial work, including the use of peer reviewing. So the reader can expect to find expertly published scientific material of general interest and general relevance.

Helmut Moritz
President, IUGG

Overview

RAYMOND JEANLOZ

Department of Geology and Geophysics, University of California, Berkeley, CA 94720-4767

PLANETARY CONTEXT

The earliest stages of planetary formation involved condensation and accumulation of matter from the solar nebula. Current thinking is that the nebular disc was hot and dynamically unstable [e.g., *Lin and Papaloizou*, 1985; *Morfill* et al., 1985; *Boss*, 1990], with the condensates aggregating rapidly into planetesimals: objects $\sim 10^3$-10^4 m diameter formed in a time period of $\sim 10^6$ years, and consisting of rock and ice grains [*Wetherill*, 1990]. Although the detailed mechanisms of coagulation are poorly understood, this general scenario has been established through a combination of observation and theory in astronomy, astrophysics, geochemistry and meteoritics [*Black and Matthews*, 1985; *Kerridge and Matthews*, 1988].

Subsequent growth of planetesimals is dominated by gravitational interactions, which define both orbital dynamics of density heterogeneities within the nebular disc--that is, the mutual gravitational attraction between planetesimals--and the accretion of matter into the proto-planetary objects. Many details of this stage can be simulated in a statistical manner through numerical computations [*Wetherill*, 1990, 1991].

It is here that the presence of Jupiter and Saturn can have a significant effect on the growth of the terrestrial planets. The giant planets exert an important gravitational influence, and can act as sources of material feeding into the growing terrestrial planetesimals. *Zharkov* [this volume] emphasizes the latter, estimating the possible composition of matter contributed from Jupiter to the planetary nuclei of Mars and Earth, as well as Saturn, Uranus and Neptune. His analysis depends on models of the internal composition and structure of the giant planets [cf. *Stevenson*, 1982; *Podolak and Reynolds*, 1985; *Podolak et al.*, 1990; *Klepeis et al.*, 1991; *Hubbard et al.*, 1991].

Evolution of the Earth and Planets
Geophysical Monograph 74, IUGG Volume 14

It is worth noting that the inclusion of proto-jovian material into the terrestrial orbits is compatible with geochemical arguments for late-stage insertion of relatively oxidized or volatile-rich material into the growing Earth [e.g., *Schmitt et al.*, 1989; *Ahrens*, 1990]. Nevertheless, the key uncertainty regards the timing of several processes that can mutually interfere with each other: growth of Jupiter to its final size, presence of a dense primordial nebula, accretion of the terrestrial planets and duration of the Sun's presumed T-Tauri episode.

In comparison with Jupiter, much more detailed and reliable modelling is possible for the internal composition of Mars. Not only are multiple samples available from this planet but, unlike the giant planets, the conditions of pressure and temperature inside Mars can be experimentally simulated with relative ease [*Jagoutz*, 1989; *Wänke*, 1991; *Pepin and Carr*, 1992; *Longhi et al.*, 1992].

Thus, *Kamaya et al.*[this volume] use laboratory studies to resolve the likely mineralogical structure of the martian mantle. They conclude that the $\sim$2000-km-thick mantle of Mars is broadly similar to the top 800 km of the Earth's mantle. Mineralogical transformations cause some of the most complex mantle structure to occur over this pressure interval, but it is unclear whether the martian core and mantle react chemically to the degree inferred for the Earth [*Knittle and Jeanloz*, 1991]. Perhaps most significant is the fact that their predictions of the internal structure of Mars are directly testable by future seismological investigations [e.g., *Golombek et al.*, 1992].

A different aspect of composition is explored by *Kaula* [this volume]. He is concerned with the contrasts evident in the tectonic styles of Venus and Earth, despite these planets being nearly identical in bulk physical properties. It is now clear from the Magellan imaging that although Venus is geologically active, plate tectonics does not describe its global dynamics in the manner found on Earth [*Solomon et al.*, 1992]. This surprising result must be explained if we are to have any confidence in our understanding of planetary tectonics.

Kaula focuses on the possible differences in volatile budgets of the two planets, primarily in the internal concentrations of water, carbon dioxide and sulfur components suggested by the available data for Venus. It is quite plausible that such differences in relatively minor constituents within Earth and Venus could have a major effect on petrology, hence on rheological and other transport properties [e.g., *Campbell and Taylor*, 1983]. Variations in the abundances of volatile constituents might therefore be responsible for the observed differences in thermal state and global tectonics. To resolve this issue will require more complete experimental determinations of the phase equilibria and thermophysical properties of the rock compositions presumed to exist in the mantle of Venus.

EARLY EARTH

Several lines of reasoning are now accepted as indicating that the Earth was hot during or shortly after accretion. Whether it was the result of giant late-stage impacts, such as the impact from which the Moon is thought to have formed [*Wetherill*, 1985; *Hartmann et al.*, 1986; *Stevenson*, 1987; *Melosh*, 1990], or the presence of a dense nebular atmosphere surrounding the growing planet [*Hayashi et al.*, 1985; *Sasaki*, 1990], current thinking is that the Earth was partly to entirely molten as it reached its final size. It therefore seems nearly inevitable that a magma ocean was present on the Hadean Earth.

The important conclusion of recent work, including that of *Abe* [this volume], is that the physical mechanisms of thermal, convective and chemical evolution of such a magma ocean are highly nonlinear [*Tonks and Melosh*, 1990; *Miller et al.*, 1991]. Therefore, both the internal differentiation of the mantle and the separation of core metal from mantle silicates involve poorly resolved processes of partial chemical equilibration [*Stevenson*, 1990]. The geochemical imprint left by these processes, perhaps the main record that we can hope to find of global differentiation, is not likely to be easily interpretable.

The subsequent differentiation of the Earth's mantle over geological history is dominated by the growth and recycling of continental crust. *Spohn and Breuer* [this volume] consider the coupled thermal and geochemical evolution of the mantle-crust system through modelling by parameterized convection. Although unable to treat the detailed dynamics that are involved, this approach allows them to follow the time evolution of the mantle-crust system in an average sense [cf. *Davies*, 1990; *Tajika and Matsui*, 1990; *Spohn*, 1991]. Thus, *Spohn and Breuer* are able to show that differentiation of the continental crust from the mantle led to a significant cooling of the interior during the Archean, followed by a more modest decrease in average mantle temperatures over time.

STATE OF THE EARTH'S INTERIOR

In order to consider the possible (or partial) geochemical equilibration associated with large-scale differentiation of the planet, it is necessary to determine phase equilibria as a function of the controlling thermodynamic variables: pressure, temperature and the fugacities of mobile components. Previous laboratory studies have emphasized the effects of temperature and pressure. However, oxygen fugacity, in particular, is considered to be significant in defining the stability of mineral phases and of various defects, as well as the valence states of Fe and other transition-metal elements within the mantle [e.g., *Hirsch and Shankland*, 1991; *Hirsch*, 1991]. It is especially important to determine how the presence of particular mineral phases buffers the oxygen fugacity of the rock; and, reciprocally, how the fugacity controls the appearance of mineral phases [e.g., *Frost*, 1991].

O'Neill et al. [this volume] present some of the first experimental work bearing on the oxygen fugacity of the Earth's deep upper mantle. They conclude that whereas the fugacity of the uppermost mantle is controlled by minor phases, the main transition-zone minerals (ringwoodite spinel, wadsleyite β-phase and majorite garnet) are able to contain substantial amounts of Fe^{3+} coexisting with the ferrous ion; the primary phases therefore buffer the oxygen fugacity of the deep upper mantle. Although more work is needed on this problem, *O'Neill et al.* reach the qualitative conclusion that the transition zone is likely to be a relatively reduced shell, compared with the more oxidizing uppermost mantle and (perhaps) lower mantle.

The most detailed information on the present state of the Earth's interior comes from seismology. Consequently, any successful model of our planet's geological evolution must be compatible with the currently observed structure of the mantle and core. The past decade's accomplishment of deducing the three-dimensional structure of the interior through seismic tomography is therefore a great achievement [e.g., *Jacobs*, 1992; *Fukao*, 1992]. It is not just that more detailed and numerous observations are available, but that the information on lateral variations in seismic velocities gives us qualitatively new insights on the heterogeneities driving--and caused by--the geodynamics of the planet.

Woodward et al. [this volume] summarize new results for the mantle, as obtained by one of the groups that pioneered global seismic tomography. Among the surprising results is that large-scale heterogeneities are systematically biased toward long, rather than short, spatial scales [*Su and Dziewonski*, 1991]. No one currently understands how this might affect the dynamics of the mantle or, for that matter, how such a biasing of the spectrum of heterogeneities came about.

Also, compatible with other recent studies [*Fukao*, 1992],

Woodward et al. find patterns of velocity anomalies suggesting that flow between the upper and lower mantle is limited. By considering the fluid-dynamical response to the inferred heterogeneities, they are then able to deduce a self-consistent model of the present convective dynamics of the Earth's mantle, from the boundary with the core to the tectonic plates at the surface.

The interpretation of seismological observations on the interior depends largely on mineral physics, which is needed to translate the velocity heterogeneities into estimated variations in composition, state (e.g., temperature, mineralogical assemblage) and physical properties at depth. *Ita and Stixrude* [this volume] and *Stixrude and Bukowinski* [this volume] present analyses pertaining to the upper mantle and the deep lower mantle, respectively. Their approach is to use a formulation grounded in basic statistical mechanics to extrapolate existing experimental measurements on the mineral equilibria and elastic properties of mantle phases [e.g., *Weiner*, 1933; *Grimvall*, 1986].

The main result of *Ita and Stixrude* is that the seismological properties of the entire upper mantle are compatible with a model of uniform bulk composition, identical with the composition inferred for the uppermost mantle ("pyrolite"-like peridotite). Others have reached similar conclusions [e.g., *Weidner*, 1985], although important questions remain in our understanding of transition-zone elasticity. Still, this result can be contrasted with the evidence indicating that the seismologically observed properties of the lower-mantle imply a difference in bulk composition from that of the upper mantle [*Jeanloz and Knittle*, 1989; *Stixrude et al.*, 1992].

Stixrude and Bukowinski give a more detailed account of recent work [*Stixrude and Bukowinski*, 1992] in which they concluded that the Mg-silicate perovskite phase is likely to be stable down to the core-mantle boundary [*Hemley and Cohen*, 1992]. This is contrary to their previous result [*Stixrude and Bukowinski*, 1990] and reinforces the possible significance of chemical reactions between the Earth's mantle and core, although there remain large uncertainties in current models of the thermal state of the lowermost mantle [*Jeanloz*, 1990; *Knittle and Jeanloz*, 1991].

TECTONICS-CLIMATE CONNECTION

The global-tectonic processes observed at the Earth's surface reflect the underlying geodynamics of the mantle. Thus, one of the most intriguing consequences of internal planetary evolution to be recently considered is the possibility that tectonics may have a significant influence on climate and the geochemistry of the ocean-atmosphere system [*Molnar and England*, 1990; *Raymo*, 1991; *Raymo and Ruddiman*, 1992; *Richter et al.*, 1992]. Uplift of mountains, which occurs in response to mantle dynamics,

causes enhanced chemical weathering and erosion, and can also influence the general atmospheric circulation around the globe [e.g., *Meehl*, 1992]. The result is a strong though indirect link between long-term tectonic and climatic evolution, as is perhaps best documented by changes that occurred during the Cenozoic: systematic cooling of the Earth's atmosphere apparently correlated with uplift of the Tibetan plateau.

François et al. [this volume] test this hypothesis--that mountain building is the main factor controlling chemical weathering--against the alternative that has been advocated, namely that surface temperature and atmospheric CO_2 content have primarily determined the rates of chemical weathering. Expanding on their previous word [*François and Walker*, 1992], they conclude from modelling the record of strontium-isotopic changes and of carbon cycling throughout the Phanerozoic that it is the tectonic model which is most successful.

Although it is not entirely resolved which is the primary driving force, climate affecting tectonics or vice-versa, it is clear that the two are closely related during the latter fraction of Earth history [*Molnar and England*, 1990; *Raymo and Ruddiman*, 1992]. As long-term changes in global climate become better understood for each of the terrestrial planets [cf. *Kasson et al.*, 1992; *Slade et al.*, 1992; *Harmon and Slade*, 1992], it may become possible to obtain a general theory of planetary evolution that involves both the interior and the external envelope.

REFERENCES

Ahrens, T. J., Earth accretion, in *Origin of the Earth* (H. E. Newsom and J. H. Jones, eds.) Oxford Univ. Press, New York, pp. 211-227, 1990.

Black, D. C. and M. S. Matthews, eds. *Protostars and Planets II*, Univ. Arizona Press, Tucson, AZ, 1985.

Boss, A. P., 3D solar nebula models: Implications for Earth Origin, in *Origin of the Earth* (H. E. Newsom and J. H. Jones, eds.) Oxford Univ. Press, New York, pp. 3-15, 1990.

Campbell, I. H., and S. R. Taylor, No water, no granites - No oceans, no continents, *Geophys. Res. Lett., 10*, 1061-1064, 1983.

Davies, G. F., Heat and mass transport in the early Earth, in *Origin of the Earth* (H. E. Newsom and J. H. Jones, eds.) Oxford Univ. Press, New York, pp. 175-194, 1990.

François, L. M., and J. C. G. Walker, Modelling the Phanerozoic carbon cycle and climate: Constraints from the Sr/Sr isotopic ratio of seawater, *Am. J. Sci., 292*, 81-135, 1992.

Frost, B. R., Introduction to oxygen fugacity and its petrologic importance, *Rev. Mineral., 25*, 1-9, 1991.

Fukao, Y., Seismic tomogram of the Earth's mantle Geodynamic implications, *Science, 258*, 625-630, 1992.

Golombek, M. P., W. P. Banerdt, K. L. Tanaka and D. M. Tralli, A prediction of Mars seismicity from surface faulting, *Science, 258*, 979-981, 1992.

Grimvall, G., *Thermophysical Properties of Materials*, North-Holland, New York, 1986.

Harmon, J. K., and M. A. Slade, Radar mapping of Mercury - Full-disc images and polar anomalies, *Science, 258*, 640-643, 1992.

Hartmann, W. K., R. J. Phillips and G. J. Taylor, eds., *Origin of the Moon*, Lunar Planetary Inst., Houston, TX, 1986.

Hayashi, C., K. Nakazawa and Y. Nakagawa, Formation of the Solar System, in *Protostars and Planets II* (D. C. Black and M. S. Matthews, eds.) Univ. Arizona Press, Tucson, AZ, pp. 1100-1153, 1985.

Hemley, R. J., and R. E. Cohen, Silicate perovskite, *Ann. Rev. Earth Planet. Sci., 20*, 553-600, 1992.

Hirsch, L. M., The Fe-FeO buffer at lower mantle pressures and temperatures, *Geophys. Res. Lett., 18*, 1309-1312, 1991.

Hirsch, L. M., and T. J. Shankland, Determination of defect equilibria in minerals, *J. Geophys. Res., 96*, 377-384, 1991.

Hubbard, W. B., W. J. Nellis, A. C. Mitchell, N. C. Holmes, S. S. Limaye and P. C. McCandless, Interior structure of Neptune - Comparison with Uranus, *Science, 253*, 648-651, 1991.

Jacobs, J. A., *Deep Interior of the Earth*, Chapman and Hall, New York, 1992.

Jagoutz, E., Sr and Nd isotopic systematics in ALHA-77005 - Age of shock metamorphism in shergottites and magmatic differentiation on Mars, *Geochim. Cosmochim. Acta, 53*, 2429-2441, 1989.

Jeanloz, R., The nature of the Earth's core, *Ann. Rev. Earth Planet. Sci., 18*, 357-386, 1990.

Jeanloz, R., and E. Knittle, Density and composition of the lower mantle, *Phil. Trans. Roy. Soc. (London), Ser. A., 328*, 377-389, 1989.

Kasson, K. R., R. N. Clayton, E. K. Gibson and T. K. Mayeda, Water in SNC meteorites - evidence for a martian hydrosphere, *Science, 255*, 1409-1411, 1992.

Kerridge, J. F., and M. S. Matthews, eds. *Meteorites and the Early Solar System*, Univ. Arizona Press, Tucson, AZ, 1988.

Klepeis, J. E., K. J. Schafer, T. W. Barbee and M. Ross, Hydrogen-helium mixtures at megabar pressures - Implications for Jupiter and Saturn, *Science, 254*, 986-989, 1991.

Knittle, E., and R. Jeanloz, Earth's core-mantle boundary: Results of experiments at high pressures and temperatures, *Science, 253*, 1438-1443, 1991.

Lin, D. N. C., and J. Papaloizou, On the dynamical origin of the Solar System, in *Protostars and Planets II* (D. C. Black and M. S. Matthews, eds.) Univ. Arizona Press, Tucson, AZ, pp. 981-1072, 1985.

Longhi, J., E. Knittle, J. R. Holloway and H. Wänke, The bulk composition, mineralogy and internal structure of Mars, in *Mars* (H. H. Kieffer, B. M. Jakosky, C. W. Snyder and M. S. Matthews, eds.) Univ. Arizona Press, Tucson, AZ, pp. 184-204, 1992.

Meehl, G. A., Effect of tropical topography on global climate, *Ann. Rev. Earth Planet. Sci., 20*, 85-112, 1992.

Melosh, H. J., Giant impacts and the thermal state of the Earth, in *Origin of the Earth* (H. E. Newsom and J. H. Jones, eds.) Oxford Univ. Press, New York, pp. 69-83, 1990.

Miller, G. H., E. M. Stolper and T. J. Ahrens, The equation of state of a molten komatiite, 2. Application to komatiite petrogenesis and the Hadean mantle, *J. Geophys. Res., 96*, 11849-11864, 1991.

Molnar, P., and P. England, Late Cenozoic uplift of mountain ranges and global climate change - Chicken or egg?, *Nature, 346*, 29-34, 1990.

Morfill, G. E., W. Tscharnuter and H. J. Völk, Dynamical and chemical evolution of the protoplanetary nebula, in *Protostars and Planets II* (D. C. Black and M. S. Matthews, eds.) Univ. Arizona Press, Tucson, AZ, pp. 493-533, 1985.

Pepin, R. O., and M. H. Carr, Major issues and outstanding questions, in *Mars* (H. H. Kieffer, B. M. Jakosky, C. W. Snyder and M. S. Matthews, eds.) Univ. Arizona Press, Tucson, AZ, pp. 120-143, 1992.

Podolak, M., and R. T. Reynolds, What have we learned from modelling giant planetary interiors? in *Protostars and Planets II* (C. D. Black and M. S. Matthews, eds.) Univ. Arizona Press, Tucson, AZ, pp. 847-872, 1985.

Podolak, M., R. T. Reynolds and R. Young, Post Voyager comparisons of the interiors of Uranus and Neptune, *Geophys. Res. Lett., 17*, 1737-1740, 1990.

Raymo, M. E., Geochemical evidence supporting T.C. Chamberlin's theory of glaciation, *Geology, 19*, 344-347, 1991.

Raymo, M. E., and W. F. Ruddiman, Tectonic forcing of late Cenozoic climate, *Nature, 359*, 117-122, 1992.

Richter, F. M., D. B. Rowley and D. J. DePaolo, Sr isotopic evolution of seawater - The role of tectonics, *Earth Planet. Sci. Lett., 109*, 11-23, 1992.

Sasaki, S., The primary solar-type atmosphere surrounding the accreting Earth: H_2O-induced high surface temperature, in *Origin of the Earth* (H. E. Newsom and J. H. Jones, eds.) Oxford Univ. Press, New York, pp. 195-209, 1990.

Schmitt, W., H. Palme and H. Wänke, Experimental determination of metal/silicate partition coefficients for P, Co, Ni, Cu, Ga, Ge, Mo, and W, and some implications for the early evolution of the Earth, *Geochim, Cosmochim, Acta, 53*, 173-185, 1989.

Slade, M. A., B. J. Butler and D. O. Muhleman, Mercury radar imaging - evidence for polar ice, *Science, 258*, 635-640, 1992.

Solomon, S. C., S. E. Smrekar, D. L. Bindschadler, R. E. Grimm, W. M. Kaula, G. E. McGill, R. J. Phillips, R. S. Saunders, G. Schubert, S. W. Squyres and E. R. Stofan, Venus tectonics: An overview of Magellan observations, *J. Geophys. Res., 97*, 13199-13255 and 16381, 1992.

Spohn, T., Mantle differentiation and thermal evolution of Mars, Mercury and Venus, *Icarus, 90*, 222-236, 1991.

Stevenson, D. J., Interiors of the giant planets, *Ann. Rev. Earth Planet, Sci., 10*, 257-295, 1982.

Stevenson, D. J., Origin of the Moon - The collision hypothesis, *Ann. Rev. Earth Planet. Sci., 15*, 271-315, 1987.

Stevenson, D. J., Fluid dynamics of core formation, in *Origin of the Earth* (H. E. Newsom and J. H. Jones, eds.) Oxford Univ. Press, New York, pp. 231-249, 1990.

Stixrude, L., and M. S. T. Bukowinski, Fundamental thermodynamic relations and silicate melting with implications for the constitution of D'', *J. Geophys. Res., 95*, 19311-19325, 1990.

Stixrude, L., and M. S. T. Bukowinski, Stability of (Mg,Fe)SiO$_3$ perovskite and the structure of the lowermost mantle, *Geophys. Res. Lett., 19*, 1057-1060, 1992.

Stixrude, L., R. J. Hemley, Y. Fei and H. K. Mao, Thermoelasticity of silicate perovskite and magnesiowüstite and stratification of the Earth's mantle, *Science, 257*, 1099-1101, 1992.

Su, W.-J., and A. M. Dziewonski, Predominance of long-wavelength heterogeneity in the mantle, *Nature, 352*, 121-126, 1991.

Tajika, E., and T., Matsui, The evolution of the terrestrial environment, in *Origin of the Earth* (H. E. Newsom and J. H. Jones, eds.) Oxford Univ. Press, New York, pp. 347-370, 1990.

Tonks, W. B., and H. J. Melosh, The physics of crystal settling and suspension in a turbulent magma ocean, in *Origin of the Earth* (H. E. Newsom and J. H. Jones, eds.) Oxford Univ. Press, New York, pp. 151-174, 1990.

Wänke, H., Chemistry, accretion and evolution of Mars, *Space Sci. Rev., 56*, 1-8, 1991.

Weidner, D. J., A mineral physics test of a pyrolite mantle, *Geophys. Res. Lett., 12*, 417-420, 1985.

Weiner, J. H., *Statistical Mechanics of Elasticity*, Wiley, New York, 1983.

Wetherill, G. W., Occurrence of giant impacts during the growth of the terrestrial planets, *Science, 228*, 877-879, 1985.

Wetherill, G. W., Formation of the Earth, *Ann. Rev. Earth Planet. Sci., 18*, 205-256, 1990.

Wetherill, G. W., Occurrence of Earth-like bodies in planetary systems, *Science, 253*, 535-538, 1991.

R. Jeanloz, Department of Geology, University of California, Berkeley, California, 94720-4767.

The Role of Jupiter in the Formation of Planets

V. N. ZHARKOV

Department of Theoretical Physics, Institute of Physics of the Earth, Academy of Sciences of the USSR, Moscow 234820, Russia

Five-layer models of Jupiter and Saturn and three-layer models of Uranus and Neptune are considered here, and shown not to support Mizuno's hypothesis that the embryos of the planets on which the accretion of gas takes place are approximately equal in mass, being 10–15 $M_\oplus$, where $M_\oplus$ is the mass of the Earth. Planetesimals expelled from Jupiter's zone after its formation exerted a significant effect on the formation of the planetary system. It is generally accepted at present that these planetesimals almost completely destroyed the feeding zone in the asteroid belt and reduced the amount of matter taking part in the formation of Mars by a factor of approximately 20. The model for the accumulation of terrestrial planets from two chemically very different components, A and B, is also connected with the influence of Jupiter.

It is suggested that the accumulation time of Mars depends on the presence of proto-Jupiter, and is limited by the time scale for growth of Jupiter. This limit has been roughly estimated to be $\sim 1.7 \times 10^7$ yr, which is approximately the time spent by the Sun in the T Tauri stage.

A new scheme for the formation of Saturn, Uranus, and Neptune is proposed. After completion of the formation of Jupiter over a period of $\sim (1\text{–}2) \times 10^7$ yr, that planet ejects a massive embryo with a mass of $\sim 5\,M_\oplus$ into the Saturn feeding zone. This nucleus initiates the formation of Saturn. After the formation of Jupiter and Saturn, massive embryos are ejected into the Uranus and Neptune feeding zones and lead to the formation of Uranus and Neptune over cosmogonically realistic time intervals.

INTRODUCTION

Jupiter, Saturn, Uranus and Neptune are fluid, convective, adiabatic planets. There are five fundamental arguments to the effect that Jupiter is a fluid body with an internal adiabatic temperature distribution. It also follows from these arguments that Jupiter is in a convective state; i.e., heat is transported from its interior by convection, and it is in a state close to hydrostatic equilibrium.

(1) If Jupiter formed as a hot body (it is hard to imagine that such a huge planet could have failed to heat up during its formation), it would not have been able to cool off during its existence, $t \sim 4.5 \times 10^9$ years since its cooling length

$$l_{cool} \sim (\chi\, t_p)^{1/2} \qquad (1)$$

is on the order of 5×10^2 km ($\chi \sim 10^{-2}\text{–}10^{-3}$ cm^2/s is the thermal diffusivity of molecular hydrogen).

(2) The heat flow from the planet's interior, as measured over broad infrared bands by the Voyager 1 and 2 spacecrafts [*Hanel et al.*, 1981, 1983], turned out to be $\sim 5.4 \times 10^3$ erg cm^{-2}/sec (the corresponding value for Saturn is $\sim 2 \times 10^3$ erg cm^{-2}/sec). Such a heat flux is about four orders of magnitude greater than the maximum heat flux which can be transported by molecular conduction [*Zharkov*, 1986].

(3) All giant planets have intrinsic magnetic fields produced in their internal, electrically conducting envelopes. These envelopes must therefore be in convective states.

(4) By considering the orbital evolution of the satellites of Jupiter, Saturn, and Uranus, *Goldreich and Soter* [1966] and *Gavrilov and Zharkov* [1977] estimated the value of the mechanical quality factor Q for these planets. It turned out that $Q_J \geq 2.5 \times 10^4$, $Q_S \geq 1.4 \times 10^4$, and $Q_U \geq 5 \times 10^3$, all of which are about two orders of magnitude greater than typical values for matter in the solid state, such as the mantles of the Earth and the terrestrial planets. Thus, these estimates can be interpreted as an indication of the **liquid** state of Jupiter, Saturn, and Uranus.

(5) The external gravitational potential for all planets has the form

$$V(r,\theta) = GM/r\, [1 - (a/r)^2\, J_2 P_2\,(\cos\theta) -$$

$$(a/r)^4\, J_4 P_4\,(\cos\theta) - (a/r)^6\, J_6 P_6\,(\cos\theta) - ...] \qquad (2)$$

i.e., it corresponds to the field of a planet in hydrostatic

Evolution of the Earth and Planets
Geophysical Monograph 74, IUGG Volume 14

equilibrium because the field of a **liquid** planet does not depend on the sign of its rotation. In (2), J_2, J_4, and J_6 are the first even gravitational moments, $P_n (\cos \theta)$ is a Legendre polynomial of the nth degree, r is the radius, θ is the polar angle or colatitude, a is the equatorial semiaxis, M is the mass of the planet, and G is the gravitational constant.

The dynamical oblateness e_D of a **liquid** planet is equal to

$$e_D = 1/2(3 J_2 + m)(1 + 3/2 J_2) + 5/8 J_4 ,\qquad (3)$$

where

$$m = \omega^2 R^3/GM = 3 \ \omega^2/4\pi\rho_0 = 3\pi/G\rho_0\tau^2 \qquad (4)$$

is the small parameter of the theory of figures, and ω, τ, R and ρ_0 are, respectively, the planet's angular velocity, rotation period, average radius and average density.

For a planet in hydrostatic equilibrium e_D is equal to the geometrical (or optical) flattening e_G

$$e_G = (a - b)/a \qquad (5)$$

where b is the polar semiaxis.

We will now explain the way in which the giant planets have been confirmed to be gas-**liquid** bodies. The critical pressure and critical temperature for hydrogen are 13 atm and 33 K, respectively. For pressures and temperatures higher than these values there is no boundary between the gaseous and **liquid** phases of molecular hydrogen. Jupiter and Saturn consist almost entirely of hydrogen, while Uranus and Neptune are covered by hydrogen mantles with a thickness of around two-tenths of a planetary radius, with the hydrogen in the supercritical state everywhere.

As a result, the gaseous atmosphere of the planet gradually becomes denser with depth, under the load of the overlying matter, and continuously transforms into a denser fluid state; there is no discrete boundary between the gaseous atmosphere and the underlying **liquid** planet. Moreover, the melting temperature of hydrogen under conditions in the interiors of Jupiter or Saturn is several times lower than the adiabat temperatures in these planets. The melting temperature of water, which is the second most significant component of the giant planets, is probably lower than the adiabat temperatures in most of Uranus and Neptune. These arguments are used as the basis for the statement that the interiors of all the giant planets are in the fluid state, with the possible exception of a small central region. The concept of the fluid state of the giant planets was developed by the author and V. P. Trubitsyn in the USSR and by W. B. Hubbard in the USA [*Hubbard et al.*, 1974].

In order to determine the temperature-pressure relationship for adiabatic models of the giant planets, one has available a selection of boundary pressures, P_1, and temperatures, T_1, which are related to each other by the adiabatic law. It is usually convenient to refer the adiabat to the surface on which $P_1 = 1$ bar. In this case, the value of T_1 is determined from a model of the planetary atmosphere.

For $P_1 = 1$ bar, the values of T_1 for all giant planets are given in Table 1. The values of the equatorial radius at the level of $P_1 = 1$ bar are denoted by a_1 and are also given in Table 1.

The problems of studying the internal structure and the chemical compositions of the giant planets are of great interest. The results obtained from giant planet models and

TABLE 1. Observational data for giant planets

Parameter	Jupiter	Saturn	Uranus	Neptune
M, g	$1.897 \ 10^{30}$ [a]	$5.68 \ 10^{29}$ [b]	$8.687 \ 10^{28}$ [x]	$10.243 \ 10^{28}$ [aa]
a_1, km	71492 ± 4 [c]	60268 ± 4 [i]	25559 ± 4 [u]	24764 [aa]
R, km	69894 [d]	58300 [e]	25270 [x]	24622 [aa]
ρ_0, g/cm^3	1.3276 [d]	0.685 [e]	1.285 [x]	1.64 [aa]
$J_2 \times 10^2$	1.4697 ± 0.0004 [d]	1.6331 ± 0.0018 [i]	0.351323 ± 0.000032 [v]	0.3539 ± 0.0010 [ff]
$J_4 \times 10^4$	-5.84 ± 0.05 [d]	-9.14 ± 0.61 [i]	-0.319 ± 0.005 [v]	[aa]
$J_6 \times 10^5$	3.1 ± 2.0 [d]	10.8 ± 5.0 [i]	$0.021 - 0.067$ [ff]	?
τ	9h 55m 29.7s [f]	10h 39m 24s $\pm$ 7s [j,k]	17.24 ± 0.01h [s,t]	16.11h [bb]
m	0.083	0.139	0.0285	0.0259
e_G	0.06487	0.095 [e]	0.0197 ± 0.0010 [w]	0.0171 [aa]
e_D	0.0651	0.094	0.0195	0.0181
T_1, K	165 ± 5 [c]	135 ± 5 [l,m]	76 ± 2 [u]	74 [bb]
E	1.67 [l]	1.78 [cc]	1.06 [dd]	2.7 [ee]
Y	0.18 ± 0.04 [g]	0.06 ± 0.05 [n]	0.26 ± 0.05 [o]	0.262 ± 0.048 [y]
C/H	2.32 ± 0.18 [h]	[p]	~20 [q]	~25 [o,q]
rel. to solar	~2 [o]	~2—6 [o]	~30 [r,u]	$\geq$60 [r]
abundance			~25 [o]	

a. Null, 1976; b. Null et al., 1981; c. Lindal et al., 1981; d. Campbell and Synnott, 1985; e. Lindal et al., 1985; f. Seidelmann and Divine, 1977; g. Gautier et al., 1985; h. Gautier et al., 1982; i. Nicholson and Porco, 1988; j. Desch and Kaiser, 1981; k. Kaiser et al., 1980; l. Hanel et al., 1981; m. Tyler et al., 1981; n. Conrath et al., 1984; o. Gautier and Owen, 1988; p. Gautier and Owen, 1984; q. Lutz et al., 1976; r. Pollack et al., 1986; s. Badenal, 1986; t. Warwick et al., 1986; u. Lindal et al., 1987; v. French et al., 1988; w. Baron et al., 1989; x. Anderson et al., 1987; y. Conrath et al., 1987; z. Smith et al., 1989; aa. Tyler et al., 1989; bb. Stone and Miner, 1989; cc. Hanel et al., 1983; dd. Pearl et al., 1990; ee. Conrath et al., 1989; ff. Zharkov and Gudkova, 1991

data (used for the construction of these models) are closely connected with several different fields of science, including the physics of high pressure, the physics and origin of the solar system, the physical and chemical processes of the protoplanetary cloud, and the calculation of the cosmochemical abundances of elements.

The flights to the planets—particularly to the giant planets—gave a new impulse to the construction of scenarios of the origin of the solar system, Earth, planets and satellites. The resulting models of the giant planets are the important new boundary conditions that are involved.

In the literature, the formation of giant planets is considered on the basis of two major hypotheses [*Pollack*, 1985; *Bodenheimer*, 1985]. According to the first hypothesis, the giant planets formed as a consequence of an instability in the gas of the solar nebula, which developed subcondensations of planetary masses that began to evolve as a unit from a very early stage. This hypothesis is in contradiction with existing models of the giant planets. In the second hypothesis, the giant planets formed by gradual accumulation of solid particles (planetesimals) into an ice-rock core of several Earth masses ($M_\oplus$), and then at a later stage the nebular gas accreted onto the core. This is the core accretion-gas capture model. The last hypothesis is frequently discussed in terms of *Mizuno's* [1980] hypothesis according to which a critical value of the core mass for all giant planets is approximately the same and equal to ~ 10–15 $M_\oplus$. But the recent models of giant planets [*Zharkov and Gudkova*, 1991] do not support Mizuno's hypothesis of approximately equal

masses ~ 10–15 $M_\oplus$ for the embryos of the giant planets on which the accretion of gas had taken place.

In *Zharkov and Kozenko* [1990], a new scheme for the formation of Saturn, Uranus and Neptune was proposed. After completion of the formation of Jupiter over a period of ~ (1–2) x 10^7 years, that planet ejects a massive embryo with a mass of ~ 5 $M_\oplus$ into the Saturn feeding zone. This nucleus initiates the formation of Saturn. After the formation of Jupiter and Saturn, ejection of massive embryos into the Uranus and Neptune feeding zones occurs. These embryos lead to the formation of Uranus and Neptune over cosmogonically realistic time intervals.

In what follows, we give an account of the role of Jupiter in the formation of the planets. In Section 2 we present observational data which are used to construct models of the giant planets, in Section 3 we consider the models of Jupiter, Saturn, Uranus and Neptune, in Section 4 we discuss the role of Jupiter in the formation of Mars and Earth, in Section 5 the formation time of Jupiter is estimated on the basis of calculating the formation time of Mars, in Section 6 we put forward a hypothesis on the role of Jupiter in the formation of the giant planets, and in Section 7 a short conclusion is given.

2. Observational Data

The observational data which have been used to construct the models are summarized in Table 1 of *Zharkov and Gudkova* [1991]. The final three rows of Table 1 contain data

TABLE 2. Abundance of substances by mass percent

Substance	MolecularWeight	Cameron [1982]		Anders and Ebihara [1982]		Anders and Grevesse [1989]	
		1	2*	1	2*	1	2*
H_2	2.016	77.03	77.21	74.12	74.28	70.31	70.59
He	4.003	20.86 4	20.90 4	23.72	23.77	27.44 1.7	27.33 1.7
Ne	20.18	0.15	0.152	0.19	0.192	0.175	0.174
N_2	28.012	—	0.094	—	0.094	—	0.110
CO	28.012	—	0.902	—	0.917	—	0.710
G		98.04	99.26	98.03	99.25	97.93	98.91
Ar	39.95	0.012	0.012	0.011	0.011	0.0102	0.0101
CH_4	16.04	0.516	—	0.5232	—	0.4083	—
NH_3	17.03	0.114	—	0.1144	—	0.1343	—
H_2O	18.02	0.768	0.189	0.806	0.218	0.9138	0.4534
I		Σ1.41	Σ0.201	Σ1.455	Σ0.229	Σ1.467	Σ0.464
SiO_2	60.09	0.174	0.174	0.163	0.163	0.151	0.151
MgO	40.32	0.124	0.124	0.117	0.117	0.109	0.109
Al_2O_3	101.96	0.013	0.013	0.011	0.011	0.011	0.011
CaO	56.08	0.01	0.01	0.009	0.009	0.009	0.009
Na_2O	61.98	0.005	0.005	0.005	0.005	0.005	0.005
FeS	87.92	0.127	0.127	0.122	0.122	0.114	0.114
FeO	71.85	0.083	0.083	0.077	0.077	0.070	0.070
Ni	58.71	0.0082	0.0082	0.008	0.008	0.007	0.007
R		Σ0.544	Σ0.544	Σ0.512	Σ0.512	Σ0.476	Σ0.476
He/H_2		0.27	0.27	0.32	0.32	0.39	0.39
I/R		2.6	0.37	2.84	0.45	3.1	1.0

*Assuming that carbon and nitrogen are present in the form of CO and N_2 in gaseous phase.

on the energetic balance of the planets (the ratio of emitted thermal to absorbed solar energy) E and the compositions of the atmospheres, specifically, the abundance of helium, Y (mass fraction of helium), and the C/H ratio relative to solar abundances.

The cosmochemical abundances proposed by *Cameron* [1982], *Anders and Ebihara* [1982], and *Anders and Grevesse* [1989] are given in Table 2 of *Zharkov and Gudkova* [1991]. New estimates with significantly increased values for the abundance of helium and the oxygen-carbon ratio give us a new insight into the internal structure of the giant planets.

The materials comprising the planets are divided into three components according to their volatility: (1) gases (H_2, He, Ne, ...)—the G component; (2) ices (CH_4, NH_3, H_2O)—the I component; (3) rocks and iron-nickel—the R component. Depending on the pressure-temperature (p,T) conditions in the protoplanetary cloud, some compounds of the I component (CH_4 and NH_3) may be present in the G component. The problem of the composition of the G and I components has undergone further development [*Fegley and Prinn*, 1989]. It is assumed that: (1) in the zone of the giant planets the proto-solar cloud contained carbon and nitrogen in the forms CO and N_2, and (2) in the proto-Jupiter and proto-Saturn clouds carbon and nitrogen existed in the forms CH_4 and NH_3 because the pressure in the gaseous phase was significantly larger.

The modeling calculations show that the formation of neither Uranus nor Neptune was accompanied by the creation of a gaseous envelope, as took place in the case of Jupiter and Saturn. Carbon exerts the most noticeable influence on G and I components. If it formed CO and was in a gaseous phase, then the mass fraction of the I component in the solid phase, which consists of I and R, and the I/R ratio would be strongly reduced. This is easy to see from Table 2, where different estimates of the solar abundances are given.

3. MODELS OF JUPITER, SATURN, URANUS AND NEPTUNE

Methods of constructing models of the giant planets are described in the books of *Zharkov and Trubitsyn* [1978] and *Hubbard* [1984], and descriptions of the models can be found elsewhere [*Hubbard and Marley*, 1989; *Podolak and Reynolds*, 1985; *Zharkov and Gudkova*, 1991]. Below we give only typical examples of the models of giant planets. A wider spectrum of models is published in *Zharkov and Gudkova* [1991].

The abundances of elements derived by *Anders and Grevesse* [1989] were used in the calculations of *Zharkov and Gudkova* [1991] (Table 2). Estimates of the helium content in the solar atmosphere and the proto-solar cloud, Y = 27.4 ± 1.7, confirm the idea that helium differentiation began not only in Saturn but in Jupiter, too.

The problem concerning the composition of dust, which consists of ice (I) and rocks (R) is of great importance. Three variants of I + R components were used: IR1 = (CH_4,

NH_3, H_2O + rocks), I/R = 3.1; IR2 = (NH_3, H_2O + rocks), I/R = 2.2, assuming that CH_4 exists in a gaseous phase; IR3 = (H_2O + rocks), I/R = 1.0, assuming that carbon and nitrogen are present in the form of CO and N_2 in gaseous phases (Table 2).

The problem of whether carbon and nitrogen exist in the gaseous or the solid state in the protoplanetary cloud is characterized by a notable uncertainty. The models with IR1 and IR3 compositions can be treated as limiting cases. Note that some quantity of carbon may be in the dust component as organic compounds, and then may transform to CH_4 in the hot hydrogen envelopes of the planets.

Recent models of Jupiter and Saturn consist of five layers: a two-layer molecular envelope, an atomic metallic envelope, and a two-layer core [*Zharkov and Gudkova*, 1991].

The condition of conservation of the solar abundance of helium in the planet constrains the structure and composition of the external core, which consists of IR components and settling helium. Standard notations for mass fractions are

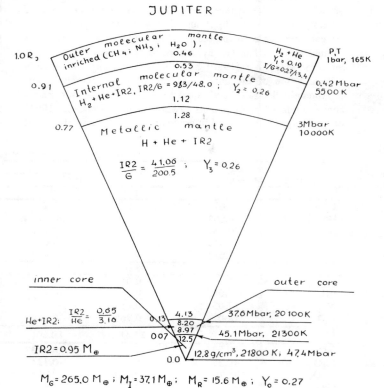

Fig. 1. Typical five-layer model of Jupiter that satisfies the observational data (second-type model). The values of pressure p, temperature T, and relative radius β, are shown at the interfaces, at the surface, and at the center. The values of the density are shown in g cm^{-3} on both sides of the interfaces. The I/G ratio in the upper envelope and the IR2/G ratio and helium content in the lower envelopes are given. The numerical values of the G, I, and R components are expressed in Earth masses ($M_\oplus$). The total mass values for the G, I, and R components are also given in Earth masses $M_\oplus$, as is the total value of helium content.

hydrogen (X), helium (Y), and the IR component (Z), with subscripts i = 1–5 denoting the layers of the planet from the surface downward (see Figure 1).

Models of Jupiter with a homogeneous concentration of helium in the hydrogen envelope $(Y_1 = Y_2 = Y_3)$ do not satisfy the condition of conservation of the solar abundance of helium in the planet, Y_0. A second type of model consists of internal hydrogen envelopes that are enriched in helium, with $Y_2 = Y_3 > Y_4$ (see Figure 1). Models constructed by *Zharkov and Gudkova* [1991] in accordance with the observational data varied, Y_1, the abundance of helium in the external envelope of Jupiter, between 0.14 and 0.22 (see Table 1). Three trial values of helium abundance were used for the internal envelopes: $Y_{2,3} = 0.23$, 0.26, and 0.28. Also, two different variants of IR, IR2 and IR3, were considered. The models of Jupiter which agree with all observational data allow the I/R ratio to be 2.4 or 1.0; i.e., it is impossible to determine the composition of dust (IR2 or IR3) in the zone of formation of Jupiter.

The CH_4 content of the atmosphere (Table 1) was used to define the enrichment of the external envelope by ices (CH_4, NH_3, H_2O) relative to solar abundances. For the construction of models of the interiors of Jupiter and Saturn the values of Z were taken to be 0.02 and 0.03 ($X_1 + Y_1 + Z_1 = 1$).

The mass of the IR core, which may be considered as an embryo of the planet, essentially depends on the distribution of helium throughout the interior of Jupiter. In the typical model shown in Figure 1, this core is small and equals only 1.5–2.0 $M_\oplus$. The resulting models [*Zharkov and Gudkova*, 1991] confirm the previous conclusion that the IR component is abundant in the hydrogen envelope of the planet ($\sim 50\ M_\oplus$, $Z_2 = Z_3 \sim 0.15$–0.20). Furthermore, the models reveal an increase in helium concentration towards the center of the planet ($Y_2 = Y_3 > Y_1$) and require some helium in the IR core. This model of Jupiter is shown in Figure 1.

The important properties of the model are enumerated as follows: (1) The upper G envelope contains an admixture of I component. (2) The lower G envelopes contain a large mass of IR2 component ($\sim 51\ M_\oplus$). This can be considered an indication that Jupiter captured and also ejected the greater part of the planetesimals from its feeding zone after accretion of its gaseous envelope. (3) The I and R components are mixed in the lower G envelopes. This feature is repeated in all giant planets. (4) The inner envelopes of the planet are enriched in helium. (5) The IR2 mass of the core is equal to 1.5–2 $M_\oplus$. (6) Some mass of helium (~ 0.5–1.0 $M_\oplus$) settles into the core during the evolution of Jupiter. The mass of the IR core, which may be considered as an embryo of the planet, essentially depends on the helium distribution in the interior of Jupiter. In the models constructed by *Zharkov and Gudkova* [1991], one of which is shown in Figure 1, this core is small and equals only 1.5–2 $M_\oplus$.

The models of Saturn are constructed in the same manner as those of Jupiter. The peculiarity of Saturn is the very low helium content of its atmosphere (Table 1). In the Saturn model the helium content in the external envelope Y_1 was taken to be in the range $Y_1 = 0.01$–0.11 and $Z_1 = 0.03$.

For Saturn one can construct models of either the first type ($Y_1 = Y_2 = Y_3$) or the second type ($Y_2 = Y_3 > Y_1$) which agree with all observational data. The distributions of density ρ, pressure p, and temperature T along the relative radius for a typical model of the second type for Saturn are shown in Figure 2. In a typical model, Saturn's core is 4–5 times larger than Jupiter's. It is equal to approximately 20–25 $M_\oplus$ and consists of helium and an IR component, the mass of which is 7 $M_\oplus$. The concentration of the IR component in the interior of Saturn is $Z_2 = Z_3 \approx 0.25$–0.30.

Zharkov and Gudkova [1991] constructed Jupiter-like models of Uranus and Neptune, with a two-layer molecular envelope. For Uranian and Neptunian models the helium content is taken as $Y = 0.26$ (Table 1), and three variants of IR compositions–IR1, IR2, and IR3–were used. We consider the IR1 type models to be the most realistic because the temperatures in the zones of formation of both planets were very low.

Typical three-layer models of Uranus and Neptune are shown in Figure 3. From model calculations, the enrichment by ices of the external envelope of Neptune is twice that of Uranus. Neptune's core has a radius approximately 1.5 times greater than that of Uranus. The compositions of the middle envelopes of both planets are similar. This confirms the

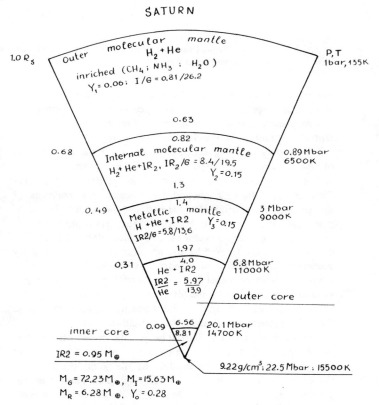

Fig. 2. Typical five-layer model of Saturn that satisfies the observational data (second-type model). The quantities shown are presented as in Fig. 1.

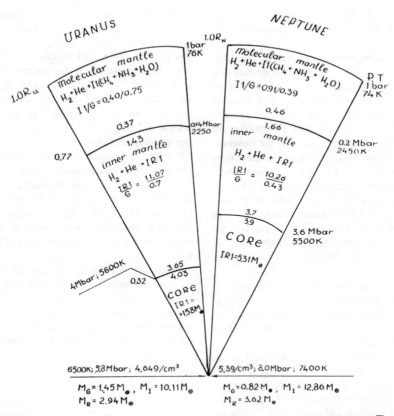

Fig. 3. Typical three-layer models of Uranus and Neptune satisfying the observational data. The quantities shown are presented as in Fig. 1.

previous conclusions (see *Zharkov and Trubitsyn* [1978]) about the huge loss of the hydrogen-helium component (~45–50 planetary masses) and the smaller G component content of Neptune in comparison with Uranus.

The models of the giant planets constrain the composition of the protoplanetary cloud, the (p-T) conditions within it, and the mechanism of formation of the planets. At present there is the following general hypothesis: the embryos of the planets were originally formed by the accumulation of planetesimals, and these embryos later gravitationally accreted their gaseous envelopes from the solar nebula. The primary cores are formed by the accumulation of rock and icy planetesimals. If the core mass has reached some critical value, it could be capable of concentrating and capturing a gaseous envelope. At the end of this process, the protoplanet captured and ejected the remaining planetesimals in the solar system and cleaned the formation zone.

Following *Mizuno's* [1980] scheme, and based on our earlier hypothesis, theoretical calculations of the formation of the giant planets and the models of the giant planets give the critical mass of embryos of all four giant planets as approximately equal to 10–15 $M_\oplus$. The results of our investigation [*Zharkov and Gudkova*, 1991] do not support Mizuno's hypothesis.

The IR core of Jupiter is ~1.5–2 $M_\oplus$, but an enormous content of IR condensate (~50–60 $M_\oplus$) was accumulated by Jupiter during and after the accretion phase. The IR core of Saturn is ~7 $M_\oplus$, i.e., noticeably larger than Jupiter's. The mass of the IR component in the hydrogen-helium envelope is twice the mass of the IR core for Saturn. Both planets are distinguished not only by their helium distribution but by the parameters cited above. So we see that the values of the critical mass of the core of these planets are very different. The previous conclusions about the loss of the G component (~10 planetary masses for Jupiter and ~15 planetary masses for Saturn) are confirmed [*Zharkov and Trubitsyn*, 1978]. This conclusion was obtained on the basis of comparing the compositions of the planets with solar abundances (Table 2).

The models of Uranus and Neptune have a small G component, so we cannot say anything about the stage of accretion during their formation. The mechanism of formation of these planets is unknown. A hypothesis suggested by *Zharkov and Kozenko* [1990] is described below. In accordance with model calculations, the IR1 cores of these planets are small and equal ~2 $M_\oplus$ for Uranus and ~5 $M_\oplus$ for Neptune.

Methane and ammonia, and therefore a significant carbon content, are not detected on the regular satellites of Jupiter.

On this basis we assume Jupiter to be depleted in carbon and thus have constructed models using IR2 and IR3 compositions.

Doubling carbon in Jupiter's atmosphere does not have any significant effect on models of its formation. Perhaps temperatures in the zone of Jupiter's formation were too high for solidification of methane. Some of the carbon in the atmosphere of Jupiter is possibly connected with the accumulation of planetesimals from the periphery of the solar system: 25 protobodies of Moon size would be sufficient. This hypothesis is based on the modern idea that protobodies crossed the solar system during the period of formation of the giant planets. The alternative explanation is that the protobodies contained a mixture of organic compounds in the zone of formation of Jupiter. This argument also applies to Saturn (the concentration of carbon in Saturn is twice that in Jupiter).

4. THE ROLE OF JUPITER IN THE FORMATION OF MARS AND EARTH

The formation of Jupiter had a great influence on the formation of the terrestrial planets. In fact, planetesimals from Jupiter's zone destroyed the seeding zone of the asteroid belt, whose mass is $\sim 10^{-3}$ $M_\oplus$, and Mars' feeding zone which, at a minimal estimate, must have contained ~ 20 times more R material than the present mass of the planet. This brings up the question of the degree of destruction of the Earth's feeding zone, which thus far has been difficult to answer quantitatively.

We can estimate the surface density of the R component in Mars' feeding zone as follows. According to the theory of planetary growth, the halfwidth of the feeding zone of a growing planet at terrestrial heliocentric distances is $\sim 0.2\, r$ (r = the radius of the planet's orbit) and is somewhat larger at the distances of the giant planets. Therefore, knowing the current mass of the planet, and in the case of the giant planet knowing also the mass of the heavy and light components on the basis of the models that have been constructed, we can estimate the initial surface density of the solid component in the planet's feeding zone from the following relation:

$$\alpha_0 = M_p/(0.8\ \xi\ \pi\ r^2) \qquad (6)$$

where M_p is the mass of the planet or its heavy component and ξ is a coefficient of order unity.

However, (6) is not applicable for evaluating the initial surface density of the dust component in the feeding zone of Mars. A substantial decrease of surface density occurs in this zone because of resonance perturbations of the giant planet during formation. Also, planetesimals from near Jupiter destroyed this zone.

We can evaluate the desired initial density by assuming that the surface density of the dust component is described by a power law and is a smooth, continuous function of the distance from the center of the solar nebula. For example,

$$\alpha(r) = \alpha_0\ r^{-3/2} \qquad (7)$$

where r is in A.U. [Safronov and Ruskol, 1982]. This can also be done by a simpler interpolation, drawing a smooth curve through the three points that determine the surface density of the R component in the zones of Venus, Earth and Jupiter using (6) [Zharkov, 1986].

The result obtained agrees with the assertion that the surface density of the dust layer in the asteroid zone may have been of the same order as in the zones of Earth and Jupiter [Ipatov, 1989]. We know the masses of Venus and Earth to high accuracy, and the mass of the silicate component of Jupiter, according to a recent model, is 16 $M_\oplus$. From Figure 4 it can be seen that $\alpha_0 \sim 9.5$ g cm^{-2} in the zone of Mars. Then, assuming no influence of Jupiter, we find that Mars must have acquired a mass estimated from (6) to be $\sim 1.2 \times 10^{28}$ g. But the actual mass of Mars is $\sim 6.4 \times 10^{26}$ g. Thus Mars was able to accumulate only about 5% of the available mass in its zone. The influence of Jupiter thus becomes appreciable, because after formation of Jupiter the feeding zone of Mars was destroyed in a cosmogonically short time.

A higher surface density of material in the feeding zone can also substantially reduce the formation time of the planet (see (8) and (9) below). The probably scenario for the formation of Jupiter shows that if Jupiter expelled about the same amount of mass as was contained in its gas envelope, then the original density of condensate in Jupiter's zone (and in Mars' zone, too) could be a factor of 2 greater, while if the mass of condensate expelled was of the order of the mass of the planet, the surface density could be a factor of 5 larger. There have also been suggestions that the total mass of solid matter in the feeding zones of the giant planets could be an order of magnitude greater than the mass of solid matter that has gone into these planets and could reach hundreds of times the mass of the Earth [Ipatov, 1989]. Therefore the solid curve in Figure 4 is a lower bound on the surface density of silicate material in the solar system, the upper bound could be considerably higher, and the distribution represented by the dashed line is not at all unlikely.

It can be seen that Jupiter could also partially disrupt the

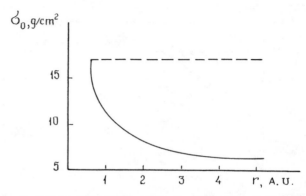

Fig. 4. Initial surface density of the R component in the protoplanetary cloud (see text).

Earth's feeding zone, and one constraint on flights of bodies of mass ~$M_{\mars}$ is the mega-impact hypothesis leading to the formation of the Moon. In that case the initial surface density in the Earth's feeding zone could have been higher, and accordingly its formation time could have been somewhat less than considered heretofore.

A model for the accumulation of terrestrial planets from two chemically very different constituents, a highly reduced component A and an oxidized component B, was postulated by *Ringwood* [1977, 1979] and *Wänke* [1981] (see also *Dreibus and Wänke* [1989]).

Based on their estimates of the bulk composition of Mars, *Dreibus and Wänke* [1987] concluded that the mixing ratio of component A:component B for Mars is 60:40, compared to a ratio of 85:15 for the Earth. Mars accreted almost homogeneously, contrary to the inhomogeneous accretion of the Earth. *Dreibus and Wänke* [1989] assumed that the oxidized component B was only added in substantial amounts after the Earth had reached about two thirds of its present mass. They also suggested that component A existed mainly at and inside the Earth's orbit, while component B dominated in the asteroid belt. In the case of the Earth, component B material was added only during a late phase of accretion, since the transfer of material from the region outside Mars' orbit required additional time. It is evident that this transfer may have been influenced by the perturbation effect of Jupiter on planetesimal orbits.

Figure 5 shows growth curves for the Earth's formation as calculated by different authors. The line parallel to the axis at $(m/M_\oplus)^{1/3} = 0.8$ distinguishes the first stage of the Earth's formation when component A was accumulated from the second stage, during which both components A and B accumulated. It is possible that the influence of Jupiter decreased the time required for the Earth's formation by a factor of two in comparison with the theoretical estimates [*Safronov*, 1972; *Vityazev et al.*, 1978; *Wetherill*, 1980]. It is also useful to remark that after perturbation by Jupiter, the feeding zones of the terrestrial planets become open systems and standard cosmological theory no longer applies.

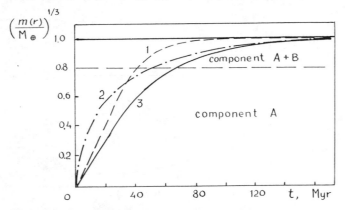

Fig. 5. Growth curves for the Earth as proposed by different authors: 1–*Safronov* [1972]; 2–*Wetherill* [1980]; 3–*Vityazev et al.* [1978] (see text).

5. FORMATION TIME FOR JUPITER

As a result of the development of models for giant planets [*Zharkov et al.*, 1974a,b; *Zharkov and Gudkova*, 1991] as well as cosmogonical concepts [*Safronov and Vityazev*, 1985; *Hayashi et al.*, 1985], the concept has arisen that giant planet formation is a four stage process, at least for Jupiter and Saturn. The dust component of the protoplanetary disk completes the first stage of planetesimal formation by settling to the central plane and decaying into clumps. The second, lengthier stage, which essentially determines the time scale for planet formation, is associated with the accumulation of planetesimals by collisions with the resultant formation of a planetary embryo. When the mass of this embryo reaches a critical value of ~2–5 $M_\oplus$, the third stage begins, with the gaseous component of the protoplanetary disk losing stability in the vicinity of the growing planet and accreting onto that planet over a cosmogonically brief time of ~10^4–10^5 years [*Hayashi et al.*, 1985]. After this, a quite brief fourth stage commences in which the final formation of the planet occurs. In this stage the powerful gravitational center which has formed empties the feeding zone by sweeping up and scattering planetesimals ejected from the feeding zone of Jupiter. These involve a mass of the same order as that of the planet itself [*Ipatov*, 1989].

We can estimate the formation time of Jupiter from two viewpoints. On the one hand, this time cannot be greater than τ_T, the time that the sun spent in the T Tauri phase, when powerful corpuscular radiation led to dissipation of the G component of the protoplanetary nebula, say τ_T ~10^7 years. On the other hand, using the theory of the growth of terrestrial planets, we can estimate the formation time of Mars which, according to what has been stated above, will also be the formation time of Jupiter. Agreement of the two estimates will indicate that the ideas on which these arguments are based are mutually consistent.

Vityazev et al. [1978] derived an approximate analytical expression to evaluate the growth time of a planet with a given initial surface density of the dust layer. It is applicable to the zones of formation of the terrestrial planets, where we can neglect the influence of gas and the ejection of bodies.

Actually, no gas remains in this zone during most of the time of formation of the terrestrial planets. The evidence for this consists of the granules in meteoritic material which show evidence of irradiation by the solar wind. These granules were fully exposed to the solar wind prior to the formation of the asteroids, the parent bodies of the meteorites. By that time there could no longer have been any gas, since one thousandth of the initial amount would have sufficed to screen the solar radiation completely.

The time for the planet to grow to 97% of its maximum possible mass given the surface density, is

$$t_{(97\%)} = 5.3 \, D, \tag{8}$$

where

$$D = (6\pi/M_\oplus)^{1/4} \frac{\delta^{3/4} r^{5/4} P}{(2\alpha_0)^{1/2}\theta^{1/4}(1+\theta)} \qquad (9)$$

Here $M_\oplus$ is the mass of the sun, r is the distance of the growing planet from the sun, P is the orbital period of the planet around the sun, δ is the density of the embryonic planet (assumed constant), α_0 is the initial surface density of the material in the planet's feeding zone, and θ is a dimensionless parameter describing the mean relative speeds of bodies in the pre-planetary swarm during accumulation (θ ~3–5).

It is not difficult to obtain the relation required to determine the growth time of Mars

$$t_{(5\%)} = 0.77 \, D \qquad (10)$$

Equation (10) also limits the time of formation of Jupiter to ~1.7 x 10^7 years, which is approximately the time spent by the Sun in the T Tauri stage. A discussion of the accumulation time of Jupiter's core on the basis of the theory of planetary accumulation is given by *Zharkov and Kozenko* [1989].

6. ON THE ROLE OF JUPITER IN THE FORMATION OF THE GIANT PLANETS

According to cosmogonic concepts, formation of the embryos of the giant planets took place in a gaseous medium and over a characteristic time t_{for} [*Hayashi et al.*, 1985]

$$t_{for} = 0.22 \frac{10^4}{f^2(1+2\theta/f)}\left(\frac{m}{10^{18}g}\right)^{1/3}\left(\frac{r}{1 A.U.}\right)^3 \qquad (11)$$

where f ~2.8 is a numerical coefficient appearing due to capture by the growing embryo of planetesimals entering the Roche sphere, θ is a numerical parameter (of the order of unity) which takes into account gravitational focusing upon collision of planetesimals with the growing planetary embryo, m is the critical mass of the planet's IR core, and r is the distance from the Sun in astronomical units. The planetary embryo formation (or growth) time is proportional to the Keplerian orbital period t_K of the planet about the Sun (t_K ~$r^{3/2}$) and inversely proportional to the surface density α_s of the dust component in the protoplanetary disk. As a result, t_{for} ~r^3 in (11). Substitution of standard parameter values in (11) yields the following values for t_{for}: ~10^9 yr for Saturn, ~10^{10} yr for Uranus, and ~10^{11} yr for Neptune. *Horedt* [1988] proposed a new model for the distribution of the condensate (I and R components) in the protoplanetary disk and carried out numerical modeling of the planetary accumulation process. As a result, somewhat lower values of t_{for} were obtained: 8 x 10^7, 3 x 10^8, 1.3 x 10^9, and 1.8 x 10^9 yr, for Jupiter, Saturn, Uranus, and Neptune, respectively.

Construction of models of the giant planets has shown that the compositions of these planets differ markedly from the solar composition—the planets enriched in the IR compo-

nent. As we have seen above, Jupiter lost ~10 planetary masses of G component during its formation. Analogous values for Saturn, Uranus, and Neptune are ~15, ~44, and ~46 planetary masses.

Current thought relates loss of the gaseous component during formation to the active state of the young Sun in its T Tauri stage for a characteristic time of ~10^7 yr. However, it is obvious that within the framework of current cosmogonic concepts, (11), reconciliation of the formation time of Saturn, Uranus, and Neptune with the characteristic duration of the young Sun's active T Tauri stage is impossible.

To avoid this difficulty, we turn to studies involving numerical modeling of the behavior of the ensemble of planetesimals [*Ip and Fernandez*, 1988; *Ipatov*, 1989]. These studies indicate that upon introducing into the ensemble of planetesimals an embryo several hundred Earth masses in size, the new object perturbs the trajectories of the planetesimals as well as sweeping up the planetesimals. As a result, many planetesimals commence to depart far from the orbit of the embryo, with this orbit deformation occurring over a cosmogonically brief time interval.

We therefore propose the following mechanism for the formation of Saturn, Uranus, and Neptune. After its accretion of gas, the proto-Jupiter embryo sweeps up planetesimals in its feeding zone and perturbs the orbits of a number of them in such a way that they begin to penetrate regions far from the Jupiter growth zone—for example, into Saturn's formation zone. Some of the larger planetesimals, close in mass to the critical mass of the Saturn embryo, having impinged upon the Saturn feeding zone over the cosmogonically brief time of ~10^4–10^5 yr, meet with a gaseous component which collects upon the embryo, thus forming a massive proto-Saturn. As was shown in the studies cited, the two powerful gravitational centers, in Jupiter and Saturn, very rapidly deform the orbits of the planetesimals, so that they begin to penetrate first the feeding zone of Uranus and then that of Neptune. Among these planetesimals are rather large ones, with masses several times that of Earth, having appreciable gaseous shells acquired in the Jupiter and Saturn feeding zones before the dissipation of gas from the protoplanetary nebula.

These large embryos for Uranus and Neptune play a dual role. On the one hand, the introduction of large embryos into the Uranus and Neptune feeding zones permits formation of both planets by the sweeping of planetesimals over a cosmogonically reasonable time period of ~10^8. On the other hand, the embryos have acquired hydrogen shells ~1–1.5 $M_\oplus$ in mass in the Jupiter and Saturn feeding zones even before the dissipation of gas. Thus, the origin of the gaseous component on Uranus and Neptune is resolved, despite the fact that formation of these planets occurred over a time interval markedly longer than the characteristic duration of the T Tauri stage.

The proposed hypothesis of giant planet formation is of course still schematic in nature and will take on more concrete features as it develops. *Hubbard and MacFarlane* [1980] called attention to the fact that at low temperatures, hydrogen compounds (the I component) of the protoplanetary disk should be enriched in deuterium, which should then be reflected in the isotopic composition of the outer G shells of the giant planets. At the present time, data have been obtained on the D/H ratio in the atmospheres of Jupiter ($2.0^{+0.6}_{-0.6} \times 10^{-5}$), Saturn ($1.6^{+1.6}_{-1.0} \times 10^{-5}$), Uranus ($7.2^{+7.2}_{-3.6} \times 10^{-5}$), and Neptune ($12^{+12}_{-8} \times 10^{-5}$) [de Bergh et al., 1989]. These data agree qualitatively with the giant-planet models and formation scenario presented in the present study. Uranus and Neptune contain a markedly higher concentration of I component in the outer G shell than do Jupiter and Saturn, and the D/H ratio in the atmospheres is correspondingly elevated.

7. CONCLUSION

In this article we have discussed the role of Jupiter in the formation of planets, and it is evident that this role was very important. But we must stress again that almost all problems connected with the effects of Jupiter are now only in an initial stage of investigation. By our request, *S. I. Ipatov* [1991] has used computer simulations to examine the role of Jupiter in the formation of Saturn, Uranus, and Neptune. He reaches the following conclusions. The results of computer runs show that even if the initial orbits of the embryos of Saturn, Uranus, and Neptune, with masses equal to several Earth masses, were highly eccentric, then the eccentricities of these growing planets could decrease and evolve to the present values. Specifically, Saturn's embryo could decrease its eccentricity due to the accretion of gas. But for the embryos of Uranus and Neptune, the decrease in eccentricity was mainly due to gravitational interactions with planetesimals. Investigation of the models taking into account the migration toward Jupiter of bodies initially located behind Saturn's orbit show that the nearly formed Saturn could migrate from Jupiter's zone. Also, embryos of Uranus and Neptune, with masses equal to a few Earth masses, could migrate from Saturn's zone, moving all the time in orbits with low eccentricities. Ipatov's work represents the first successful test of the hypothesis considered here.

Acknowledgments. The author is grateful to Prof. Raymond Jeanloz for valuable comments. Preparation of this paper was partially supported by NASA.

REFERENCES

Anders, E., and M. Ebihara, Solar system abundances of the elements, *Geochim. Cosmochim. Acta, 46*, 2363-2380, 1982.

Anders, E., and N. Grevesse, Abundances of the elements: Meteoritic and solar, *Geochim. Cosmochim. Acta, 53*, 197-214, 1989.

Anderson, J. D., J. K. Campbell, R. A. Jacobson, D. N. Sweetnam, A. H. Taylor, A. J. R. Prentice, and G. L. Tyler, Radio science with Voyager 2 at Uranus: results on masses and densities of the planet and five principal satellites, *J. Geophys. Res., 92*, 14877-14883, 1987.

Badenal, F., The double tilt of Uranus, *Nature, 231*, 809-810, 1986.

Baron, R. L., R. G. French, and J. L. Elliot, The oblateness of Uranus at the 1-mbar level, *Icarus, 78*, 119-130, 1989.

Bodenheimer, P., Evolution of the giant planets, in *Protostars and Plan-*

ets II, edited by D. C. Black and M. S. Matthews, The University of Arizona Press, Tucson, 873-894, 1985.

Cameron, A. G. W., in *Essays in Nuclear Astrophysics*, edited by C. Barues et al., Cambridge University Press, 23, 1982.

Campbell, J. K., and S. P. Synnott, Gravity field of the Jovian system from Pioneer and Voyager tracking data, *Astron J., 90*, 364-372, 1985.

Conrath, B. J., D. Gautier, K. A. Hanel, and J. S. Hornstein, The helium abundance of Saturn from Voyager measurements, *Astrophys. J., 282*, 807-815, 1984.

Conrath, B., D. Gautier, R. Halen, G. Lindal, and A. Marten, The helium abundance of Uranus from Voyager measurements, *J. Geophys. Res., 92*, 15003-15010, 1987.

Conrath, B., F. M. Flasar, R. Havel, V. Kunde, W. Maguire, I. Pearl, I. Pirraglia, R. Samuelson, P. Gierasch, A. Weir, B. Bezard, D. Gautier, D. Cruikshank, L. Horn, R. Springer, and W. Shaffer, Infrared observations of the Neptunian system, *Science, 246*, 1454-1459, 1989.

deBergh, C., B. L. Lutz, T. Owen, and J. P. Maillard, Monodeuterated methane in the outer solar system, its detection and abundance on Neptune, 1989, *Astrophys. J., 355*, 661-666, 1990.

Desch, M. O., and M. I. Kaiser, Voyager measurements of the rotation period of Saturn's magnetic field, *Geophys. Res. Lett., 8*, 253-256, 1981.

Dreibus, G., and H. Wänke, Volatiles on Earth and Mars: A comparison, *Icarus, 71*, 225-240, 1987.

Dreibus, G., and H. Wänke, Supply and loss of volatile constituents during the accretion of terrestrial planets, in *Origin and Evolution of Planetary and Satellite Atmosphere*, edited by S. K. Atreya et al., The University of Arizona Press, Tucson, 268-288, 1989.

French, R. G., J. L. Elliot, L. A. Kangas, K. I. Meech, and M. E. Ressler, Uranian ring orbits from Earth-based and Voyager occultation observations, *Icarus, 73*, 349-378, 1988.

Gautier, D., B. Conrath, M. Flaser, R. Hanel, V. Kunde, A. Chedin, and N. Scott, The helium abundance of Jupiter from Voyager, *J. Geophys. Res., 86*, 8713-8720, 1985.

Gautier, D., B. Bezard, A. Marten, J. P. Baluteau, N. Scott, A Chedin, V. Kunde, and R. Hanel, The C/H ratio in Jupiter from the Voyager infrared investigation, *Astrophys. J., 257*, 901-912, 1982.

Gautier, D., and T. Owen, Observational constraints on models for giant planet formation, in *Protostars and Planets II*, edited by D. C. Black and M. S. Matthews, The University of Arizona Press, Tucson, 832-846, 1985.

Gautier, D., and T. Owen, The composition of outer planet atmospheres, in *The Origin and Evolution of Planetary and Satellite Atmospheres*, edited by T. Gehrels, University of Arizona Press, Tucson, 1988.

Gavrilov, S. V. and V. N. Zharkov, Love numbers of the giant planets, *Icarus, 32*, 443-449, 1977.

Goldreich, P., and S. Soter, Q in the solar system, *Icarus, 5*, 375-389, 1966.

Hanel, R. A., B. J. Conrath, I. W. Herath, V. G. Kunde, and J. A. Pirraglia, Albedo, internal heat, and energy balance of Jupiter – Preliminary results of the Voyager infrared investigation, *J. Geophys. Res., 86*, 8705-8712, 1981.

Hanel, R. A., B. J. Conrath, V. G. Kunde, J. C. Peale, and J. A. Pirraglia, Albedo, internal heat flux, and energy balance of Saturn, *Icarus, 53*, 262-285, 1983.

Hayashi, C., K. Nakazawa, and Y. Nakagava, Formation of the solar system, in *Protostars and Planets II*, edited by D. C. Black and M. S. Matthews, The University of Arizona Press, Tucson, 1100-1153, 1985.

Horedt, G. P., Evolutionary models of the planets, *Astron. Astrophys., 202*, 284-294, 1988.

Hubbard, W. B., *Planetary Interiors*, Van Nostrand-Reinhold Co., New York, 1984.

Hubbard, W. B., V. P. Trubitsyn, and V. N. Zharkov, Significance of gravitational moments for interior structure of Jupiter and Saturn, *Icarus, 21*, 147-151, 1974.

Hubbard, W. B., and J. J. MacFarlane, Theoretical predictions of deuterium abundances in the Jovian planets, *Icarus, 44*, 676-682, 1980.

Hubbard, W. B., and M. S. Marley, Optimized Jupiter, Saturn and Uranus interior models, *Icarus, 78*, 102-118, 1989.

Ipatov, S. I., Migration of the planetesimals during the last stages of giant planet accumulation, *Astron. Vestn., 23*, 27-38, 1989.

Ipatov, S. I., Orbital evolution of growing giant planet embryos moving initially in highly eccentrical orbits, *Pis'ma Astron. Zh., 17*, 268-280, 1991 (Sov. Astron. Lett., in press).

Ip, W. H., and J. A. Fernandez, Exchange of condensed matter among the outer and terrestrial protoplanets and the effect on surface impact and atmospheric accretion, *Icarus, 74*, 47-61, 1988.

Kaiser, M. L., D. Desch, J. W. Warwick, and J. B. Pearce, Voyager detection of nonthermal radio emission from Saturn, *Science, 209*, 1238-1240, 1980.

Lindal, G. F., D. N. Sweetnam, and V. R. Eshleman, The atmosphere of Saturn: An analysis of the Voyager radio occultation measurements, *Astron. J., 90*, 1136-1146, 1985.

Lindal, G. F., J. R. Lyons, D. N. Sweetnam, V. R. Eshleman, D. P. Hinson, and G. L. Tyler, The atmosphere of Uranus: Results of radio occultation measurements with Voyager 2, *J. Geophys. Res., 92*, 14987-15001, 1987.

Lindal, G. F., G. E. Wood, G. S. Levy, J. D. Anderson, D. N. Sweetnam, H. B. Hotz, B. J. Buckles, D. P. Holmes, P. E. Doms, V. R. Eshelman, G. L. Tyler, and T. A. Croft, The atmosphere of Jupiter: An analysis of Voyager occultation measurements, *J. Geophys. Res., 86*, 8721-8727, 1981.

Lutz, B. L., T. Owen, and R. D. Cess, Laboratory band strengths of methane and their application to the atmospheres of Jupiter, Saturn, Uranus, Neptune, and Titan, *Astrophys. J., 203*, 541-551, 1976.

Mizuno, H., Formation of the giant planets, *Progr. Theoret. Phys., 64*, 544-557, 1980.

Nicholson, P. D., and C. C. Porco, A new constraint of Saturn's zonal gravity harmonics from Voyager observations of an eccentric ringlet, *J. Geophys. Res., 93*, 10209-10224, 1988.

Null, G. W., Gravity field of Jupiter and its satellites from Pioneer 10 and Pioneer 11 tracking data, *Astron. J., 81*, 1153-1161, 1976.

Null, G. W., E. L. Lau, E. D. Biller, and J. D. Anderson, Saturn gravity results obtained from Pioneer 11 tracking data and Earth-based Saturn satellite data, *Astron. J., 86*, 456-468, 1981.

Pearl, J. S., B. J. Conrath, R. A. Hanel, J. A. Pirraglia, and A. Counstenis, The albedo, effective temperature and energy balance of Uranus, as determined from Voyager IRIS data, *Icarus, 84*, 12-28, 1990.

Podolak, M., and T. Reynolds, What have we learned from modeling giant planet interiors? in *Protostars and Planets II*, edited by D. C. Black and M. S. Matthews, The University of Arizona Press, Tucson, 847-872, 1985.

Pollack, J. B., Formation of the giant planets and their satellite ring systems: An overview, in *Protostars and Planets II*, edited by D. C. Black and M. S. Matthews, The University of Arizona Press, Tucson, 791-831, 1985.

Pollack, J. B., K. Rages, K. H. Baines, J. T. Bergstrahl, D. Wenkert, and E. Danielson, Estimates of the bolometric albedos and radiation balance of Uranus and Neptune, *Icarus, 65*, 442-466, 1986.

Ringwood, A. E., Composition of the core and implications for origin of the Earth, *Geochem. J., 11*, 111-135, 1977.

Ringwood, A. E., *On the Origin of the Moon*, Springer Verlag, New York, 1979.

Safronov, V. S., *Evolution of the Protoplanetary Cloud and Formation of the Earth and the Planets*, Nauka Press, Moscow (in Russian). Trans. NASA TTF-677 1972.

Safronov, V. S., and E. L. Ruskol, On the origin and initial temperature of Jupiter and Saturn, *Icarus, 49*, 284-296, 1982.

Safronov, V. S., and A. V. Vityazev, Origin of the solar system, *Sov. Sci. Rev. Astrophys. Space Phys., 4*, 1-89, 1985.

Seidelmann, P. K., and N. Divine, Evaluation of Jupiter longitudes in system III (1965), *Geophys. Res. Lett., 4*, 65-68, 1977.

Smith, B. A., L. A. Soderblom, D. Banfield, C. Barnet, A. T. Basilevsky, R. F. Beebe, K. Bollinger, J. M. Boyce, A. Brahic, G. A. Briggs, R. H. Brown, C. Chyba, S. A. Collins, T. Colvin, A. F. Cook II, D. Crip, S. K. Croft, D. Cruikshank, J. N. Cuzzi, G. E. Danielson, M. E. Davies, E. De Jong, L. Dones, J. Godfrey, J. Goguen, I. Grenier, V. R. Haemmerle, H. Hammel, C. J. Hansen, C. P. Helfenstein, C. Howell, G. E. Hunt, A. P. Ingersoll, T. V. Johnson, J. Kargel, R. Kirk, D. I. Kuehn, S. Limaye, H. Masursky, A. McEwen, D. Morrison, T. Owen, W. Owen, J. B. Pollack, C. C. Porco, K. Rages, P. Rogers, D. Rudy, C. Sagan, J. Schwartz, E. M. Shoemaker, M. Showalter, B. Sicardy, D. Simonelli, J. Spenser, L. A. Sromovsky, C. Stoker, R. G. Strom, V. E. Suomi, S. P. Synott, R. J. Terrile, P. Thomas, W. R. Thompson, A. Verbiscer, and J. Veverka, Voyager 2 at Neptune: Imaging science results, *Science, 246*, 1422-1449, 1989.

Stone, E. C., and E. D. Miner, The Voyager 2 encounter with the Neptunian System, *Science, 246*, 1417-1421, 1989.

Tyler, G. L., V. R. Eshleman, J. D. Anderson, G. S. Levy, G. F. Lindal, G. E. Wood, and T. A. Croft, Radio science investigations of the Saturn system with Voyager 1: Preliminary results, *Science, 212*, 201-206,1981.

Tyler, G. I., D. N. Sweetnam, J. D. Anderson, S. E. Borutzki, J. K. Campbell, V. R. Eshleman, D. L. Gresh, E. M. Gurola, D. P. Hinson, N. Kawashima, E. R. Kursinski, G. S. Levy, C. F. Lindal, J. R. Lyons, E. A. Marouf, P. A. Rosen, R. A. Simpson, and G. E. Wood, Voyager radio science observations of Neptune and Triton, *Science, 246*, 1466-1473, 1989.

Vityazev, A. V., G. V. Pechernikova, and V. S. Safronov, Limiting masses, distances and times for the accumulation of the planets of the terrestrial group, *Sov. Astron., AJ., 22*, 60-63, 1978.

Wänke, H., Composition of terrestrial planets, *Phil. Trans. R. Soc. London, A303*, 287-302, 1981.

Warwick, J. W., D. R. Evans, J. H. Romig, C. B. Sawyer, M. D. Desch, M. L. Kaiser, J. K. Alexander, T. D. Carr, D. H. Staelin, S. Gulkis, R. L. Poynter, A. Aubier, A. Boischot, Y. Leblanc, A. Lecacheux, B. M. Petersen, and Pl Zarka, Voyager 2 radio observations of Uranus, *Science, 233*, 102-106, 1986.

Wetherill, G. W., Formation of the terrestrial planets, *Ann. Rev. Astron. Astrophys., 18*, 77-113, 1980.

Zharkov, V. N., *Interior Structure of the Earth and Planets*, Trans. from the Russian by W. B. Hubbard and R. A. Masteler. Harwood Acad. Publ., Chur, Switzerland, 1986.

Zharkov, V. N., and T. V. Gudkova, Models of giant planets with a variable ratio of ice to rock, *Ann. Geophysicae, 9*, 357-366, 1991.

Zharkov, V. N., and A. V. Kozenko, Formation time of Jupiter, *Pis'ma Astron. Zh., 15*, 745-749, 1989 (Sov. Astron. Lett., 15 (4), 322-324, 1989).

Zharkov, V. N., and A. V. Kozenko, On the role of Jupiter in formation of giant planets, *Pis'ma Astron. Zh., 16*, 169-173, 1990 (Sov. Astron. Lett., 16 (1), 73-74, 1990).

Zharkov, V. N., A. B. Makalkin, and V. P. Trubitsyn, Models of Jupiter and Saturn. II. Structure and composition, *Astron. Zh., 51*, 1288-1297, 1974a (Sov. Astron. 768, 1975).

Zharkov, V. N., and V. P. Trubitsyn, *Physics of Planetary Interiors*, edited by W. B. Hubbard, Pachart, Tucson, 1978.

Zharkov, V. N., V. P. Trubitsyn, I. A. Tsarevskii, and A. B. Makalkin, Equations of state of cosmochemical substances and the structure of the planets, *Izv. Akad. Nauk SSSR Fiz. Zemli, 10*, 3-14, 1974b (Izv. Acad. Sci. USSR, Phys. Earth, 10, 1975).

V. N. Zharkov, Department of Theoretical Physics, Institute of Physics of the Earth, Academy of Sciences of the USSR, Moscow 123810, Russia

High Pressure Phase Transitions
in a Homogeneous Model Martian Mantle

Noriko Kamaya, Eiji Ohtani, Takumi Kato and Kosuke Onuma

Institute of Mineralogy, Petrology and Economic Geology,
Tohoku University, Sendai, Japan

The mineralogical structure of a model Martian mantle has been estimated from high pressure, high temperature experiments. The sequence of phase assemblages with increasing depth in the Martian interior is summarized as follows : olivine + pyroxene + plagioclase (+ minor phases) in the mantle at depths shallower than 230 km; olivine + pyroxene + garnet from 230 to 1280 km; garnet + modified spinel and/or spinel ($(Mg,Fe)_2SiO_4$) + magnesiowustite in the Martian transition zone from 1280 to 1800 km; garnet + Mg-perovskite + Ca-perovskite + magnesiowustite in the Martian lower mantle from 1800 to 2020 km. The Martian core is assumed to be a mixture of iron-nickel alloy and iron sulfide in this model. This model satisfies the observed mean density and moment of inertia estimate of 0.365 (Reasenberg, 1977; Kaula and Asimow, 1991) for Mars.

Introduction

The chemical composition of Mars can be estimated to some extent on the basis of geophysical data obtained by many spacecrafts launched to date. Many authors have suggested an "Fe-rich mantle model" because of the mean density and the moment of inertia of Mars [Anderson, 1972; Ringwood, 1977; McGetchin and Smyth, 1978; Morgan and Anders, 1979; Goettel, 1983].

Knowledge of the high pressure and temperature phase transitions of an Fe-rich mantle composition is indispensable for estimating the structure of the Martian interior correctly. Patera and Holloway [1982] clarified the phase relations of the model mantle composition of Mars estimated by Morgan and Anders [1979] at pressures up to 2 GPa.

The purpose of this work is to study the phase relations in the Martian mantle experimentally to high pressures, thereby providing a new model of the Martian interior which satisfies both geochemical and geophysical constraints.

Experimental Procedure

We used a piston cylinder high pressure apparatus for experiments up to 3 GPa, a DIA6 cubic anvil high pressure apparatus for experiments at 5 and 6.4 GPa, and an MA8 multiple anvil high pressure apparatus above 12.5 GPa.

The cell assembly used for the piston cylinder apparatus consists of a graphite tube heater, a glass insulation tube and a talc pressure medium. A Pt-Pt13%Rh thermocouple was used to measure run temperatures. The cell assembly of the DIA6 apparatus consists of a graphite tube heater, a zirconia thermal insulation tube and a pyrophyllite pressure medium. The starting material was packed directly into the graphite tube heater. A W3%Re-W25%Re thermocouple was used to measure the run temperatures. The cell assembly used in the MA8 experiments is essentially similar to those described by Kato and Kumazawa [1985] and Ohtani [1987]. The cell assembly consisted of a graphite sample capsule, a magnesia pressure medium, TiC electrodes, and a twin plate heater made of a mixture of tungsten carbide and diamond powder. Magnesia pressure media containing about 10 wt.% CoO were used for runs above 20 GPa. This pressure medium shows better thermal insulation compared to that made of pure magnesia, and is suitable for the high temperature runs. A W3%Re-W25%Re thermocouple was used to measure the run temperatures.

In all experiments, the pressure was applied first, and then the temperature was increased to the desired value and held constant. The sample was quenched by turning off the electric power after being held at a constant pressure and temperature for a desired time. The experimental products were analyzed by using an X-ray powder diffractometer, a Debye-Scherrer camera with

Evolution of the Earth and Planets
Geophysical Monograph 74, IUGG Volume 14

TABLE 1. Compositions of Starting Material

	MA	MA-OL
SiO2 (wt.%)	42.1	43.0
Al2O3	6.5	10.1
MgO	30.2	23.2
FeO	16.0	15.9
CaO	5.3	7.7

the Gandolfi attachment, and an electron probe micro-analyzer.

Two starting compositions MA and MA-OL were used in this work (see Table 1). The mantle composition of Mars estimated by Morgan and Anders [1979] was simplified to the five component system CaO–MgO–FeO–Al2O3–SiO2 for both starting materials. The composition MA-OL, which contains 10wt.% olivine (see Table 1), was used in order to identify minor phases in the system, such as pyroxene and garnet. A mixture of the five reagents CaCO3, Fe2O3, MgO, Al2O3 and SiO2 was melted in a graphite container at one atmosphere and quenched into a glass. The starting materials thus synthesized consisted of glass and a small amount of olivine and magnetite.

The experimental P–T conditions nearly fit the geotherm in Mars for the convection model suggested by Johnston and Toksoz [1977] (see below, Figure 3). The temperature at a depth of 230 km in the mantle is 1000 °C, and at 1800 km depth, at the upper–lower mantle boundary, it is 1820 °C. Although this temperature distribution has no thermal boundary layer at the core–mantle boundary, it may be reasonable because of weak convection and low heat flux in the core as suggested by the very weak magnetic field. The experimental conditions were in the range 1–20 GPa and 1000–1790 °C. The run durations were in the range 1 minute at 20 GPa to 66.3 hrs at 1 GPa. The diffraction peaks of pyroxene in the run products using the MA-OL starting composition were easily identified compared to those using the MA starting composition. Most pyroxene and garnet grains synthesized have diameters smaller than 20 μ m.

EXPERIMENTAL RESULTS AND DISCUSSION

Phase transformations in the Martian interior

The experimental conditions and run products are summarized in Table 2. A small amount of melt was observed in the runs conducted at 1–3 GPa and 1000 °C. The partial melting may be caused by the effect of a small amount of water in the charge. The phase assemblage at these conditions was olivine + clinopyroxene + orthopyroxene + glass. Garnet appears and the phase assemblage changes to olivine + pyroxene + garnet at a pressure between 1.5 and 3 GPa at 1000 °C.

TABLE 2. Experimental Results

No.	P(GPa)	T(°C)	Duration	Results
1	1.0	1000	10 hr	Ol+Cpx+Opx+Gl
2	1.0	1000	66.3 hr	Ol+Cpx+Opx+Gl
3	1.5	1000	24 hr	Ol+Cpx+Opx+Gl
4	3.0	1000	5 hr	Ol+Px+Gt+Gl
5	5.0	1100	2 hr	Ol+Px(+Gt)
6	6.4	1200	2 hr	Ol+Px(+Gt)
7	12.5	1582	19 min	Ol+Gt(+Px)
*8	12.5	1590	23 min	Ol+Gt+Px
9	14.4	1672	22 min	Ol+Gt(+Px)
*10	14.4	(1650)	41 min	Ol+Gt+Px
11	18.0	1600	16 min	MS+Gt+MW
12	18.0	1780	2 min	MS+Gt+MW
13	20.0	1790	1 min	MS or Sp+Gt+MW

*; MA-OL composition was used.
Abbreviations are Ol, olivine; Cpx, clinopyroxene; Opx, orthopyroxene; Px, pyroxene; Gl, glass; Gt, garnet; MS, modified spinel; MW, magnesiowustite; Sp, spinel.

Olivine transforms to modified spinel at a pressure between 14.4 and 18 GPa at around 1700 °C. The transformation of pyroxene to garnet starts at around 10 GPa and is complete at 16 GPa where magnesiowustite appears in the same pressure range. Magnesiowustite, majorite, and spinel coexist stably above 18 GPa in the MA composition. The compositions of these phases are

TABLE 3. Representative Chemical Compositions of Minerals Coexisting at 18GPa 1600 °C.

	Modified Spinel-3	Magnesio-wustite-1	Garnet-5
SiO2 (wt.%)	37.62	0.11	47.90
Al2O3	0.13	0.26	10.79
FeO	23.00	56.70	9.87
MgO	38.49	42.81	23.28
CaO	0.03	0.12	8.75
Total	99.25	100.00	100.58
	(O=4.000)	(O=1.000)	(O=12.000)
Si	0.989	0.001	3.438
Al	0.004	0.003	0.913
Fe	0.506	0.423	0.593
Mg	1.509	0.570	2.490
Ca	0.001	0.001	0.673
Total	3.009	0.998	8.106

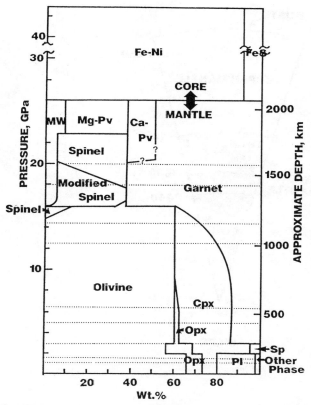

Fig. 1. Phase transformations in the Martian interior on the basis of the present experimental data. Dotted lines show experimental conditions. Pl, plagioclase; Opx, orthopyroxene; Cpx, clinopyroxene; MW, magnesiowustite; Mg–Pv, Mg–perovskite; Ca–Pv, Ca–perovskite.

given in Table 3. It is worth noting that magnesiowustite occurs in the transition zone in this Martian composition.

Figure 1 illustrates the phase transformations in the Martian interior on the basis of the present experimental data. We assumed the simplest model of Mars in which the Martian mantle is chemically homogeneous, although there is a possibility that the Martian transition zone and the lower mantle are chemically distinct from the upper mantle. The run products made at 1–3 GPa were partially molten, so the estimation of the subsolidus mineral assemblage in this pressure range was made by a CIPW norm calculation. The mineral assemblages above 3 GPa were estimated from the experimental products. The weight fraction of each phase was calculated from the mass balance of the experimental products. The transformation patterns of each mineral above 20 GPa were estimated by the phase relations in spinel [Ito and Katsura, 1989] and garnet [Irifune et al., 1989].

The sequences of the phase transitions in the Martian mantle are summarized as follows. The assemblage of olivine, plagioclase, clinopyroxene and orthopyroxene in the shallow mantle reacts to form a mixture of olivine, clinopyroxene, orthopyroxene and spinel

$((Mg,Fe)Al_2O_3)$ at 2 GPa. At 3 GPa, the assemblage changes to olivine, clinopyroxene, garnet and orthopyroxene. Then orthopyroxene gradually dissolves into clinopyroxene up to 10 GPa. The Ca–poor clinopyroxene dissolves into garnet and disappears at 16 GPa. Magnesiowustite also appears at this pressure after the complete dissolution of pyroxene into garnet. Olivine transforms to modified spinel at 16.5 GPa. The transformation from modified spinel to spinel occurs over the range of 17.5 to 20 GPa. Spinel decomposes to Mg–perovskite and magnesiowustite at 23 GPa. Ca–perovskite forms by exsolution from garnet at 20 GPa.

The Martian core is assumed to be a mixture of iron–nickel alloy as was proposed by Morgan and Anders [1979]. The radius of the core can be calculated to be 1350 km from the density distribution of the mantle in this study (Figure 3) and the observed data on the mean density of Mars. The depth of the core–mantle boundary is about 2040 km, which corresponds to about 26 GPa.

Figure 2 shows the model of the Martian interior estimated from the phase transformation behavior of the Martian mantle (Figure. 1). In this Martian model the Martian mantle consists of three layers, the upper mantle (–1280 km depth), the transition zone (1280–1800 km depth), and the lower mantle (1800–2040 km depth). The mineral assemblage of each layer is similar to those of the Earth's mantle, but the thicknesses are quite different. The Martian "upper mantle" and the "transition region" are thicker and the "lower mantle" is thinner than those of the Earth.

Density and Seismic Wave Velocity Profiles of the Martian Interior

A density profile of the Martian interior is illustrated in Figure 3. Densities of minerals are calculated using the Birch–Murnaghan equation of state, and the parameters used are listed in Table 4. The temperature profile in the Martian interior suggested by Johnston and Toksoz [1977] was adopted for the present calculation (see also Fig.3).

The density increase up to 230 km depth is caused by a change in the mineral assemblage, from plagioclase lherzolite through spinel lherzolite to garnet lherzolite. The density gradually increases from 230 km to 1280 km because of compression and dissolution of clinopyroxene into garnet. Between 1280 km and 1800 km there are density jumps corresponding to the phase transitions in olivine. Olivine transforms to modified spinel and further to spinel in this region. Magnesiowustite and Ca–perovskite also appear in the same region. The largest density jump occurs at 1800 km depth because of the decomposition of spinel into Mg–perovskite and magnesiowustite. The core–mantle boundary is located at a depth around 2040 km.

Although seismic data are absent for the Martian interior, the seismic velocity profiles may be calculated using a method proposed by Okal and Anderson [1978]

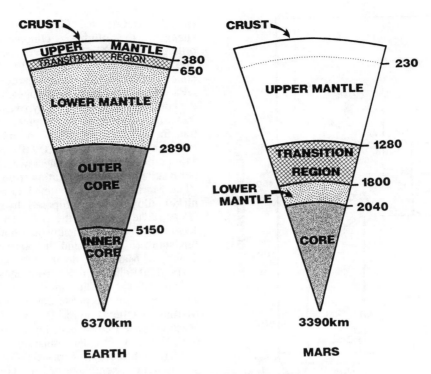

Fig. 2. The internal structures of the Earth and Mars.

for the Martian model constructed in this work. The seismic velocities, Vp and Vs, were calculated from the seismic parameter ϕ and the poisson ratio σ. The seismic parameter $\phi = Vp^2 - 4Vs^2/3$ is calculated by Anderson's [1967] seismic equation of state,

$$\phi = \phi_0 \left(\frac{\rho}{M}\right)^{\frac{1}{E}} \qquad (1)$$

where ρ is the density, M is the mean atomic weight, and ϕ_0 and E are parameters taken from various Earth models. The poisson ratio $\sigma = (Vp^2 - 2Vs^2)/(2(Vp^2 - Vs^2))$ is determined from the Earth model C2 [Anderson and Hart, 1976] for the Martian mantle and data for pure iron for the Martian core. The seismic velocities are given by the following equations:

$$Vp = \sqrt{\frac{3\phi(1-\sigma)}{(\sigma+1)}} \qquad (2)$$

$$Vs = \sqrt{\frac{3\phi(1-2\sigma)}{(2\sigma+2)}} \qquad (3)$$

In this model, the core is assumed to be solid. This assumption is based on the weak magnetic field of Mars, which has been reported to be about 3×10^{-4} of the terrestrial magnetic field [Arvidson et al., 1980].

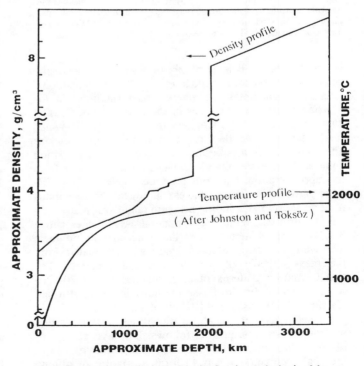

Fig. 3. The density profile in the Martian interior on the basis of the present experimental data.

TABLE 4. Parameters Used for Calculating the Density Profile of Martian Interior

	Olivine	Modified Spinel	Spinel	Orthopyroxene	Clinopyroxene	Garnet
α $(10^{-6}K^{-1})$	26.2①	20.6①	18.6+2.7×X①	27①	27①	18①
$d\alpha/dT(10^{-8}/K^2)$	–	1.7①	–	–	–	–
$\rho_0(g/cm^3)$	3.222+1.182×X②	3.472+1.24×X②	3.548+1.3×X②	3.204+0.799×X②	3.277+0.38×X②	3.562+0.758×X②
$K_{0S}(GPa)$	129.1+8.4×X①	167①	213–16×X①	104②	113+7×X②	(113.47+2.44×X/(0.61–0.005×X)④①
δ_S	4+1.5×X①	3①	3.5①	6①	6①	6.3①
γ_G	1.25①	1.3①	1.35①	1.1④	1.1④	1.1①
K_{0S}'	5.2①	4①	4.8①	5②	4.5②	4.5①
α (0,T)	$(0.3052\times10^{-4}+0.8504\times10^{-8}\times T-0.5824\times T^{-2})\times(1-X)+(0.266\times10^{-4}+0.8736\times10^{-8}\times T-0.2487\times T^{-2})\times X$③	–	$(0.2367\times10^{-4}+0.5298\times10^{-8}\times T+0.5702\times T^{-2})\times(1-X)+(0.2455\times10^{-4}+0.3591\times10^{-8}\times T-0.3703\times T^{-2})$③	$(0.1391\times10^{-4}+2.544\times10^{-8}\times T+0.1282\times T^{-2})\times(1-X)+(0.183\times10^{-4}+1.413\times10^{-8}\times T-0.19\times T^{-2})\times X$③	$(0.1391\times10^{-4}+2.544\times10^{-8}\times T+0.1282\times T^{-2})\times(1-X)+(0.183\times10^{-4}+1.413\times10^{-8}\times T-0.19\times T^{-2})\times X$③	$(0.2338\times10^{-4}+0.5706\times10^{-8}\times T-0.4924\times T^{-2})\times(1-X)+(0.1808\times10^{-4}+1.182\times10^{-8}\times T-0.5442\times T^{-2})\times X$③

	Ca–Perovskite	Mg–Perovskite	Anorthite	Magnesio-wustite	Iron	FeS	Nickel
α $(10^{-6}K^{-1})$	20①	20①	15⑤	3.10+4.5×X①	38⑧	38⑧	38⑧
$d\alpha/dT(10^{-8}/K^2)$	1.6①	1.6①	0.0204⑤⑥	1.9①	2⑧	2⑧	2⑧
$\rho_0(g/cm^3)$	4.13②	4.014+1.07×X②	2.76⑦	3.583+2.28×X②	7.875⑥	5.4⑩	8.91⑥
$K_{0S}(GPa)$	227②	266②	91.2⑤	162.7+17×X①	170⑨	120⑩	170⑩
δ_S	3①	3①	–	3①	–	–	–
γ_G	1.3①	1.3①	0.7⑤	1.5+0.07×X①	–	–	–
K_{0S}'	3.9②	3.9②	–	4.21–0.5×X①	–	–	–
K_{0T}'	–	–	4	–	4	4	4
$dK/dT(10^{-4})$	–	–	–	–	–131⑨	–131⑨	–131⑨

X; atomic ratio of Fe/(Mg+Fe), T; temperature (K)

① Lees et al. (1983)
② Duffy and Anderson (1989)
③ Watanabe (1982)
④ Jeanloz and Thompson (1983)
⑤ Birch (1952)
⑥ Carmichael (1984)
⑦ Birch (1961)
⑧ Skinner (1966)
⑨ Simmons and Wang (1971)
⑩ King and Ahrens (1973)

The calculated seismic velocity profile of the Martian interior is given in Figure 4. Seismic velocities gradually increase in the upper mantle, jump at the top of lower mantle, and then decrease at the core–mantle boundary.

Discussion

Geophysical properties for the present model of the Martian interior have to be compared with the geophysical data of Mars. The geophysical constraints used to make the present model are the total mass and the radius of Mars [Arvidson et al., 1980]. The moment of inertia factor of Mars is the most important property for constraining the model.

Determination of the moment of inertia factor of Mars is complicated due to a significant departure from hydrostatic equilibrium of the Tharsis Plateau. Three

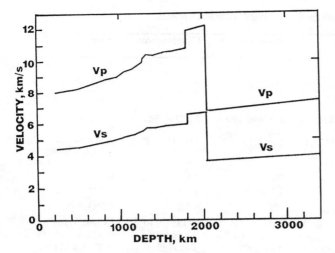

Fig. 4. Seismic velocity profiles in the present Martian model.

values of the moment of inertia factor have been presented to date: 0.376–0.372 [Binder and Davis, 1973], 0.365 [Reasenberg, 1977] and 0.345 [Bills, 1989]. Ringwood [1979] supported the estimation of Binder and Davis [1973]. He argued that the Tharsis Plateau was likely to be partially compensated at a depth of 100 km [Ringwood, 1979] although Reasenberg assumed that it was not compensated. Moreover Ringwood pointed out that the stress for supporting the Tharsis Plateau in Reasenberg's model was too large to explain the long term strength of the Martian mantle. Arvidson et al. [1980] suggested that Mars is not in hydrostatic equilibrium from a comparison between a dynamic flattening expected by a hydrostatic equilibrium and a geometric flattening of Mars. He supported Reasenberg's estimation of the moment of inertia factor from this analysis. Kaula [1979] and Kaula and Asimow [1991] obtained the same value by a different procedure of calculation. They evaluated that Bills' estimate, 0.345, is too small.

A moment of inertia factor of the present Martian model can be calculated using the density profile of the model. The moment of inertia factor is given as C/MR^2, where C is the moment of inertia, M is the mass, and R is the mean radius. The moment of inertia C of a sphere is defined by the following equation:

$$\frac{C}{MR^2} = \frac{8}{3}\pi \int_0^R r^4 D(r) dr \qquad (4)$$

where r is the distance from the center and D(r) is the density in the Martian interior. D(r) cannot be represented by a simple equation, but it is given in Figure 3. Using the density profiles of Mars given in Figure 3, we calculated the moment of inertia factor of our model to be 0.364. This moment of inertia factor is

very close to 0.365 as proposed by Reasenberg [1977] and Kaula and Asimow [1991].

Dreibus and Wanke [1985] suggested that the shergottite parent body, SPB, corresponds to Mars. The SPB mantle has a density greater than that of the present MA mantle because of its higher FeO content, although it has lower CaO and Al_2O_3 contents compared to the MA mantle. The core of SPB which is enriched in sulfur has a density lower than that of MA. Because of its smaller density contrast between the mantle and core, the moment of inertia factor of SPB is calculated to be 0.367 which is greater than that of MA.

CONCLUSIONS

The phase transformations of an Fe–rich mantle model composition of Mars were studied up to 20 GPa. The model of the Martian interior based on these results indicates that the Martian upper mantle and the transition region are thicker and the lower mantle is thinner than those of the Earth.

The density and seismic wave velocity profiles have been obtained on the basis of the phase transformation sequences in the Martian interior. The present model gives a moment of inertia factor of 0.364, which is consistent with 0.365 as proposed by Reasenberg [1977] and Kaula and Asimow [1991].

Acknowledgment. We would like to appreciate Professors K. Aoki of Tohoku University and H. Tanaka of Yamagata University for providing a chance for EPMA analysis. We wish to thank C. B. Agee, C. Herzberg and D. C. Rubie for review and valuable suggestions. This work was supported by a grant–in–aid of Ministry of Education, Science, and Culture of Japanese Government.

REFERENCES

Anderson, D. L., A seismic equation of state, *Geophys.J. Roy. Astron. Soc., 13,* 9–30, 1967.

Anderson, D. L., The internal composition of Mars, *J. Geophys Res., 77,* 789–795, 1972.

Anderson, D. L., and R. S. Hart, An Earth model based on free oscillations and body waves, *J. Geophys. Res., 81,* 1461–1475, 1976.

Arvidson, R. E., K. A. Goettel, and C. M. Hohenberg, A post–Viking view of Martian geologic evolution, *Rev. Geophys. Space Phys., 18,* 565–603, 1980.

Bills, B. G., The moment of inertia of Mars, *Geophys.Res. Lett., 16,* 385, 1989.

Binder, A. B., and D. R. Davis, Internal structure of Mars, *Phys. Earth Plan. Int., 7,* 477–485, 1973.

Birch, F., Elasticity and constitution of the Earth's interior, *J. Geophys.Res., 57,* 227–286, 1952.

Birch, F., The velocity of compressional waves in rocks to 10 kilobars, part 2, *J. Geophys. Res., 66,* 2199–2224, 1961.

Carmichael, R. S., *Handbook of Physical Properties of Rocks, 3,* 340pp., CRC PRESS, 1984.

Dreibus, G., and H. Wanke, Mars, a volatile–rich planet, *Meteoritics, 20,* 367–381, 1985.

Duffy, T. S., and D. L. Anderson, Seismic velocities in mantle minerals and the mineralogy of the upper mantle, *J. Geophys. Res., 94,* 1895–1912, 1989.

Goettel, K. A., Present constraints on the composition of the mantle of Mars, *Year Book Carnegie Inst. Washington, 82,* 363–366, 1983.

Irifune, T., J. Susaki, T. Yagi, and H. Sawamoto, Phase

transformations in diopside CaMgSi$_2$O$_6$ at pressures up to 25GPa, *Geophys. Res. Lett., 16,* 187–190, 1989.

Ito, E., and T. Katsura, A temperature profile of the mantle transition zone, *Geophys. Res. Lett., 16,* 425–428, 1989.

Jeanloz, R., and A. B. Thompson, Phase transitions and Mantle discontinuities, *Rev. Geophys. Space Phys., 21,* 51–74,1983.

Johnston, D. H., and M. N. Toksoz, Internal structure and properties of Mars, *Icarus, 32,* 73–84, 1977.

Kato, T., and M. Kumazawa, Incongruent melting of Mg$_2$SiO$_4$ at 20 GPa, *Phys. Earth Planet. Inter., 41,* 1–5, 1985.

Kaula, W. M., The moment of inertia of Mars, *Geophys. Res. Lett., 6,* 194–196, 1979.

Kaula,W. M.,and P. D. Asimow, Tests of random density models of terrestrial planets, *Geophys. Res. Lett., 18,* 909–912, 1991.

King, D. A., and T. J. Ahrens, Shock compression of iron sulphide and the possible sulphur content of the Earth's core, *Nature Phys. Sci., 243,* 82–84, 1973.

Lees, A.C., M.S.T. Bukowinski, and R. Jeanloz, Reflection properties of phase transition and compositional change models of the 670–km discontinuity, *J. Geophys. Res., 88,* 8145–8159, 1983.

McGetchin, T.R., and J. R. Smyth, The mantle of Mars; some possible geological implications of its high density, *Icarus, 34,* 512–536, 1978.

Morgan, J. W., and E. Anders, Chemical composition of Mars, *Geochim. Cosmochim. Acta, 43,* 1601–1610, 1979.

Ohtani, E., Ultrahigh–pressure melting of a model chondritic mantle and pyrolite compositions,in *High–Pressure Research in Mineral Physics,* edited by M.H. Manghnani and Y. Syono, TERRA PUB/Amer. Geophys. Union, Tokyo/Washington D.C., 87–93, 1987.

Okal, E. A., and D. L. Anderson, Theoretical models for Mars and their seismic properties, *Icarus, 33,* 514–528, 1978.

Patera, E.S., and J.R. Holloway, Experimental determinations of the spinel–garnet boundary in a Martian mantle composition, *Proc. Lunar Planet. Sci.Conf. 14th,* in *J. Geophys. Res., 87,* A31–A36, 1982.

Reasenberg, R. D., The moment of inertia and isostasy of Mars, *J. Geophys.Res., 82,* 369–375, 1977.

Ringwood, A. E., Composition and Origin of the Earth, *Publication No. 1299, Res. School of Earth Sciences,* Australian National Univ., Canberra, 65pp, 1977.

Ringwood, A. E., *Origin of the Earth and Moon,* SpringerVerlag, 295pp., 1979.

Simmons, G., and H. Wang, *Single Crystal Elastic Constants and Calculated Aggregate Properties: a Handbook, 2nd edition,* M. I. T. PRESS, Cambridge, 370 pp., 1971.

Skinner, B. J., Thermal expansion, in *Handbook of Physical Constants, 97,* Geol. Soc. Amer. Mem., 75–96, 1966.

Watanabe, H., Thermochemical properties of synthetic high–pressure compounds relevant to the earth's mantle,in *High–Pressure Research in Geophysics,* edited by S. Akimoto and M. H. Manghnani, CAPJ/REIDEL, Tokyo/Dordrecht • Boston • London, 441–464, 1982.

N. Kamaya, E. Ohtani, T. Kato and K. Onuma, Institute of Mineralogy, Petrology and Economic Geology, Tohoku University, Sendai, Japan.

Compositional Evolution of Venus

WILLIAM M. KAULA

University of California, Los Angeles, CA 90024, U.S.A.

Venus is without water in its outer parts; not only in the atmosphere, but in the upper mantle, to account for the high ratio of correlated gravity to topography. The upper mantle stiffness required by this high ratio leads to a regionalization of magmatism, and thus a lower rate of crustal formation. There is still sufficient heat in places, however, for secondary differentiations of the crust. From several indicators, Venus appears to have retained in its outer parts appreciable carbon dioxide and sulfur. But all hypotheses proposed to date have their difficulties; more needs to be understood about the physics and chemistry of several magma types.

INTRODUCTION

Venus is of particular interest as the planet by far the closest to the Earth in bulk properties — mass, mean density, and distance from the Sun — but evolving quite differently in secondary properties. Two of these secondary properties — the high temperature and perpetual cloud cover of the atmosphere— have led to Venus being poorly sampled chemically. But this meager sampling, plus detailed observations by Pioneer, Venera, and Magellan of properties that are not chemical but influenced by composition, furnish incontrovertible evidence that Venus is quite different from Earth, and thus a basis for fruitful speculation about how Venus evolved to this state.

Considerations relevant to Venus's composition can be divided into four chategories:

- o direct observations of composition;
- o indirect evidences of composition;
- o origin circumstances (mainly planetesimal infall);
- o evolutionary circumstances (mainly mantle convection).

DIRECT OBSERVATIONS OF COMPOSITION

Direct observations of chemistry fall into two categories: sensing of volatiles, remotely and in situ; and sensing of the solid surface, by x-ray fluorescence and gamma ray spectrometry from landers.

Remote Sensing of Volatiles

Observations of Venus from the Earth (surface or satellite) are frustrated by the perpetual cloud barrier, and, with one outstanding exception, have been superseded by *in situ* observations from spacecraft, probes, and landers. This outstanding exception is of a temporal change: upper atmosphere sulfur dioxide, SO_2, from EUV and haze. From these data, there has been inferred a remarkable variability. Both haze and EUV showed peaks in SO_2 in 1959 and 1978. The latter was followed by a marked decline of a factor of ten

Evolution of the Earth and Planets
Geophysical Monograph 74, IUGG Volume 14
Copyright 1993 by the International Union of Geodesy and Geophysics and the American Geophysical Union.

in the years 1978-1984 [Esposito et al., 1988]. Since 1984, the SO_2 has leveled off, showing negligible variation [Na et al., 1990].

The evident explanation for this sporadicity is that Venus is like the Earth in having major outburts of pyroclastic volcanism (such as El Chichon and Pinatubo) on a decade timescale. Models of SO_2 dispersal indicate that the 1978 outburst on Venus was remarkably big by terrestrial standards [Klose et al., 1992]. The alternative of an atmospheric phenomenon has not been modelled, for lack of an adequate physical basis; it is hard to see even an atmosphere as massive as Venus's sustaining such large decade-scale oscillations. In any case, the existence of SO_2 in the upper atmosphere of Venus is important in requiring that Venus has pyroclastic volcanism, and hence volatiles in its upper mantle to drive such volcanism [Prinn & Fegley, 1987].

In Situ Sensing of Volatiles

Both the Pioneer Venus and Venera spacecraft sent probes with mass spectrometers into the atmosphere. These measurements led to some remarkable surprises, particularly in the abundances of inert gases, in which Venus appears to be like the gas-rich carbonaceous chondrites, rather than the gas-poor Earth. Table 1 gives the main results.

The low radiogenic argon [40]Ar indicates that the greater abundance of Venus in primordial argon [36+38]Ar cannot be due to greater efficiency in outgassing. In fact, Venus is less efficient, probably for the simple reason that it has negligible erosion, while that on Earth is sufficient to remove an average of more than 1 km per 100 My [Howell and Murray, 1986]. Hence the high primordial gas abundances in Venus must be attributed to differences in origin circumstances. These differences are discussed in the section thereon below.

X-ray Fluorescence

X-ray fluorescence experiments were carried on three of the Soviet landers. These measurements were applied to samples from cores penetrating about a centimeter deep below the surface [Surkov et al., 1983, 1984]. The results are given in Table 2, with typical Earth mid-ocean ridge (MORB) and alkali basalt values for comparison.

TABLE 1. ATMOSPHERE + CRUST VOLATILE ABUNDANCES
Ratio Venus / Earth, in proportion to planetary mass

Volatile	Pioneer	Venera
^{36}Ar	72	120
^{20}Ne	20	
^{84}Kr	3	70
Xe	<30	
CO_2		1/2
N_2		2
H_2O		10^{-5}
Atmospheric ^{40}Ar		1/4
K/U		1

From *Donahue and Pollack [1983].*

The uncertainties in Table 2 vary appreciably, from about 3-4 percent for SiO_2 and MgO to 0.6 percent or less for K_2O. Clearly above the noise level is the most striking difference of the Venus measurements: the presence of volatiles— the high sulfur abundances and the detection of chlorine. Also striking is the high K_2O abundance at the Venera-13 site. The other differences— lower CaO and SiO_2, higher MgO— are within the range of not-too-scarce Earth rocks. The main uncertainty about the x-ray fluorescence is whether it is representative of the bedrock. Although the erosion is slight, the effects of chemical weathering over several 100 My could penetrate deeper than the cores.

Gamma Ray Spectrometry

Gamma ray spectrometers were carried on seven of the Soviet landers [Surkov et al., 1987]. The results are shown in Figure 1. The range in K_2O, from 0.3 percent at Venera-10 and -14 sites to 4.0 percent at Venera-8 and -13 sites, is well established, as well as the correlation of Th and U abundances therewith. The K/U and K/Th ratios in Figure 1 are about the same as Earth's, indicating a similar density of radiogenic heat sources. The range in abundances over a factor of twenty from *both* x-ray and gamma data is strong evidence that the Venus crust has had secondary differentiation, as discussed below.

TABLE 2. ABUNDANCES OF OXIDES IN THE SURFACE OF VENUS
Weight Percentages

Oxide	Venera 13	Venera 14	Vega 2	MORB	Alkali Basalt
SiO_2	45	49	46	50	45
TiO_2	1.6	1.2	0.2	0.7	2
Al_2O_3	16	18	16	16	13
FeO	9	9	8	8	13
MgO	11	8	12	9	12
CaO	7	10	8	14	10
Na_2O	?	?	?	2	3
K_2O	4	0.2	0.1	0.1	0.7
SO_3	1.6	0.9	4.7	0	0
Cl	0.3	0.4	<0.3		

From *Hess and Head [1990].*

INDIRECT EVIDENCES OF COMPOSITION

I include in this category a variety of observations, some clearly relevant to composition, others arguably so:

o planetary mean density;
o ridge altimetry;
o gravity: altimetry ratio;
o persistence of large craters;
o tectonics of topographic highs;
o intrinsic radar reflectivity;
o sinuous rilles;
o pancake volcanoes.

Planetary Mean Density

Venus has a mean density, reduced to uniform pressure, about 2.5 percent less than the Earth's. Of the various conjectures in the past as to the reason for this difference [Ringwood and Anderson, 1977; Anderson, 1980; Goettel et al., 1981], that of a deeper basalt: eclogite transition seems ruled out by the findings (see below) that Venus has a limited amount of crust: perhaps no more than the Earth. Still viable are the hypotheses that Venus has a higher oxidation level or a lower Fe: Mg ratio than Earth.

Ridge Altimetry

A systematic study was made of Pioneer Venus altimetry of all elevated features except Ishtar Terra by Kaula and Phillips [1981]. In this study, the fall-off with height from a ridge was necessarily referred to the square root of distance, $s^{1/2}$, rather than time, $t^{1/2}$, as is customary in analysis of Earth ocean rises. However, the technique was tested out on Earth, and gave results on the high side for implied rate-of-spreading and heat delivery, as would be expected from the effects of off-rise volcanism. On Venus, the analysis obtained an implied heat delivery of less than 15 percent of the total, compared to 70 percent on Earth [Sclater et al., 1980], primarily because of insufficient ridge length and secondarily because of excessive rate-of-drop-off. Another important difference is that heights of Venus ridges do not constitute anywhere near as tight a distribution about a mean height as do Earth's ocean rises. Hence Venus cannot have a mode of mantle convection that provides a generous supply of material to preexisting rifts, a circumstance that has implications for heat delivery, and thence magmatism, as commented in the last section below.

Gravity: Topography Ratios

Another important finding from the Pioneer Venus project is that the gravity has a systematic correlation with topographic elevation, and has a ratio thereto implying a depth of compensation in excess of 150 km for either significant features [Smrekar and Phillips, 1992] or spherical harmonic coefficients [Kaula, 1990b]. Venus is unique among the terrestrial planets in this property [Kaula, 1992]. It thus is unavoidable that Venus has a much more viscous upper mantle than does Earth, to shove the compensation so deep [Kiefer et al., 1986; Phillips, 1990]. The only plausible hypothesis to date to explain this extraordinary stiffness is that the upper mantle of Venus is much drier than Earth's, because there is no water being recycled from an ocean [Kaula, 1990c]. However, quantification of this model requires an appreciable extrapolation from laboratory experi-

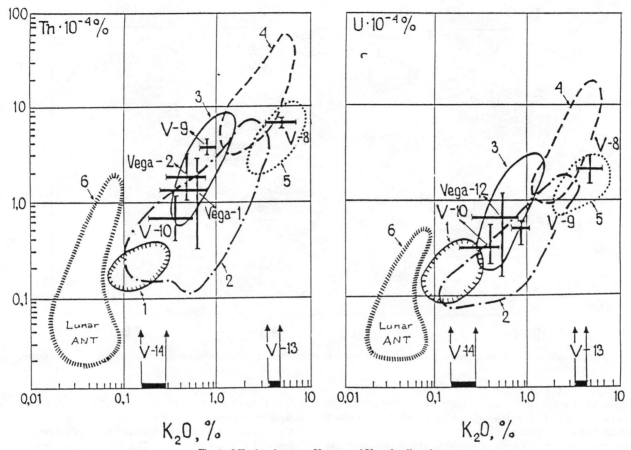

Fig. 1. LIL abundances at Venera and Vega landing sites.

ments [Karato et al., 1986; Karato, 1989]. Also, experiments have not been done on the rheological effects of carbon dioxide comparable to those done for the effects of water.

Persistence of Large Craters

A study by Grimm and Solomon [1988] analyzed depth: diameter ratios of craters with diameters from 30 to 140 km. These ratios were rather high, indicating a slow isostatic adjustment, which in turn sets a minimum on the effective viscosity at shallow depths. The available experimental data on crustal rocks [Shelton & Tullis, 1981; Caristan, 1982] indicate that at Venus temperatures, with plausible temperature gradients, these minima are reached at depths of less than 20 km, which thus becomes an estimate of crustal thickness. This carries appreciable implication for the rates of crustal creation and recycling on Venus, as discussed below. However, experimental difficulties make the available rheological data on crustal rocks dubious; new work is needed. But, in any case, the more extensive data obtained by Magellan have confirmed that the depth: diameter ratios of craters on Venus average more than twice as much as on Earth [Sharpton and Edmunds, 1991], so there is a marked *difference* between the two planets in effective viscosity at depths up to a few tens of kilometers.

Another study by Grimm and Solomon [1987] used craters to limit the rate of volcanism, which covers the craters, to less than 2 km^3/yr. This result has been refined from Magellan imagery to less than 1 km^3/yr [Phillips et al., 1992]. This rate is moderately less than the Earth (some commentaries that get a much higher rate for Earth are based on the erroneous assumption that *all* oceanic crust formation is extrusive; actually it is almost 90 percent intrusive [see e.g., Lister, 1980], like the continents [Carmichael et al., 1974]).

So it appears that Venus has both about the same volume of crust as Earth and a similar rate of volcanism, which still leaves quite uncertain the rate of plutonism.

Tectonics of Topographic Highs

Venus also appears similar to the Earth in that most deformation and large scale volcanism are associated with topographic highs. I concentrate on the five regions that have extensive parts with altitude more than 4 km above mean planetary radius: see Figure 2.

The lowest areas, such as Atalanta Planitia and Sedna Planitia, have a lot of small scale features, created perhaps over more than 1 Ga. But they do not show broad features manifesting major influence from mantle convection, and seem analogous to Earth's lowest features, the ocean basins (even though Venus does not have sea floor

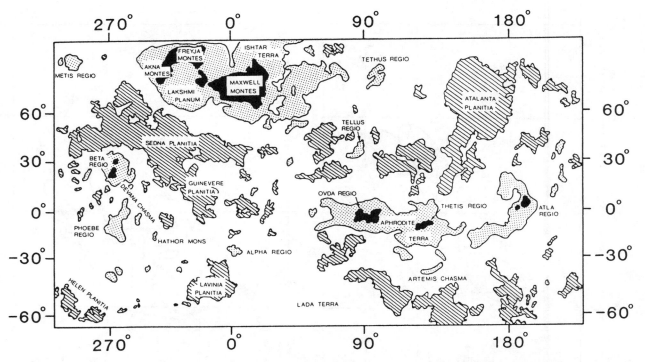

Fig. 2. Altimetry of Venus. Ranges of elevation with respect to mean planetary radius: cross-hatched, <0 km; blank, 0-1 km; stippled, 1-3 km; black, >3 km.

spreading like Earth). But the marked negative gravity anomaly at Atalanta indicates that it may be an incipient down-welling [Bindschadler et al., 1992].

Two of the high regions are characterized by a lot of volcanism and very high gravity anomalies, as well as high topography: Atla Regio and Beta Regio, both of whose geoid highs are 30 percent more than the greatest on Earth: see Figure 3. Their tectonic deformation is mainly rifting: see Figure 4. These features are clearly major uplifts by mantle convection, perhaps plumes from the core boundary [Bindschadler et al., 1992]. Hence their chemistry should reflect either, or both, of chemistry at depth and much magmatic processing— probably different from the water-dependent processes conjectured to be significant in major features on Earth.

Two more of the high regions are strongly deformed: Ovda Regio and Thetis Regio: see Figure 5. They do show some volcanism as well, but it seems auxiliary. However, they do not have strong gravity signals: see Figure 6. In this respect, they are modest compared to most ocean-under-continent subduction zones on Earth, suggesting that there is not the same deep recycling and associated differentiation. Probably a better analogue, if one must be had, is continent-under-continent thrusting, as in Asia. It should be mentioned, however, that the extent to which Ovda Regio and Thetis Regio are convergent features is still debated [Phillips et al., 1991; Bindschadler et al., 1992; Solomon et al., 1992].

The final high region, Ishtar, is quite varied in its characteristics, containing both a clearly convergent major feature, Maxwell Montes, and an extensive area of volcanism, Lakshmi Planum. Ishtar Terra is clearly influenced by both significant mantle variations and

resistant blocks preexisting from the past: "cratons." It also appears to demonstrate that, despite the stiffness of the upper mantle, there can occur marked spatial variations in Venus's flow on the scale of a few 100 km: see Figure 7. The gravity signal over Ishtar is positive— appreciably more than over Ovda— but the high altitude of the Pioneer Spacecraft at its latitude makes it impossible to infer details.

Intrinsic Radar Reflectivity

Most (but not all) high regions on Venus are extraordinarily bright to the radar: see Figure 8, which covers part of the some area as Figure 7, but whose brightness range is less distorted by Maxwell Montes. This highland brightness arises partly from roughness, but analyses show that a major part of it must come from a high dielectric constant: high enough that it is satisfied only by pyrrhotite ($Fe_{0.877}S$) among familiar minerals. But the chemistry that would stabilize pyrrhotite is not understood, so this is a *faute de mieux* hypothesis. Variability of the brightness height from 2.5 to 4.8 km suggests that weathering effects affected by winds are significant in producing conductive secondary mineral assemblages. The principal high feature not markedly conductive is the volcano Maat Mons (see Figures 3 and 4). [Klose et al., 1992].

Thus the brightness of the highlands on Venus are suggestive of appreciable sulfur abundance in Venus's outer parts, but cannot be said to be conclusive as to composition, since the processes to fix so much pyrrhotite— or other compound of high dielectric constant— have not been identified, let alone quantified.

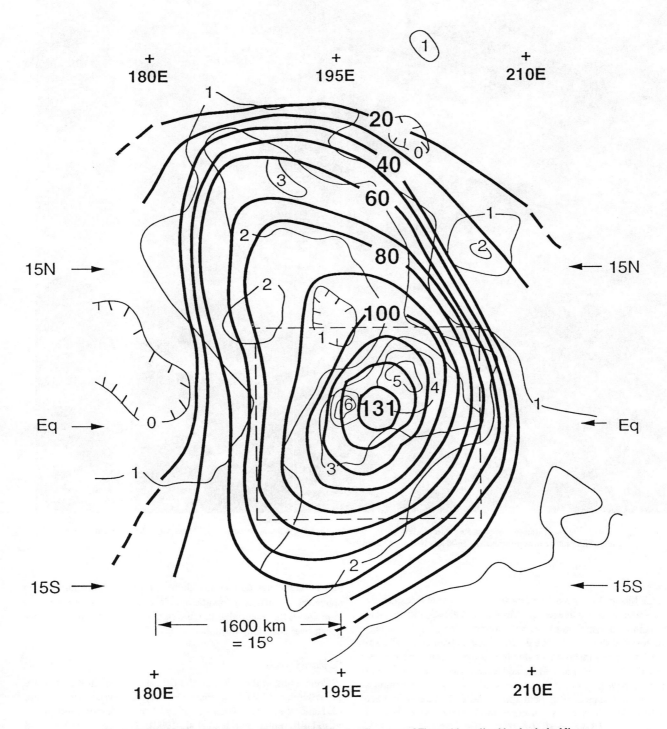

Fig. 3. Geoid (in m) and topography (in km) of Atla Regio. Coverage of Figure 4 is outlined by the dashed line.

Fig. 4. Magellan radar imagery of Atla Regio, including Maat Mons (center left), 6.5 km elevation, and Ozza Mons (upper right), 5 km elevation. The picture is approximately 1600 km N-S x 1850 km E-W.

Sinuous Rilles

Like the Moon, Venus has long, meandering channels despite the lack of water as an erosive agent. More than 200 relic channel and valley landform complexes have been found in the Magellan imagery, such as that shown in Figure 9. One of these channels is 6800 km long. Materials to make the channels that have been explored are ultramafic silicate melts, sulfur, and carbonate lavas. It is not understood how to prevent the freezing of a silicate melt within a much shorter distance. But both sulfur and carbonate lavas have water-like viscosities at temperatures moderately above those of Venus's surface. [Baker et al., 1992].

Again, the sinuous rilles are a suggestive, but not conclusive, evidence of appreciable volatiles in the outer parts of solid Venus. In

this instance, the doubts are about the physical mechanisms to concentrate a sufficient abundance of low melting-point material and to maintain the material within the right temperature range to erode basalt over such long distances.

Pancake Domes

Contrasting to the rilles are the pancake domes, in that they are quite round, and look to be created by very viscous lavas: see Figure 10. Hence they are evidence of highly siliceous material, and thus of crustal differentiation [Head et al., 1992]. The site of the Venera-8 landing, at which high LIL's were measured (Figure 1), is inferred to be in a pancake dome area [Basilevsky et al., 1991].

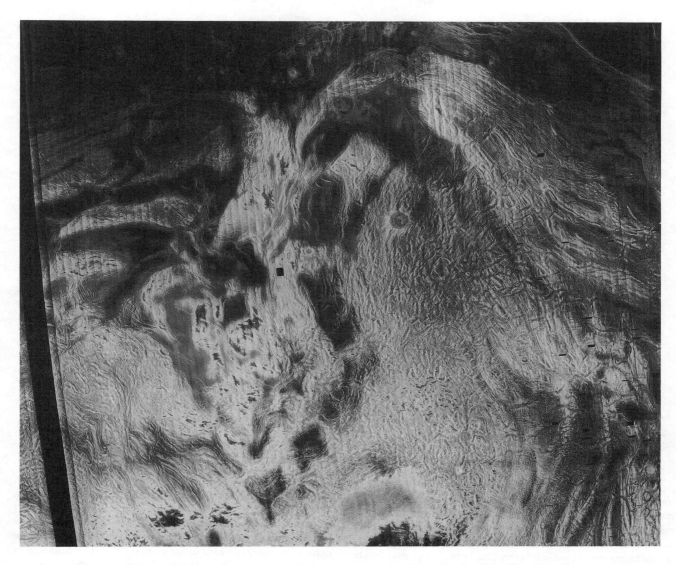

Fig. 5. Magellan radar imagery of Ovda Regio. The drop from the SW corner to the NE corner is about 4 km. The picture is approximately 1600 km N-S x 1850 km E-W.

ORIGIN CIRCUMSTANCES

It is now concurred that the terrestrial planets were formed by infall of planetesimals that were rather big in the terminal phases [Wetherill, 1990]. Such a model appears necessary to explain the extraordinary differences in primary properties among the terrestrial planets.

But only two of the marked differences in secondary properties between Venus and Earth must be attributed to the formation phase of their evolutions: rotation and primordial inert gases (Venus's lack of a moon could be due to its rotation: if there were any satellite, it would spin in from tidal friction). These differences are so big as to rule out anything but a catastrophic difference between the two planets in origin circumstances. Most obvious, the largest impact in

Venus was much smaller than the largest impact in Earth, as first hypothesized by Cameron [1983]. Monte Carlo models of planetary formation from planetesimals indicate that marked differences in rotation and energy deposition between the two planets, while of considerably less than fifty percent probability, can plausibly occur [Kaula, 1990a].

A great difference in size between the largest impact into the Earth and the largest into Venus is quite plausible physically. Indeed, given formation of planets from planetesimals, the burden is on those who argue to the contrary to hypothesize a way of preventing it. But the consequences of great impacts and the subsequent magma ocean are still very imperfectly understood. Intuitive notions that the great

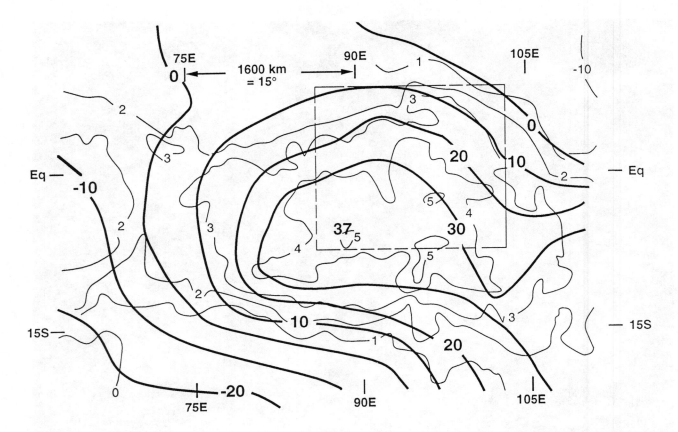

Fig. 6. Geoid (in m) and topography (in km) of Ovda Regio. Coverage of Figure 5 is outlined by the dashed line.

impact sorts out abundances according to volatility, and that any magma ocean would lead to stratification of the mantle, could be mistaken. At present, the main resort is to great computer modellings of impacts [Benz et al., 1989; Cameron and Benz, 1991], while study of a turbulent magma ocean is barely underway [Tonks and Melosh, 1990].

But these complications may be avoidable for the one matter of concern in this paper: the implication of the great retention of inert gases by Venus (Table 1 above) for the retention of other volatiles. Gaining a lot of [36+38]Ar implies gaining a lot of H_2O and CO_2—much more of the latter than now in Venus's atmosphere. One might conjecture that ionizing effects on water and carbon dioxide consequent on a great impact would lead to their loss while the inert gases are retained. But it is the Earth that had the greater impact, so this effect should there work even more to retain inert gases in preference to active.

Hence the problem is how to dispose of a lot of H_2O and CO_2, as implied by the inert gas retentions. The solar fluxes that have prevailed through nearly all solar system history dispose of only a modest amount of H_2O, and none of the CO_2. Hence we must find other discriminants between the active and inert gases. Two have

been suggested that are applicable to H_2O, but none applicable to CO_2.

A discriminant proposed by Kasting et al [1983, 1984] is that the Sun was indeed much more active than it is now—enough to heat the upper atmosphere of Venus so that when H_2O was photodissociated, the hydrogen was blasted off hydrodynamically, rather than being lost by subsequent escape processes that are too slow to be effective. Here the essential discriminant is that H_2O is subject to photo-dissociation, while the inert gases are not. The unknown in this hypothesis is the level of solar activity.

A second discriminant proposed by Zhang and Zindler [1988] is that H_2O was retained in Venus's mantle because it is much more soluble in magmas: see Table 3.

Differences in solubility have been succesful in explaining inert gases on Earth [Zhang and Zindler, 1989]. Outgassing apparently depends on a lot more than merely bringing the volatiles within a few tens of kilometers of the surface, as has been assumed in some geophysical studies. Thus, to exsolve water from a melt, it must be brought within 1.3 km of the surface. The hypothesis is quite plausible; its main unknown is how much a big planetesimal is outgassed upon impact, and how much is carried with its solid

Fig. 7. Magellan radar imagery of south central Ishtar Terra. The drop from the crest of Maxwell Montes, 64 N, 5 E, to the radar dark area at 61 N, 2 E, a distance of 360 km, is about 9 km. The scarp running E-W at 62 N between 345 E and 5 E has drops of 3 km in 60 km, despite its smooth appearance. The dark area in the NW is Lakshmi Planum, including the caldera Sacajewea at 59 N, 337 E. The bright area in the west is Danu Montes; the rough terrain SE thereof is Clotho Tessera. The picture is approximately 1600 km N-S x 1900 km E-W.

constituents into the planetary interior. Understanding retention versus outgassing on impact is still at the experimental stage with small bodies [Ahrens, 1991]. The hypothesis also requires that the retained water be deep in Venus's mantle, because of the constraint that the upper mantle be highly viscous [Kaula, 1990b]. The combination of a dry upper mantle and wet lower mantle in Venus would have major implications for the nature of its convection.

Both the foregoing hypotheses imply extraordinary abundances of carbon dioxide and sulfur in Venus's mantle. What they would do to its rheology is unknown. At low pressures, the petrological effects of CO_2— in the absence of H_2O— are mainly to raise the melting temperature and to decrease the proportion of SiO_2 in the melt [Hess and Head, 1990].

EVOLUTIONARY CIRCUMSTANCES

Mantle convection must exist on Venus; otherwise, it would be impossible to support the highlands (Figure 2) with such deep apparent depths of compensation. Mantle convection thus dominates the long term thermal and compositional evolution of Venus, as it does the Earth's. But the size and extent of the highlands on Venus suggest that mantle convection is not as vigorous as on Earth; perhaps only one-third as much. But even so convection on Venus

Fig. 8. Magellan radar imagery of central Ishtar Terra. The rough area on the east not appearing in Figure 7 is Fortuna Tessera.

is probably similar to that on Earth in that it gets quite close to the surface: within kilometers. Any systematically layered convective system with a uniform thermal boundary layer on top would have excessive interior temperatures because of the steep temperature gradients associated with conductive transfer across the top layer and any internal boundary layers. A real heterogeneous planet evolves to a more complex system that is more efficient. That on the Earth entails extraordinary lateral variations in heat transfer [Kaula, 1983]. Venus's convective system very likely also has lateral variability, but of a different character from Earth's.

Mantle convection in Venus differs from that on Earth most obviously because of the lower water content. Lower stress levels may also be significant. The effective viscosity— i.e., stress: strain rate ratio— has a strongly nonlinear inverse dependence on stress. Thus lower stress levels make a material appear stiffer, although in

Venus this effect is probably not as important as the dryness.

Convection in the Earth can be described as "concentrated": i.e., a major portion of the material and heat coming to the surface— 70 to 90 percent— is through the ocean rises and sea floor spreading. It has long been recognized that this process requires a driver, which is generally thought to be the subduction zones: the plunging plates pull the oceanic lithosphere behind them, as well as strongly influencing the pattern of mantle flow. But perhaps even more important to the unique style of convection in the Earth is the asthenosphere: secondarily as a means of allowing the oceanic lithosphere to move over the interior, but primarily as a means to move a tremendous amount of material and heat to the rises, which must draw on a zone extending thousands of kilometers to the side. This process is sometimes called "passive upwelling"; a more heuristic term is "sink-driven."

But without a thick asthenosphere, there cannot be large scale

Fig. 9. Magellan radar imagery of northern Itzapapalotl Tessera, including a sinous rille most prominent at 76.8 N, 337 E, but extending at least 150 km SE and 100 km SWS, then W, where it blends into the dark plain.

lateral transfer in the upper mantle. Flows will be more regional, and the delivery of heat and material to the crust and surface more scattered; a "distributed" convection. The evident idea is that plumes would be relatively more important. Plumes are certainly the mode suggested by fluid dynamical models. But fluid dynamical models are based on materials much simpler than real materials, and an actual planet may have a more complex "multi-cell" system, as sketched in Figure 11. This is likely to be true for Venus, which does not have an energy source from the core, and may not have significant downthrusts similar to the subducted slabs on Earth, as suggested by the mild gravity signal at Ovda Regio (Figure 6). Hence it may be driven by sources distributed throughout the mantle. The small size of most coronae — two thirds are less than 300 km in diamter [Stofan et al., 1992] — supports this idea of distributed sources.

The upper mantle stiffness implied by the gravity: topography ratio, together with the variations in surface character within a few hundred kilometers shown by Magellan imagery (Figure 7), argue strongly for "distributed" convection; more so than the theoretical conjectures in the preceding paragraph. But, germane to the subject of this paper, most important is the implication for magmatism of such a flow system. Because it does not concentrate heat like the Earth's flow system, magmatism will be much more in the nature of many low percent partial melts, rather than a few large percentage melts, and have a lower global volume rate. Also lowering the amount of magmatism is the higher melting temperature in the absence of H_2O [Hess and Head, 1990]. The most significant consequence of less magmatism would be a lower rate of crustal formation. A lower rate would explain the apparently low volume of crust, thus alleviating

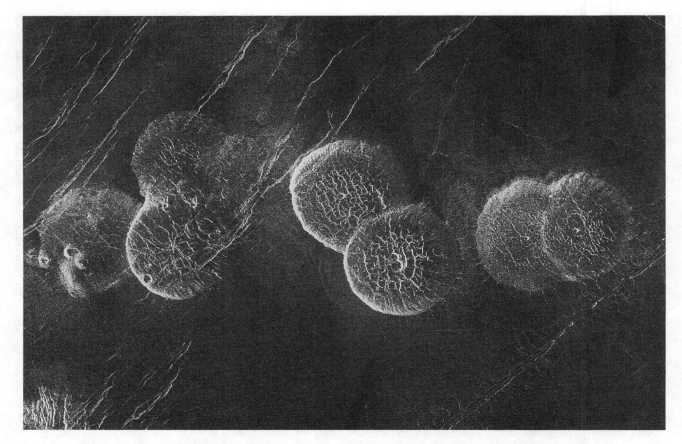

Fig. 10. Magellan radar imagery of seven pancake domes, averaging 25 km in diameter and 0.75 km in height, at 30 S, 12 E.

the need to find mechanisms to recycle crust, emphasized by *Kaula* [1990b].

CONCLUSIONS

The inert gas abundances appear to require a much higher abundance of CO_2 than apparent in the atmosphere. They also entail a high abundance of H_2O. This water cannot be in the upper mantle; whether it is in the lower mantle or early lost to space through vigorous solar activity is unsure. There are several secondary evidences of abundant other volatiles in the upper mantle of Venus, all of them suggestive but far from conclusive, because essential

Table 3: Solubilities of Gases in Magmas

Ratio of Solubilities S	Measured value
$S(H_2O) / S(CO_2)$	60
$S(H_2O) / S(N_2)$	310
$S(H_2O) / S(Ar)$	500

From *Zhang and Zindler [1988]*.

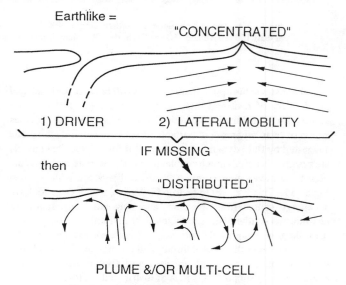

Fig. 11. A sketch of styles of mantle convection.

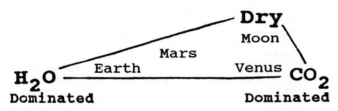

Fig. 12. Volatile control of petrological character.

processes are not yet defined. These include the long sinuous rilles, the low SiO_2 and high SO_3 from x-ray fluorescence, the high radar reflectivity of the highlands, and the sporadicity of upper altitude SO_2 abundance. The Venera 8 and 13 landing sites, as well as the pancake domes, indicate that there has been secondary differentiation in the crust of Venus. In the absence of water, this differentiation is quite different from the Earth's; probably more like the early differentiations on the moon creating norite, but not necessarily so, in view of Venus having volatiles other than H_2O. Ternary diagrams have become fashionable in comparing planets; Figure 12 is an offer of such, which we make isosceles to suggest that the absence of H_2O is more important to the petrology than is the presence of CO_2.

But most important is that the regional nature of mantle convection imposed by the stiffness of the upper mantle and the higher melting temperatures due to lack of water have led to a much lower rate of crustal formation than on Earth; there is lacking tremendous concentration of heat as at ocean rises.

Acknowledgments. This paper has benefited from comments by Duane L. Bindschadler, Roger J. Phillips, and John A. Wood. This research is supported by NASA grant NAGW-2085.

REFERENCES

Ahrens, T. J., Bounds on the Earth's volatile content from accretion, In *Chemical Evolution of the Earth and Planets*, E. Takahashi, ed., American Geophysical Union, Washington, submitted, 1992.

Anderson, D. L., Tectonics and composition of Venus, *Geophys. Res. Let.* 7, 101-102.

Baker, V. R., G. Komatsu, T. J. Parker, V. C. Gulick, J. S. Kargel, and J. S. Lewis, Channels and valleys on Venus: preliminary analysis of Magellan data. *J. Geophys. Res.* 97, submitted, 1992.

Basilevsky, A. T., M. A. Ivanov, and O. V. Nikolayeva, Venera-8 landing site: preliminary analysis of Magellan imagery, *Lun. Plan. Sci. Conf.* 22, 57-58.

Bindschadler, D. L., G. Schubert, and W. M. Kaula, Coldspots and hotspots: global tectonics and mantle dynamics of Venus, *J. Geophys. Res.* 97, submitted, 1992.

Cameron, A. G. W., Origin of the atmospheres of the terrestrial planets, *Icarus* 56, 195-201, 1983.

Cameron, A. G. W. and W. Benz, The origin of the moon and the single impact hypothesis IV, *Icarus* 92, 204-216, 1991.

Caristan, Y., The transition from high temperature creep to fracture in Maryland diabase, *J. Geophys. Res.* 87, 6781-6790, 1982.

Carmichael, I. S. E., F. J. Turner, and J. Verhoogen, *Igneous Petrology*, McGraw-Hill, New York, 739 pp., 1974.

Donahue, T. M. and J. B. Pollack, Origin and evolution of the atmosphere of Venus, In *Venus*, D. M. Hunten, L. Colin, T. M. Donahue, and V. I. Moroz, eds., Univ. Arizona Press, 1003-1036, 1983.

Esposito, L. W., M. Copley, R. Eckert, L. Gates, A. I. F. Stewart, and H. Worden, Sulfur dioxide at the Venus cloud tops, 1978-1986, *J. Geophys. Res.* 93, 5267-5276, 1988.

Fegley, B. and R. G. Prinn, Estimation of the rate of volcanism on Venus from reaction rate measurements, *Nature* 337, 55-58, 1989.

Goettel, K. A., J. A. Shields, and D. A. Decker, Density constraints on the composition of Venus, *Proc. Lunar Planet. Sci. Conf.* 12B, 1507-1516, 1981.

Grimm, R. E. and S. C. Solomon, Limits on modes of lithospheric heat transport on Venus from impact crater density, *Geophys. Res. Let.* 14, 538-541, 1987.

Grimm, R. E. and S. C. Solomon, Viscous relaxation of impact crater relief on Venus: constraints on crustal thickness and thermal gradient, *J. Geophys. Res.* 94, 11,911-11,929, 1988.

Head, J. W., and 7 others, Venus volcanism: initial analysis from Magellan data, *Science* 252, 276-288, 1991.

Head, J. W. and 6 others, Volcanism: pancake domes and other flows of apparent high viscosity, *J. Geophys. Res.* 97, submitted, 1992.

Hess, P. C. and J. W. Head, Derivation of primary magmas and melting of crustal materials on Venus: some preliminary petrogenetic considerations, *Earth, Moon, and Planets* 50/51, 57-80, 1990.

Howell, D. G. and R. W. Murray, A budget for continental growth and denudation. *Science* 233, 446-449, 1986.

Karato, S., Defects and plastic deformation in olivine. In *Rheology of Solids and of the Earth*, S. Karato and M. Toriumi, eds., Oxford, 176-208, 1989.

Karato, S., M. S. Paterson, and J. D. Fitzgerald, Rheology of synthetic olivine aggregates: influence of grain size and water, *J. Geophys. Res.* 91, 8151-8176, 1986.

Kasting, J. F. and J. B. Pollack, Loss of water from Venus. I Hydrodynamic escape of hydrogen, *Icarus* 53, 479-508, 1983.

Kasting, J. F., J. B. Pollack, and T. P. Ackerman, Response of the Earth's atmosphere to increases in solar flux and implications for loss of water from Venus, *Icarus* 57, 335-355, 1984.

Kaula, W. M., Minimal upper mantle temperature variations consistent with observed heat flow and plate velocities, *J. Geophys. Res.* 88, 10,323-10,332, 1983.

Kaula, W. M., Differences between the Earth and Venus arising from origin by large planetesimal infall, In *Origin of the Earth*, H. Newsome and J. Jones, eds., Oxford, 45-57, 1990a.

Kaula, W. M., Venus: a contrast in evolution to Earth, *Science* 247, 1191-1196, 1990b.

Kaula, W. M., Mantle convection and crustal evolution on Venus, *Geophys. Res. Lett.* 17, 1401-1403, 1990c.

Kaula, W. M., Properties of the gravity fields of terrestrial planets, In *Gravity Field Determination from Space and Airborne Measurements*, O. L. Colombo, ed., Springer-Verlag, accepted, 1992.

Kaula, W. M. and R. J. Phillips, Quantitative tests for plate tectonics on Venus, *Geophys. Res. Let.* 8, 1187-1190, 1981.

Kiefer, W. S., M. A. Richards, B. H. Hager, and B. G. Bills, A dynamical model of Venus's gravity field, *Geophys. Res. Let.* 13, 14-17, 1986.

Klose, K. B., J. A. Wood, and A. Hashimoto, Mineral equilibria and the high radar reflectivity of Venus mountain-tops, *J. Geophys. Res.* 97, submitted, 1992.

Lister, C. R. B., Qualitative models of spreading-center processes, including hydrothermal penetration, In *Oceanic Ridges and Arcs: Geodynamical Processes*, M. N. Toksoz, S. Uyeda, and J. Francheteau, eds., Elsevier, 143-158, 1980.

Na, C. Y., L. W. Esposito, and T. E. Skinner, International ultraviolet Explorer observation of Venus SO_2 and SO, *J. Geophys. Res.* 95, 7485-7491, 1990.

Phillips, R. J., Convection-driven tectonics on Venus, *J. Geophys. Res.* 95, 1301-1316, 1990.

Phillips, R. J., R. E. Grimm, and M. C. Malin, Hot-spot evolution and the global tectonics of Venus, *Science* 252, 651-658, 1990.

Phillips, R. J., and 6 others, Impact crater distribution on Venus: implications for planetary resurfacing history, *J. Geophys. Res.* 97, submitted, 1992.

Prinn, R. G., and B. E. Fegley, Jr., The atmospheres of Venus, Earth, and Mars: a critical comparison, *Ann. Rev. Earth Planet. Sci.* 15, 171-212, 1987.

Ringwood, A. E. and D. L. Anderson, Earth and Venus: a comparative study, *Icarus* 30, 243-253, 1977.

Sclater, J. G., C. Jaupart, and D. Galston, The heat flow through oceanic and continental crust and the heat loss of the Earth, *Rev. Geophys. Space Phys.* 18, 269-311, 1980.

Sharpton, V. I. and M. S. Edmunds, Depth-diameter data for large impact structures on Venus: implications for crater modification, *EOS Trans. AGU* 72 (44) *Suppl.* 289, 1991.

Shelton, G. and J. Tullis, Experimental flow laws for crustal rocks, *EOS Trans. AGU* 62, 396, 1981.

Smrekar, S. E. and R. J. Phillips, Venusian highlands: geoid to topography ratios and their implications, *Earth Planet Sci. Let.*, accepted, 1992.

Solomon, S. C., and 10 others, Venus tectonics: an overview of Magellan observations, *J. Geophys. Res.* 97, submitted, 1992.

Stofan, E. R. and 6 others, Global distribution and characteristics of coronae and related features on Venus: implications for origin and relation to mantle processes, *J. Geophys. Res.* 97, submitted, 1992.

Surkov, Y. A., and 5 others, Determination of elemental composition of rocks on Venus by Venera 13 and 14, *J. Geophys. Res.* 88, A481-494, 1983.

Surkov, Y. A., and 4 others, New data on the composition, structure, and properties of Venus rock obtained by Venera 13 and Venera 14, *J. Geophys. Res.* 89, B393-402, 1984.

Surkov, Y. A., and 5 others, Uranium, Thorium, and Potassium in the Venusian rocks at the landing sites of Vega 1 and 2, *J. Geophys. Res.* 92, E537-540, 1987.

Tonks, H. B. and H. J. Melosh, The physics of crystal settling and suspension in a turbulent magma ocean, In *Origin of the Earth*, H. E. Newsom and J. H. Jones, eds., Oxford, 151-174, 1990.

Wetherill, G. W., Formation of the Earth, *Ann. Rev. Earth Plan. Sci.* 18, 205-256, 1990.

Zhang, Y. and A. Zindler, Did Venus ever have an equivalent surface water mass of the terrestrial oceans? *EOS Trans. AGU* 69, 1294, 1988.

Zhang, Y. and A. Zindler, Noble gas constraints on the evolution of the Earth's atmosphere, *J. Geophys. Res.* 94, 13,719-13,737, 1989.

William M. Kaula, University of California, Los Angeles, CA 90024

Thermal Evolution and Chemical Differentiation of the Terrestrial Magma Ocean

YUTAKA ABE[1]

Water Research Institute, Nagoya University
Nagoya, Japan

Release of gravitational energy during the formation stage of the Earth results in global melting and formation of a magma ocean. It is believed that melt-solid separation in the magma ocean causes chemical differentiation of the proto-mantle. However, near-chondritic relative abundances of refractory lithophile elements in the upper mantle may suggest an undifferentiated mantle.

Chemical differentiation of the magma ocean is complicated by various processes, such as impact stirring, convective mixing, cooling and solidification. To clarify the chemical evolution of the proto-mantle I examine physical processes in the magma ocean by using a one-dimensional two phase flow model. I also calculate trace-element transport in the magma ocean by taking into account the partitioning of elements between solid and melt phases. It is shown that melt-solid separation in the magma ocean is seriously affected by vigorous convection. Melt-solid separation is suppressed by very efficient convective mixing at high melt fractions. However, it proceeds rapidly at intermediate melt fractions (20 ~ 30%), at which large viscosity prevents efficient convective mixing and high permeability results in efficient separation of melt and solid. Hence, chemical differentiation of the proto-mantle should proceed at intermediate melt fractions, if solid and melt phases have different density and composition. Although melt-solid separation changes the abundance ratio between compatible and incompatible elements, both the abundance ratio among compatible elements and that among incompatible elements are kept unchanged. This result suggests that apparent chondritic relative abundances of refractory lithophile elements in the upper mantle does not necessarily rule out differentiation of the proto-mantle.

INTRODUCTION

Release of gravitational energy is expected to cause global melting and formation of a magma ocean during accretion of the Earth [e.g., Safronov, 1978; Hayashi et al., 1979; Kaula, 1979; Coradini et al., 1983; Davies, 1985; Abe and Matsui, 1985, 1986; Matsui and Abe; 1986a, 1986b; Sasaki and Nakazawa, 1986; Zahnle et al., 1988; Melosh, 1990]. The blanketing effect of the proto-atmosphere results in melting of the surface of the growing Earth [Hayashi et al., 1979; Abe and Matsui, 1985, 1986; Matsui and Abe; 1986a, 1986b; Zahnle et al., 1988]. A surface magma ocean is efficiently formed by this mechanism when the size of the Earth-forming planetesimals are small [Abe, 1988; Rintoul and Stevenson, 1988]. Deposition of heat by planetesimal impacts also heats up the subsurface layers of the proto-

Earth and results in a subsurface magma ocean [Safronov, 1978; Kaula, 1979; Coradini et al., 1983; Davies, 1985]. High efficiency of heat burial is expected when the Earth-forming planetesimals are relatively large. In particular, if a giant impact occurs, a completely molten deep magma ocean should be produced [Melosh, 1990]. Formation of the magma ocean triggers iron-silicate separation and formation of the core. Then core formation proceeds synchronously with accretion of the Earth, and a deep partially molten magma ocean is sustained by gravitational energy released by core formation [Sasaki and Nakazawa, 1986]. Thus, from a theoretical point of view, it seems difficult to accrete the Earth without making any kind of magma ocean.

One should note, however, that the initial state of a magma ocean depends on its cause. The magma ocean formed by giant impacts should be very deep and hot; the initial temperature distribution should be adiabatic and above the liquidus at least in the upper part. To the contrary, proto-atmosphere blanketing and core formation result in many cases in only a partially molten magma ocean.

It was believed that chemical differentiation should have proceeded in the mantle owing to gravitational separation of solid and melt, if the Earth had experienced extensive

---

[1] Now at Department of Earth and Planetary Physics, Faculty of Science, University of Tokyo, Tokyo, JAPAN.

Evolution of the Earth and Planets
Geophysical Monograph 74, IUGG Volume 14

melting. Recently Kato et al. [1988] argued that the hypothesis of a deep magma ocean which extends to the lower mantle conflicts with observed trace element abundances in the upper mantle. They experimentally determined element partitioning between silicate perovskite, garnet and melt. According to their results, separation of Mg-perovskite from melt causes variations of Lu/Hf, Sc/Sm and Hf/Sm ratio in the residual melt. Hence, if the initial composition of the mantle was nearly chondritic, melt-solid separation in the deep magma ocean should cause the relative abundances of these elements to deviate from chondritic values. To the contrary, the present upper mantle posses near-chondritic relative abundances of Ca, Al, Sc, Yb, Sm, Zr and Hf, suggesting either no melt-solid separation or complete rehomogenization of initial layering in the deep mantle. Chemical layering might be destroyed by subsequent mantle convection to some extent. However, it would be difficult to rehomogenize the initial layering completely, because chemical heterogeneity of small scale would likely not be completely homogenized by solid state convection. Thus, they concluded that the mantle did not experience extensive melting during the formation of the Earth.

However, the intuitive idea of melt-solid separation in the magma ocean is not obviously correct. Melt-solid separation processes in the magma ocean are very complicated because of the following factors: (1.) Mode of melt-solid separation depends on melt fraction: it is controlled by crystal settling or flotation in the melt at high melt fraction. On the other hand, it is controlled by permeable flow of melt at low melt fraction. (2.) Turbulent convection expected in the magma ocean disturbs or prevents the melt-solid separation. Recently, Tonks and Melosh [1990] showed that convective mixing suppresses the melt-solid separation in some cases. A similar conclusion was also obtained by Miller et al. [1991b]. (3.) Magma ocean solidifies at the last stage of accretion. If solidification proceeds very rapidly, the magma ocean might be "quenched" in an undifferentiated state. (4.) Growth of the Earth results in (a) addition of undifferentiated matter to the magma ocean, (b) secular increase of pressure caused by increase of the Earth's mass, and (c) stirring by planetesimal impacts.

The purpose of this study is to examine various physical processes which affect the thermal and chemical evolution of the magma ocean by using a simple one-dimensional numerical model. At first, we examine the elementary processes which affect the melt-solid separation in a partially molten material, namely, melt-solid separation rate in an undisturbed fluid layer, convective mixing and impact stirring. Then, we examine the thermal evolution of the magma ocean by taking into account heat transport by convection and melt-solid separation. Finally, we examine the chemical evolution of the magma ocean based on trace element transport and partitioning between solid and melt phases.

ELEMENTARY PROCESSES OF MELT-SOLID SEPARATION

At first we consider elementary processes of melt-solid separation in the magma ocean. We neglect melting/solidifica-

tion for the first step. Namely, we treat melt-solid separation as if treating sedimentation of sand particles in water. Based on this assumption, we can discuss melt-solid separation processes separately from heat transport processes. Coupled processes of melt-solid separation and heat transport will be discussed in the next section.

Under the assumption of no melting/solidification, melt-solid separation is treated merely as a mass-transfer process in a two-phase mixture. Let us consider mixture of solid and melt phases, with solid-phase density ρ_s and melt-phase density ρ_m. The average density of the mixture ρ is given by

$$\frac{1}{\rho} = \frac{1}{\rho_s}(1 - \phi) + \frac{1}{\rho_m}\phi \qquad (1)$$

where ϕ is the melt fraction (mass fraction of melt). The masses of solid and melt phases per unit volume of mixture (or spatial density of solid and melt phase) ρ_s^* and ρ_m^*, respectively, are given by

$$\rho_s^* \equiv (1 - \phi)\rho = \frac{\rho_s \rho_m (1 - \phi)}{\rho_s \phi + \rho_m (1 - \phi)},$$
$$\rho_m^* \equiv \phi\rho = \frac{\rho_s \rho_m \phi}{\rho_s \phi + \rho_m (1 - \phi)} \qquad (2)$$

The mass of each phase is conserved, because we assumed no melting/solidification to occur. Mass conservation of each phase in a one dimensional system is given by

$$\frac{\partial \rho_m^*}{\partial t} + v_m \frac{\partial \rho_m^*}{\partial z} = -\rho_m^* \frac{\partial v_m}{\partial z} \qquad (3)$$
$$\frac{\partial \rho_s^*}{\partial t} + v_s \frac{\partial \rho_s^*}{\partial z} = -\rho_s^* \frac{\partial v_s}{\partial z} \qquad (4)$$

where v_m and v_s are velocities of melt and solid phases averaged over a small domain in the mixture (Figure 1). z

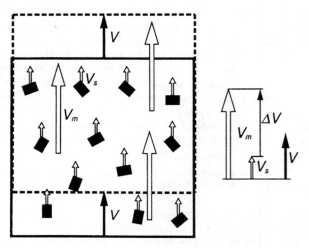

Fig. 1.   Relationship among the velocities of local barycenter, v, melt phase, v_m, and solid phase, v_s.

is the vertical coordinate (z increases in upward direction). We define the velocity of the local barycenter of the mixture, v, and vertical mass flux of melt, J_m, relative to the local barycenter as follows

$$v \equiv \frac{\rho_m^*}{\rho_m^* + \rho_s^*} v_m + \frac{\rho_s^*}{\rho_m^* + \rho_s^*} v_s = \phi v_m + (1 - \phi) v_s \quad (5)$$

$$J_m \equiv \rho_m^*(v_m - v) = \rho \phi (1 - \phi)(v_m - v_s) \quad (6)$$

J_m characterizes the melt-solid separation processes. It will be evaluated for various cases in the following sections.

By using v and J_m, v_m and v_s is written as follows:

$$v_m = v + \frac{J_m}{\rho \phi}, \quad v_s = v - \frac{J_m}{\rho(1 - \phi)} \quad (7)$$

Substituting (7) into (3) and (4), we obtain,

$$\frac{\partial \rho_m^*}{\partial t} + v \frac{\partial \rho_m^*}{\partial z} = -\rho_m^* \frac{\partial v}{\partial z} - \frac{\partial J_m}{\partial z} \quad (8)$$

$$\frac{\partial \rho_s^*}{\partial t} + v \frac{\partial \rho_s^*}{\partial z} = -\rho_s^* \frac{\partial v}{\partial z} + \frac{\partial J_m}{\partial z} \quad (9)$$

For simplicity, we assume that both the solid and melt phases are incompressible: namely, ρ_m and ρ_s are constant. Then (8) and (9) are transformed into the following.

$$\frac{\partial \phi}{\partial t} + v \frac{\partial \phi}{\partial z} = -\frac{1}{\rho} \frac{\partial J_m}{\partial z} = \frac{\rho_s \rho_m}{\rho(\rho_s - \rho_m)} \frac{\partial v}{\partial z} \quad (10)$$

This is the governing equation of the melt-solid separation for a one dimensional, incompressible system without melting or solidification.

We define the characteristic time scale, τ, of melt-solid separation as the time scale required for a fractional change in melt or solid. Consider a partially molten layer of thickness L with uniform initial melt fraction ϕ_0. We define the characteristic time scale of separation as the shorter one of the e-folding time of melt fraction or solid fraction change in the lower half of the layer. Then, τ is given by

$$\tau \equiv \frac{\rho L}{2 J_m} \text{MIN}[\phi_0, 1 - \phi_0] \quad (11)$$

where J_m value is evaluated at the center of the layer.

Gravitational Separation

When the melt fraction is large, the relative motion of solid and melt is expected to be controlled by crystal settling or flotation. Then the separation would be controlled by Stokes' law. On the other hand, when the melt fraction is small, mass flux is controlled by the permeable flow of melt in the solid matrix. In both cases, however, mass flux due to gravitational separation of solid and melt, J_{gm}, is written in the same form:

$$J_{gm} = \frac{a^2 g(\rho_m - \rho_s) \rho_s \rho_m}{\eta_m \rho} F(\phi) \quad (12)$$

where a is the size of the solid grain, η_m is the viscosity of melt, and g is the gravitational acceleration. $F(\phi)$ is a function of melt fraction. $F(\phi)$ depends on the flow law and is given as follows:

$$F(\phi) = \begin{cases} \dfrac{1}{1000} \dfrac{\rho_s^3 \phi^3}{[\rho_m + (\rho_s - \rho_m)\phi]^2 \rho_m (1 - \phi)} \\ \qquad \text{for } \phi < \dfrac{\rho_m}{11.993 \rho_s + \rho_m} \quad (13a) \\[2ex] \dfrac{5}{7} \dfrac{\rho_s^{5.5} \rho_m \phi^{5.5}(1 - \phi)}{[\rho_m + (\rho_s - \rho_m)\phi]^{6.5}} \\ \qquad \text{for } \dfrac{\rho_m}{11.993 \rho_s + \rho_m} < \phi < \dfrac{\rho_m}{0.29624 \rho_s + \rho_m} \quad (13b) \\[2ex] \dfrac{2}{9} \dfrac{\rho_m \rho_s \phi (1 - \phi)}{[\rho_m + (\rho_s - \rho_m)\phi]^2} \\ \qquad \text{for } \dfrac{\rho_m}{0.29624 \rho_s + \rho_m} < \phi \quad (13c) \end{cases}$$

Formulae (13a) and (13b) are derived from the permeable-flow law based on the permeability formula given by Blake-Kozeny-Carman and Rumpf-Gupta, respectively. The original formula for permeability is given as a function of the volume fraction of solid phase, $c \equiv (1 - \phi)/[1 + (\rho_s/\rho_m - 1)\phi]$. Since the numerical constant in (13a) is given by McKenzie [1984] based on melting experiments, (13a) is expected to be a good approximation for silicate melt at melt fractions smaller than about 0.07. Formula (13b) is recommended for $c = 0.3 \sim 0.65$ by Dullien [1979]. Formula (13c) is derived from Stokes' law. These formulae are combined in (13) so that $F(\phi)$ is continuous.

As shown in (12), J_{gm} is proportional to the density difference and square of grain size, and inversely proportional to the viscosity of melt. We use the following parameter values as standard: $a = 1$ mm, $(\rho_s - \rho_m)/\rho_m = 0.05$, $\rho_m = 4000$ kg/m^3, and $\eta_m = 100$ Pas (viscosity of basalt melt). (Standard parameter values used in this study are shown in Table 1). Calculated values of J_{gm} are shown in Figure 2 as a function of melt fraction. Since we choose a relatively high value of melt viscosity as the standard (estimated viscosity of pyrolite melt is about $0.1 \sim 1$ Pas), J_{gm} values calculated from the standard parameters are expected to give a lower bound for the mass flux in a magma ocean. The dependence of mass flux on the melt fraction is very large, as seen in Figure 2. A slight change of melt fraction results in order-of-magnitude changes in mass flux.

Figure 3 shows the characteristic time scale of melt-solid separation, τ_g, as a function of melt fraction and depth of the

TABLE 1. Standard Parameter Values

symbol	value	meaning
a	1mm	grain size
ρ_s	4200kg/m^3	density of solid
$\Delta \rho$	$\rho_s - \rho_m$	density difference
η_s	10^{21} Pas	viscosity of solid
η_m	100Pas	viscosity of melt
C_p	1000J/kgK	specific heat
q_r	0	heat production rate
Δh	4×10^5 J/kg	difference of specific enthalpy
Δv	$\rho_m^{-1} - \rho_s^{-1}$	difference of specific volume
α	10^{-5} K^{-1}	thermal expansion coefficient
k_0	4W/mK	heat conductivity

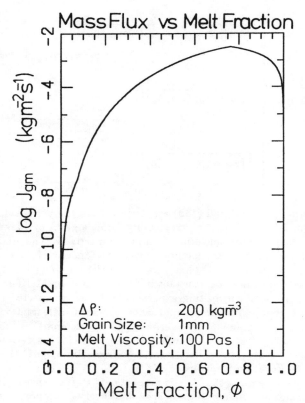

Mass Flux vs Melt Fraction

$\Delta \rho$:   200 kgm^{-3}
Grain Size: 1mm
Melt Viscosity: 100 Pas

Fig. 2. Mass flux due to gravitational separation, J_{gm}, is shown as a function of melt fraction, ϕ. Parameters used in the calculation are shown in Table 1.

magma ocean. When the melt fraction is larger than about 0.1, melt-solid separation affects a layer thicker than about 1000 km within 10^8 years. Since J_{gm} is underestimated in our case, a realistic separation time would be even shorter. Since we assumed no melting or solidification to occur, this result may not be directly applicable to a real magma ocean. However, it suggests that melt-solid separation and chemical differentiation of the mantle are expected to occur, if a magma ocean is kept undisturbed during accretion of the Earth.

Convective Mixing

We can approximate the convective mixing by turbulent diffusion in a vigorously convecting magma ocean. The vertical mass flux of melt due to convective mixing is given by Fick's law:

$$J_{cm} = -\kappa_c \rho \frac{\partial \phi}{\partial z} \qquad (14)$$

where κ_c is the eddy diffusivity for convective mass transport. Based on mixing-length arguments and Reynolds' analogy, we assume that κ_c is the same as the eddy diffusivity, κ_h, for heat transport, which is given by mixing length theory as follows [Vitense, 1953; Sasaki and Nakazawa, 1986]:

$$\kappa_h = \begin{cases} 0 \qquad\qquad\qquad\qquad\quad \text{for } g[(\frac{\partial T}{\partial z})-(\frac{\partial T}{\partial z})_s]<0 \quad (15a) \\[2ex] \frac{\alpha g l^4}{18\nu}\left[(\frac{\partial T}{\partial z})-(\frac{\partial T}{\partial z})_s\right] \\[1ex] \qquad\qquad \text{for } g[(\frac{\partial T}{\partial z})-(\frac{\partial T}{\partial z})_s]>0 \text{ and } \kappa_h<\frac{9}{8}\nu \quad (15b) \\[2ex] \sqrt{\frac{\alpha g l^4}{16}\left[(\frac{\partial T}{\partial z})-(\frac{\partial T}{\partial z})_s\right]} \\[1ex] \qquad\qquad \text{for } g[(\frac{\partial T}{\partial z})-(\frac{\partial T}{\partial z})_s]>0 \text{ and } \kappa_h>\frac{9}{8}\nu \quad (15c) \end{cases}$$

where κ is thermal diffusivity, ν is kinematic viscosity, α is the thermal expansion coefficient at constant pressure, and l is the mixing length. Formula (15a) implies no eddy diffusion at stable stratification. Formula (15b) is based on the formulation given by Sasaki and Nakazawa [1986], which is derived for a viscous fluid and is valid at relatively low eddy diffusivity. Formula (15c) is based on the formulation by Vitense [1953], which is derived for an inviscid fluid and is used for highly turbulent convection.

Mixing length theory was also used in previous work for estimating turbulent velocity [Tonks and Melosh, 1990; Miller et al., 1991b]. These estimates are essentially equivalent to the eddy diffusivity given by (15c), which is derived for an inviscid fluid based on consideration

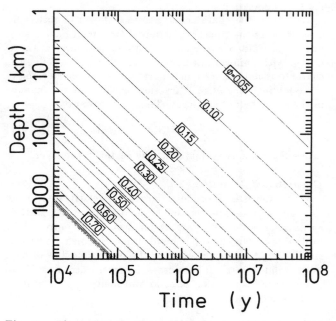

Fig. 3. Characteristic time scale of melt-solid separation, τ_{gm}, in a quiet magma ocean as a function of depth and initial melt fraction. Standard values shown in Table 1 are used for calculation. When melt fraction is higher than about 0.1, characteristic separation time in a 100km-depth magma ocean is shorter than 10^7 years.

of energy exchange between gravitational and kinematic energy [Vitense, 1953]. This formalism often appears in the astrophysical literature. Although it is appropriate highly turbulent less viscous fluid, this formulation is not adequate for highly viscous fluids because it neglects viscous drag in the estimation of velocities. (Note that viscosity does not appear in (15c)). Hence, it gives a significant overestimate of eddy diffusivity and turbulent velocity for highly viscous fluids in which viscous drag plays an important role. Instead, we use (15b) for highly viscous fluids. Formula (15b) is derived based on considering the balance between the buoyancy force and viscous drag operating on a fluid parcel [Sasaki and Nakazawa, 1986].

We can eliminate temperature gradient by combining (15) and (22) derived below. Then the eddy diffusivity is given as a function of heat flux J_q:

$$
\kappa_h = \begin{cases} 0 & \text{for } \frac{J_q}{\rho C_p \kappa} + (\frac{\partial T}{\partial z})_s < 0 \quad (16a) \\[2ex] \frac{\kappa}{2}\left\{\sqrt{\frac{2\alpha|g|l^4}{9\kappa\nu}\left[\frac{J_q}{\rho C_p \kappa} + (\frac{\partial T}{\partial z})_s\right] + 1} - 1\right\} & \\[1ex] & \text{for } \frac{J_q}{\rho C_p \kappa} + (\frac{\partial T}{\partial z})_s > 0 \text{ and } \kappa_h < \frac{9}{8}\nu \quad (16b) \\[2ex] \left[\frac{J_q \alpha |g| l^4}{16\rho C_p}\right]^{1/3} & \\[1ex] & \text{for } \frac{J_q}{\rho C_p \kappa} + (\frac{\partial T}{\partial z})_s > 0 \text{ and } \kappa_h > \frac{9}{8}\nu \quad (16c) \end{cases}
$$

In the above estimate of κ_h we ignore the effect of solid particles on the convective motion. This assumption results in overestimation of the eddy diffusivity, because the existence of dense solid particles tends to suppress convective motion [Koyaguchi et al., 1990]. Thus, the convective mass flux estimated below gives the upper bound.

The mixing length l and kinematic viscosity ν are the most important parameters which control the eddy diffusivity. We choose the distance from the boundary as the mixing length l. Then, the calculated heat flux agrees well with experimental results over a wide range of Rayleigh numbers. Hence we can expect that the estimated eddy diffusivity is also a good approximation as long as convective motion is less affected by solid particles.

Viscosity of partially molten material is estimated based on an empirical formula for lava viscosity [Marsh, 1981] and Voigt's viscosity limit for mixtures:

$$\eta = \text{MIN}\left[c\eta_s + (1-c)\eta_m, \eta_m(1-Ac)^{-2.5}\right] \quad (A = 1.67)$$

Rewrite η as a function of ϕ.

$$\eta = \text{MIN}\left\{\frac{(1-\phi)\rho_m \eta_s + \rho_s \phi \eta_m}{(1-\phi)\rho_m + \rho_s \phi}, \; \eta_m\left[\frac{\rho_m + (\rho_s - \rho_m)\phi}{[1 - 1.67(1-\phi)]\rho_m + (\rho_s - \rho_m)\phi}\right]^{2.5}\right\} \quad (17)$$

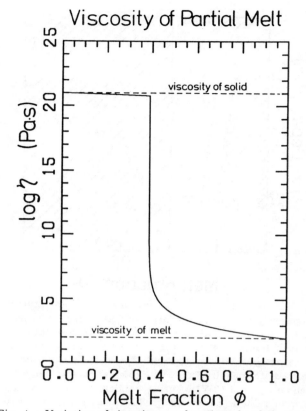

Fig. 4. Variation of viscosity as a function of melt fraction. Critical melt fraction at which viscosity jump occurs is about 0.4 for this case.

where η_s is the viscosity of solid particles. We used $\eta_s = 10^{21}$ Pas as a standard value. The estimated viscosity is shown in Figure 4 as a function of melt fraction ϕ. A very large viscosity jump (up to 19 orders of magnitude) occurs at $\phi \approx 0.4$. It is well known that partially molten materials show such a drastic change of viscosity at intermediate melt fraction [e.g., Arzi, 1978]. In the following, we call the melt fraction at which the viscosity jump occurs the 'critical melt fraction', ϕ_c. The critical melt fraction is ≈ 0.4 in our model. Such a drastic change is essentially caused by a change of solid-particle connectivity at ϕ_c. Since solid particles are mutually connected at $\phi < \phi_c$, the viscosity of the mixture is controlled by that of the solid particles. On the other hand, at $\phi > \phi_c$, the viscosity is controlled by that of the melt, because solid particles are separated. In real system, ϕ_c should depend on various parameters, such as crystal shape, crystal size distribution and strain rate.

The mass flux of melt in a vigorously convecting melt-solid mixture is given by $J_m = J_{gm} + J_{cm}$. Hence, we can calculate solid-melt separation in a vigorously convecting layer by using (10), (12) and (14). The results (not shown here) agree well with experimental results at large melt fraction [Martin and Nokes, 1988].

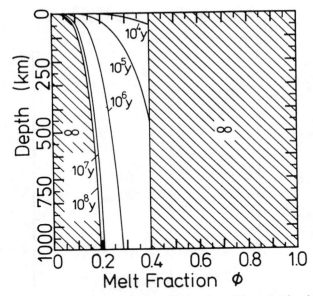

Fig. 5. Characteristic time scale of melt-solid separation in a vigorously convecting magma ocean as a function of melt fraction and depth. At high melt fraction ($\phi >\sim$ 40%) and low melt fraction ($\phi <\sim$ 20%) convective mixing suppress the separation. However, rapid separation is expected at intermediate melt fraction. The calculation was made for given heat flux of 10W/m^2 using standard parameter values shown in Table 1 (but we assumed no phase change between melt and solid to occur).

We can also estimate the characteristic time scale of separation by using (11) and $J_m = J_{gm} + J_{cm}$. Figure 5 shows the characteristic separation time in a vigorously convecting layer as a function of melt fraction and depth of the magma ocean. In this calculation, J_q values are fixed at 10W/m^2, which is the estimated heat flux during core formation (formation time $\approx 10^8\text{y}$). Although vigorous convection prevents melt-solid separation at a melt fraction smaller than 0.1~0.2 or larger than ~0.4, separation proceeds rapidly at intermediate melt fractions. This behavior is caused by the variation of partial-melt viscosity with ϕ. When the melt fraction is higher than ϕ_c (≈ 0.4 in this case), efficient convective mixing occurs, owing to the very low viscosity, and melt-solid separation is suppressed. This is consistent with the recent work by Tonks and Melosh [1990] and Miller et al. [1991b]. With decreasing ϕ, eddy diffusivity decreases drastically at ϕ_c, because of the viscosity jump which is controlled by the connectivity of solid particles, as noted before. In contrast, the mass flux of gravitational separation, which is essentially controlled by connectivity of the melt phase, does not show such a drastic change at ϕ_c (see for example Figure 2). Hence, rapid separation occurs at $\phi < \phi_c$. A further decrease of ϕ results in decreased melt connectivity. The separation rate slows down, and convective mixing again surpasses gravitational separation at $\phi < \sim 0.2$.

Figure 6 shows the characteristic time scale of separation for increased heat flux; $J_q = 1000\text{W/m}^2$ for 6a and 10^5W/m^2 for 6b. Since the efficiency of convective mixing increases with increasing heat flux, the range of melt fraction at which melt-solid separation proceeds is narrower at higher heat flux. However, even in the case of $J_q = 10^5\text{W/m}^2$, melt-solid separation proceeds at a melt fraction just below the critical melt fraction. In addition, high heat flux continues only for a very short time span in the real magma ocean. If heat flux is sustained by energy release due

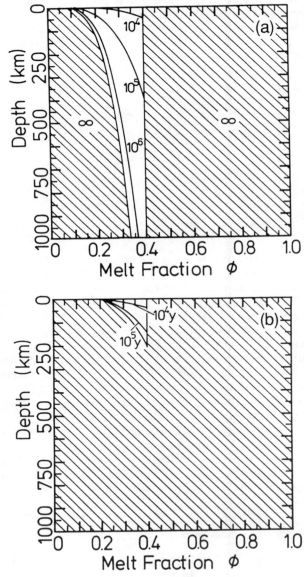

Fig. 6. Characteristic time scale of melt-solid separation in a vigorously convecting magma ocean at increased heat flux. (a) heat flux, $J_q = 100\text{W/m}^2$, (b) $J_q = 1000\text{W/m}^2$. See also Figure 5.

to core formation, duration times of heat fluxes as high as 10^5W/m^2 are only about 10^4 years because such a high heat flux requires a very short core-formation time. Moreover, the calculated characteristic time should be overestimated, because gravitational mass flux is underestimated and the eddy diffusivity is overestimated in this study, as discussed before. Hence, melt-solid separation should proceed more easily in the real magma ocean.

Thus, gravitational separation is expected to occur at the melt fraction just below the critical melt fraction ϕ_c, even if the magma ocean is vigorously convecting.

Impact Stirring

Planetesimal impacts stir up the magma ocean and interrupt the gravitational separation process. Unlike convective mixing, impact stirring cannot be approximated by diffusion, because planetesimal impacts are intermittent and the intervals between impacts are not short compared to the characteristic time of melt-solid separation. Instead, we consider that the time interval of impact stirring gives an upper bound for the duration time of separation. In the following we estimate the time interval of impact stirring.

If we assume that the proto-planet increases its radius at a constant rate, the time interval of impact stirring at the depth d, $\tau_s(d)$, is approximately given by the following equation

$$\tau_s(d) = \frac{S_s(d)}{\pi(r_E - d)^2}\tau_{acc} \qquad (18)$$

where r_E is the radius of the planet, τ_{acc} is the accretion time, and $S_s(d)$ is the total area stirred by planetesimal impacts at the depth d during accretion. Depth and area stirred by an impact depend on the size of the impacting planetesimals: a larger planetesimal stirs a greater depth into the mantle. Suppose the impact of a planetesimal (mass m) stirs a region of depth d and area s. Then the minimum mass of planetesimals which stirs the region of depth d, $m_c(d)$, is given by reversing the relationships. The total area, $S_s(d)$, disturbed during accretion at depth d, is given by the sum of $s(m)$ for all impacting planetesimals larger than $m_c(d)$ during accretion.

We assume that the planetesimal distribution follows a power law. Then the size distribution is specified by two parameters: the largest planetesimal mass, m_{ix}, and the power law index, q. We consider two mechanisms of impact stirring. One is the penetration of impacting planetesimals into the proto-planet, and the other is crater formation. We estimate $\tau_s(d)$ for both processes and choose the shorter one.

According to Kieffer and Simonds [1980], the penetration depth, d_p, is given by

$$d_p = 2r_i \frac{V_i}{V_s}, \qquad (19)$$

where r_i is the radius of the planetesimal, and V_i and V_s are the impact velocity and shock-wave velocity, respectively. When planetesimals and the proto-mantle have similar

material properties, V_s is given by $V_s = C_0 + sV_i/2$, where C_0 and s are the bulk sound velocity and equation of state parameter, respectively. In the following, we assume $C_0 = 3000\text{m/s}$ and $s = 1.5$. The area stirred by planetesimal penetration, s_p, is approximately given by the cross section of the planetesimal, πr_i^2.

Crater depth, d_e, and area, s_e, can be estimated based on the crater scaling law given by Holsapple and Schmidt [1982] as follows:

$$d_e = fr_e$$
$$s_e = \pi r_e^2 \qquad (20)$$
$$\pi fr_e^3 = B(\frac{m_i}{\rho})^{1-\beta/3}(\frac{g}{V_i^2})^{-\beta}$$

where r_e is the radius of the crater, m_i is the mass of the planetesimal, and f is the ratio of the crater depth to the radius. B and β are nondimensional parameters determined from either experiments or theoretical consideration. We consider three sets of B and β values: one set is estimated based on cratering experiments on quartz sand [Holsapple and Schmidt, 1982] and two sets are given by Davies [1985] as the limit of "energy scaling" and "momentum scaling".

Figure 7 shows the time interval of the impact stirring, τ_s, as a function of depth, d. Three cases with different cratering parameters ("quartz sand", "momentum scaling"

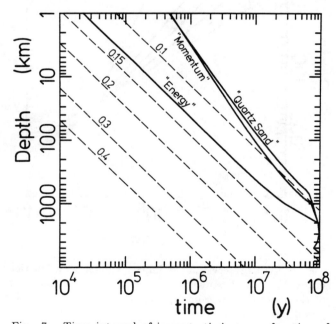

Fig. 7. Time interval of impact stirring as a function of depth, z. Three cases with different cratering parameters ("quartz sand", "momentum scaling" and "energy scaling") are shown. Thin broken lines indicate the characteristic time of the gravitational separation, τ_{gm}. Numbers attached to lines are melt fraction, ϕ.

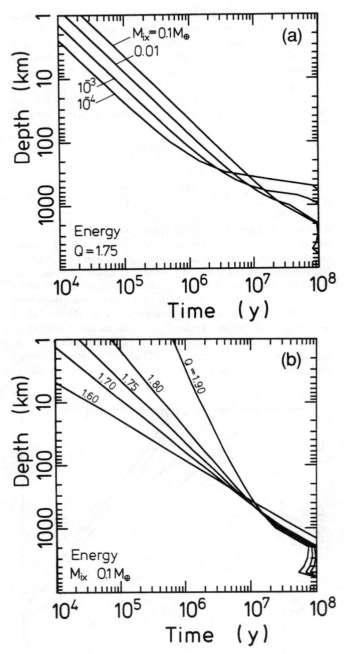

Fig. 8. Time interval of the impact stirring as a function of depth, z. (a) dependence of impact interval on the mass of the largest planetesimals, m_{ix}, (b) dependence on the power law index, q. In both cases "energy scaling" is used for calculation. See also Figure 7.

and "energy scaling") are shown. In the calculation of Figure 7 we assumed $m_{ix} = 0.1 M_E$, $q = 1.75$ and $\tau_{acc} = 1 \times 10^8$ y. Thin broken lines indicate the characteristic times of melt-solid separation, τ_{gm}. Numbers attached to lines are

melt fraction ϕ. If we adopt cratering parameters of "quartz sand" and "momentum scaling", $\tau_s(d)$ is always larger than $\tau_{gm}(d)$ at $\phi \approx 0.1$. For the case of "energy scaling", $\tau_s(d)$ is comparable to $\tau_{gm}(d)$ at $\phi \approx 0.13$. Figure 8 shows the dependence of τ_s on m_{ix} and q values. With decreasing m_{ix}, the impact interval decreases in the shallower part ($d \leq 100$ km). However, even if m_{ix} is as small as $10^{-4} M_E$ (~ 600 km in diameter), the characteristic separation time is shorter than the impact interval at $\phi > 0.2$. Similarly, even if we change the q values, the impact interval does not decrease significantly. This implies that significant gravitational separation proceeds within the impact interval, if the melt fraction is larger than about 0.2. In other words, interruption of the gravitational separation by planetesimal impact occurs only if the melt fraction is smaller than about 0.2.

As discussed in the previous section, melt-solid separation is already suppressed by vigorous convection at $\phi < 0.2$. Hence, we can conclude that impact stirring is less efficient than convective mixing. Moreover, planetesimal impacts may contribute to separation rather than stirring, because planetesimal impacts heat up the magma ocean and increase the melt fraction. When impact heating results in more than 20% melting, gravitational separation proceeds before the next impact, thus the impact accelerates the separation.

THERMAL EVOLUTION OF MAGMA OCEAN

In the previous section we considered only dynamical aspects of magma ocean evolution, namely the processes of separation and mixing assuming no melting or solidification. In the real magma ocean, however, the melt fraction is controlled by the thermal state of the magma ocean. Hence, the differentiation and thermal processes are inevitably coupled. In this section we discuss the thermal evolution of a magma ocean by taking into account melt-solid separation, and convective heat and mass transport using a one-dimensional model.

For simplicity, we ignore the growth of the Earth, namely addition of undifferentiated material, pressure increase, and impact stirring. This assumption is appropriate, if we consider the evolution of a magma ocean formed by a giant impact. We assume spherical symmetry and local thermal equilibrium between melt and solid phases. Then the equation of energy balance at constant pressure is given as follows:

$$[C_p + \Delta h(\frac{\partial \phi}{\partial T})_p]\frac{\partial T}{\partial t} = -4\pi \frac{\partial}{\partial m}r^2[J_q + \Delta h(J_{gm} + J_{cm})] + \Delta v|g|J_m + q_r \qquad (21)$$

where T is temperature, r is the distance from the center of the planet, m is mass coordinate ($m = \int_0^r 4\pi r'^2 dr'$), and q_r is heat production rate. Also, Δh and Δv are the differences of specific enthalpy and specific volume between melt and solid phases, respectively, and J_q, J_{gm} and J_{cm} are heat flux, mass flux of melt due to gravitational separation and mass flux of melt due to convective mixing, respectively.

The melt fraction ϕ is calculated from $\phi = (T -$

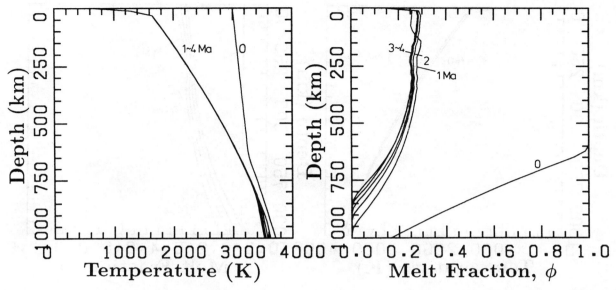

Fig. 9.   Time evolution of temperature and melt fraction distribution for case A. Profiles are shown at 1 million year interval.

$T_{sol})/(T_{liq} - T_{sol'})$, where T_{liq} and T_{sol} are the liquidus and solidus temperatures, respectively. T_{liq} and T_{sol} are the same as those used by Abe and Matsui [1986], which are based on estimates of Ohtani [1983].

Taking into account convective heat transport, the heat flux J_q is given by

$$J_q = -\rho C_p \kappa_h [(\frac{\partial T}{\partial r}) - (\frac{\partial T}{\partial r})_s] - \rho C_p \kappa (\frac{\partial T}{\partial r}) \quad (22)$$

where $(\partial T/\partial r)_s$ is the adiabatic temperature gradient, and κ and κ_h are thermal diffusivity and eddy thermal diffusivity, respectively. The eddy diffusivity, κ_h, is given by (15), and the mass fluxes J_{gm} and J_{cm} are given by (12) and (14), respectively.

For simplicity, we assume blackbody radiation at the surface and that no atmosphere is present. To solve (21) we adopt a finite difference numerical scheme with 100 equal-mass grid points. The upper portion of the grid is further divided into sublayers in order to resolve the very thin thermal boundary layer. The thickness of the smallest sublayer is 1 mm. We use a modified Euler-backward method of time stepping to stabilize the calculation. Numerical values of the parameters used in this study are summarized in Table 1.

We calculate the evolution of a magma ocean for two initial conditions. In case A, we consider the evolution from an adiabatic temperature profile with high surface temperature (3000 K). The uppermost part (depth, $d < 600$ km) is completely molten at first. This case is chosen to approximate a magma ocean just after a giant impact. In case B, we consider the evolution from a state of partial melting with constant melt fraction $\phi_0 = 0.1$. This case is chosen to approximate a magma ocean formed by the blanketing effect

of a proto-atmosphere [Abe and Matsui, 1986]. The results are shown in Figures 9 and 10.

Figure 9 shows the time evolution of vertical profiles of temperature and melt fraction for the case A. Because of low viscosity at high melt fraction, efficient convective heat transport cools the magma ocean rapidly during the first 1 Ma. However, once the melt fraction reaches the critical value, $\phi_c \approx 0.4$, convective motion slows down owing to the drastic increase of viscosity. Then cooling of the magma ocean is controlled by heat transport due to melt-solid separation. Since the melt-solid separation rate decreases rapidly with decreasing melt fraction (see Figure 2), cooling slows down at the melt fraction $0.2 \sim 0.3$. Hence, the melt fraction remains at $0.2 \sim 0.3$ for more than 4 Ma in the region shallower than 600 km.

Since we assumed blackbody radiation at the surface and no atmosphere, the heat flux at the surface given in our calculation is an upper bound. In other words, the calculation gives an upper bound for a realistic cooling rate of the magma ocean. Nevertheless, cooling is found not to be fast enough to 'quench' the magma ocean and to prevent melt-solid separation. This is because the drastic increase of viscosity slows down convective heat transport at intermediate melt fractions. Moreover, melt-solid separation is the factor controlling the cooling and solidification rate of the magma ocean at intermediate to low melt fractions.

The results for case B are shown in Figure 10. Although the melt fraction is initially low ($\phi = 0.1$), it increases to 0.3 within about 1 Ma, and its distribution pattern is quite similar to that of case A. At low melt fractions, high viscosity and the slow separation rate of melt and solid result in low efficiency of heat transport. Then, heat transported

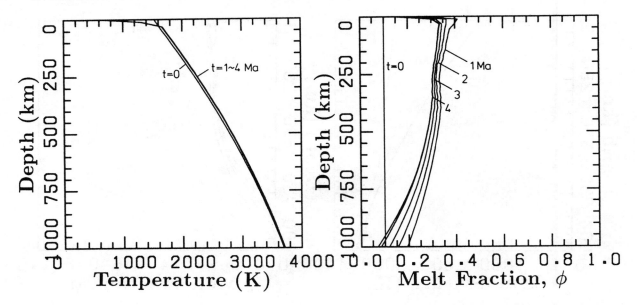

Fig. 10.   Time evolution of temperature and melt fraction distribution for case B. Profiles are shown at 1 million year interval.

from the deep interior accumulates at shallower levels and the melt fraction increases. Once the melt fraction exceeds ~ 0.3, efficient melt-solid separation occurs, which transports heat and cools the magma ocean. Thus, melt fraction increases until efficient melt-solid separation occurs. Hence, even in this case, it seems difficult to prevent melt-solid separation during the evolution of the magma ocean.

The cooling time of the magma ocean, which is controlled by melt-solid separation as discussed above, is not short. On the other hand, the thermal states of cases A and B become quite similar to each other within a relatively short time period (about 1 Ma), in spite of the large difference in their initial conditions. Thus, the thermal relaxation time of the magma ocean is rather short if no atmospheric blanketing is operating. We may conclude that the thermal evolution of the magma ocean is relatively insensitive to the initial thermal state.

CHEMICAL DIFFERENTIATION IN THE MAGMA OCEAN

We assume equilibrium partitioning of elements between two phases. Then, the equation of mass conservation of each element is given by

$$\frac{\partial \omega_i}{\partial t} = -4\pi \frac{\partial}{\partial m} r^2 \Big[\frac{1 - K_i}{K_i + (1 - K_i)\phi} \omega_i J_{gm} - \rho \kappa_c \frac{\partial \omega_i}{\partial r} \Big] \quad (23)$$

where ω_i is the mass fraction of element i. K_i is the Nernst type partition coefficient of element i between melt and solid phases. K_i is defined by

$$K_i \equiv \frac{(\text{concentration of element i in solid phase})}{(\text{concentration of element i in melt phase})}. \quad (24)$$

$K_i > 1$ for a compatible element i and $K_i < 1$ for an incompatible element i.

For simplicity, we assume that partition coefficients are independent of pressure, temperature and composition, and that the initial distribution of elements is uniform in the magma ocean. We consider five elements with different partition coefficients; three incompatible elements ($K_i = 0.01$, 0.1 and 0.5) and two compatible elements ($K_i = 10$ and 100). Figures 11 and 12 show the time evolution of trace element distributions for cases A and B, respectively. As shown in Figure 11, incompatible elements are depleted to $\sim 10\%$ of the initial concentration in deeper regions ($d > 50$ km) and enriched by 10 times at shallower levels ($d < 50$ km). On the hand, compatible elements are depleted to less than 0.1% of the initial concentration in shallower depths ($d < 400$ km) and slightly enriched in deeper regions ($d > 400$ km). Although the element distribution pattern is completely different between compatible elements ($K_i = 10$ and 100) and incompatible elements ($K_i = 0.5$, 0.1 and 0.01), it is quite similar within each group in spite of large differences in K_i values. Thus, the element distribution does not directly reflect differences in partition coefficients. In particular, if K_i is larger than ~ 10 or smaller than ~ 0.1, the resulting element distribution pattern is almost independent of the K_i values. In other words, the abundance ratios of elements are not changed by melt-solid separation when their K_i values are larger than about 10 or smaller than about 0.1. In addition, both compatible and incompatible elements are depleted in the depth range of 50 km $< d <$ 400 km.

Figure 12 shows the distribution pattern of elements for case B. It should be noted that the resulting pattern of ele-

Element Transportation

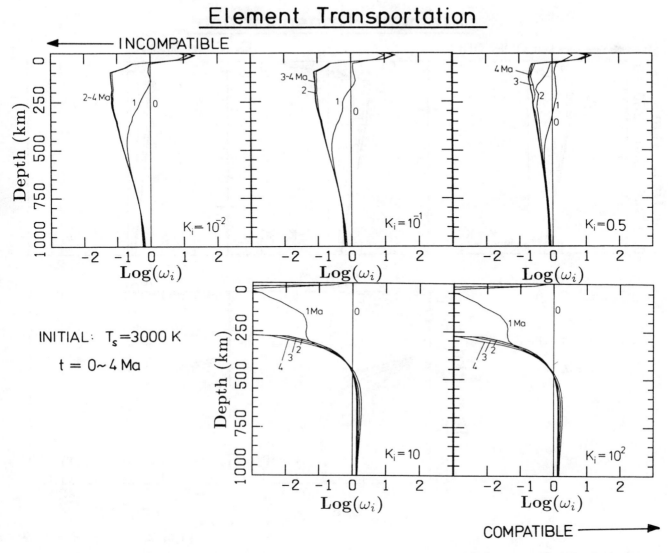

Fig. 11.   Time evolution of trace element concentration for case A. Concentration of elements with different partition coefficients are shown at 1 million year interval.

ment distribution is quite similar to that of case A, in spite of large differences in the initial melt fraction. This is because no chemical differentiation proceeds during the initial, high melt-fraction stage of case A owing to efficient convective mixing. In other words, chemical differentiation is controlled by melt-solid separation just below the critical melt fraction, as suggested in the previous section. Thus the initial condition is also less important for chemical differentiation, once the melt fraction approaches ϕ_c. As shown above, thermal evolution is less sensitive to initial conditions. Hence, the chemical evolution appears to be rather insensitive to the initial condition.

DISCUSSION AND CONCLUSION

We examined various processes which control the thermal and chemical evolution of the terrestrial magma ocean. The results are summarized as follows:

(1) Gravitational separation of melt and solid is a very rapid process. Significant separation is expected to occur when a partially molten magma ocean with melt fraction larger than about 0.1 is kept undisturbed for $\sim 10^8$ years.

(2) Planetesimal impacts contribute to separation of melt and solid, rather than stirring, if impacts result in

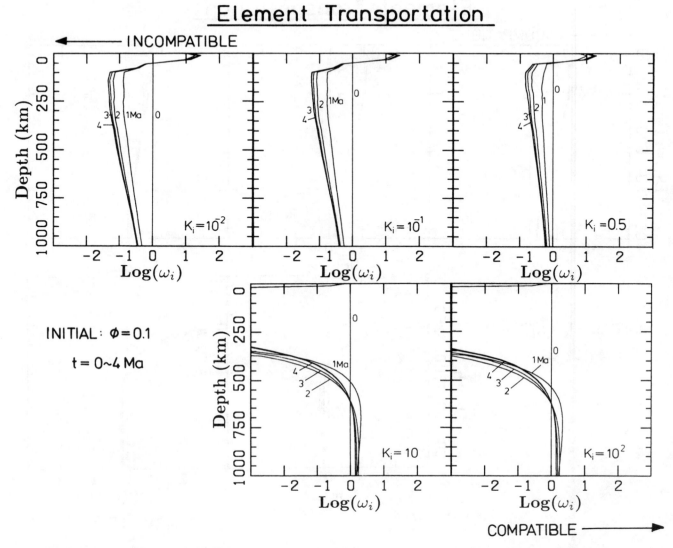

Fig. 12.   Time evolution of trace element concentration for case B. Concentration of elements with different partition coefficients are shown at 1 million year interval.

more than ~ 20% melting and the accretion time is ~ 10^8 years.

(3) Vigorous convection disturbs the melt-solid separation and prevents the development of chemical layering when the melt fraction is higher than the critical melt fraction ϕ_c (30~40%).   Just below the critical melt fraction, however, gravitational separation proceeds rapidly.   At even lower melt fractions, gravitational separation might also be suppressed by convection.

(4) Because of the large viscosity variation in partial melt the magma ocean is kept just below the critical melt fraction, at which melt-solid separation proceeds most efficiently.

(5) Hence, it is difficult to keep the proto-mantle undifferentiated, if density and compositional differences exist between solid and melt.

(6) Chemical differentiation is controlled by the compositional difference between melt and solid at intermediate melt fractions just below the critical values, rather than the values at the liquidus or solidus.

(7) The resulting element distribution, however, does not always reflect differences in partition coefficients. Moreover, it is insensitive to K_i values larger than about 10 or smaller than about 0.1.

(8) Thermal and chemical evolution is rather insensitive to the initial condition.

One of the most important results of this study is the result (3). One may think this conflicts with previous studies by Tonks and Melosh [1990] and Miller et al. [1991b], which suggested that melt-solid separation was suppressed by convective mixing. They obtained their results by comparing velocities of turbulent convection and melt-solid separation. Although we used a somewhat different method, this is not the cause of the apparent difference in results because our eddy diffusivity is essentially equivalent to their turbulent velocity and we also compare velocities in our calculation, implicitly.

Actually, the conflict between our results and those of Tonks and Melosh [1990] is just superficial. Tonks and Melosh [1990] considered only the high melt-fraction case and showed that melt-solid separation was suppressed by turbulent convection. However, they did not consider the intermediate melt-fraction case, at which we showed rapid separation to occur. We also showed no separation to occur within the melt fraction range considered by Tonks and Melosh [1990]. Therefore, no conflict exists between our results and those of Tonks and Melosh.

Miller et al. [1991b] considered intermediate to low melt-fraction cases as well as the high melt-fraction case. However, their results indicate essentially no melt-solid separation. This difference is caused by the difference in estimating convective velocity. Miller et al. [1991b] estimated the velocity of turbulent convection by using mixing length theory for non-viscous fluids. Their turbulent velocity is equivalent to the eddy diffusivity given by (15c). This formula gives a significant overestimate of turbulent viscosity and eddy diffusivity for highly viscous fluids as noted before. Since the viscosity of partial melt is very high at intermediate to low melt fractions, they have likely overestimated convective mixing over the melt fraction range that we observed rapid separation. This is the reason why they obtained different results from ours.

Melt-solid separation and convective mixing critically depend on the rheological properties of partial melt. With decreasing melt fraction, both convective mixing rate and melt-solid separation rates decrease owing to increase in viscosity and decrease in permeability. The drastic increase of viscosity at the critical melt fraction results in a drastic decrease in the convective mixing rate. On the other hand, permeability does not show such a drastic change at the critical melt fraction and the melt-solid separation rate remains high just below the critical melt fraction. This is the reason why rapid separation occurs just below the critical melt fraction. Hence, whether or not melt-solid separation proceeds depends on the melt-fraction dependency of viscosity and permeability at intermediate melt fractions. If permeability decreases significantly at melt fractions higher than the critical value, melt-solid separation should be suppressed by convective mixing.

At this moment, both viscosity and permeability are poorly constrained at intermediate melt fractions. We estimated these quantities separately based on empirical formulae ((17)

and (13)). Nevertheless, we can expect the result (3) is rather general. Permeability is controlled by the connectivity of melt. In contrast, the viscosity of partial melt is primarily controlled by the connectivity of solid particles at intermediate melt fractions. In other words, the critical melt fraction is the melt fraction at which the connectivity of the solid fraction changes drastically. It is obvious from simple geometrical considerations that a drastic change of solid-phase connectivity occurs at higher melt fractions than does a significant change in melt-phase connectivity. Then, a viscosity jump or drastic stiffening should occur while permeability remains high. Hence, melt-solid separation just below the critical melt fraction should be a general feature of the magma ocean evolution, and the result (3) holds for cases with viscosity profiles different from ours.

Another important result of this study is the result (7), caused by repeated melting and solidification in a vigorously convecting magma ocean. This result indicates that the relative abundance of elements might be kept unchanged, even if significant element transport occurs, when K_i values are smaller than about 0.1 or larger than about 10. It suggests that nearly chondritic relative abundances of these elements may not be used as evidence against chemical differentiation in the proto-mantle.

We assumed that the solid phase is always denser than the melt phase in this study. However, the possibility of a density cross-over both in the upper and lower mantle has been suggested [e.g., Agee and Walker, 1988; Ohtani, 1985; Miller et al, 1991a]. If a density cross-over exists in the mantle, the evolution of the magma ocean might be significantly affected. For example, the possibility of extensive homogenization is suggested by a preliminary calculation, which is not shown here, when the solid phase is lighter than the melt phase. However, it is difficult to examine the consequences of density cross-over, at present. A density cross-over is caused by composition (such as Fe/Mg ratio) differences between solid and melt phases, as well as by compressibility differences [e.g., Jeanloz, 1990]. Hence, a consistent model of magma-ocean evolution with a density cross over requires modeling of major-element transport and the dependence of material properties on composition. In particular, we have to take into account the composition dependence of melting temperature in the model. Both experimental and theoretical work is required to establish the distribution of composition-dependent melting temperatures. Moreover, the effect of solid particles on convective motion, which is ignored in this study, will be very important if a density cross-over exists. Further investigation is required to clarify the effect of a density cross-over on the evolution of the Earth.

Acknowledgment.   The author express his thanks to H. Yurimoto for his valuable comments on the usage of partition coefficients and interpretation of the results, and E. Takahashi, E. Ohtani and D. J. Stevenson for stimulating discussions. He appreciates R. Jeanloz, H. J. Melosh and G. H. Miller for their helpful review comments and valuable suggestions. Calculation was done in FACOM M-780 at the Nagoya University Computation Center.

This work was partially supported by Grant-in-Aid for Scientific Research on Priority Areas (No.02246102) and Grant-in-Aid for Co-operative Research (A) (No.02302033).

REFERENCES

Abe, Y., Conditions required for sustaining a surface magma ocean, *Proc. 23rd ISAS Lunar Planet Symp.*, 225–231, Inst. Space Astron. Sci., Sagamihara, 1988.

Abe, Y., Gravitational differentiation, convective mixing and impact stirring in the early earth, *Proc. 23rd ISAS Lunar Planet Symp.*, 226–231, Inst. Space Astron. Sci., Sagamihara, 1990.

Abe, Y. and T. Matsui, The formation of an impact-generated H_2O atmosphere and its implications for the early thermal history of the earth, *J. Geophys. Res.*, *90, supple.*, c545–559, 1985.

Abe, Y. and T. Matsui, Early evolution of the earth: accretion, atmosphere formation and thermal history, *J. Geophys. Res.*, *91, supple.*, e291–302, 1986.

Agee, C. B. and D. Walker, Mass balance and phase density constraints on early differentiation of chondritic mantle, *Earth Planet. Sci. Lett*, *90*, 144–156, 1988.

Arzi, A., Critical phenomena in the rheology of partially melted rocks, *Tectonophysics*, *44*, 173–184, 1978.

Coradini, A., C. Federico and P. Lanciano, Earth and Mars: early thermal profiles, *Phys. Earth Planet. Int.*, *31*, 145–160, 1983.

Davies, G. F., Heat deposition and retention in a solid planet growing by impacts, *ICARUS*, *63*, 45–68, 1985.

Dullien, F. A. L., *Porous media fluid transport and pore structure*, Academic Press, New York, 1979.

Hayashi, C., K. Nakazawa and H. Mizuno, Earth's melting due to the blanketing effect of the primordial dense atmosphere, *Earth Planet. Sci. Lett.*, *43*, 22–28, 1979.

Holsapple, K. A. and R. M. Schmidt, On the scaling of crater dimensions 2. impact processes, *J. Geophys. Res.*, *87*, 1849–1870, 1982.

Jeanloz, R., Thermodynamics and evolution of the Earth's interior: high-pressure melting of silicate perovskite as an example, *Proceedings of The Gibbs Symposium*, 211-226, American Mathematical Society, 1990.

Kato, T , A. E. Ringwood and T. Irifune, Experimental determination of element partitioning between silicate perovskites, garnets and liquids: constraints on early differentiation of the mantle, *Earth Planet. Sci. Let.*, *89*, 123–145, 1988.

Kaula, W. M., Thermal evolution of earth and moon growing by planetesimal impacts, *J. Geophys. Res.*, *84*, 999–1008, 1979.

Kieffer, S. W. and C. H. Simonds, The role of volatiles and lithology in the impact cratering process, *Rev. Geophys. Space Sci.*, *18*, 143–181, 1980.

Koyaguchi, T., M. A. Hallworth, H. E. Huppert and S. J. Sparks, Sedimentation of particles from a convecting fluid, *Nature*, *343*, 447–450, 1990.

Marsh, B. D., On the crystallinity, probability of occurrence, and rheology of lava and magma, *Contrib. Mineral. Petrol.*, *78*, 85–98, 1981.

Martin, D. and R. Nokes, Crystal settling in a vigorously convecting magma chamber, *Nature*, *332*, 534–536, 1988.

Matsui, T. and Y. Abe, Evolution of an impact-induced atmosphere and magma ocean on the accreting Earth, *Nature*, *319*, 303–305, 1986a.

Matsui, T. and Y. Abe, Impact-induced atmospheres and oceans on Earth and Venus, *Nature*, *322*, 526–528, 1986b.

McBirney, A. R. and T. Murase, Rheological properties of magmas, *Ann. Rev. Earth Planet. Sci.*, *12*, 337–357, 1984.

McKenzie, D., The generation and compaction of partially molten rock, *J. Petrol.*, *25*, 713–765, 1984.

Melosh, H. J., Giant impacts and the thermal state of the early earth, in *Origin of the earth*, 69–83, Oxford, New York, 1990.

Miller, G. H., E. M. Stolper and T. J. Ahrens, The equation of state of a molten komatiite, 1. shock wave compression to 36 GPa. *J. Geophys. Res.*, *96*, 11831–11848, 1991a.

Miller, G. H., E. M. Stolper and T. J. Ahrens, The equation of state of a molten komatiite, 2. application to komatiite petrogenesis and hadean mantle. *J. Geophys. Res.*, *96*, 11849–11864, 1991b.

Ohtani, E., Melting temperature distribution and fractionation in the lower mantle, *Phys. Earth Planet. Int.*, *33*, 12–25, 1983.

Ohtani, E., The primordial terrestrial magma ocean and its implication for stratification of the mantle, *Phys. Earth Planet. Int.*, *38*, 70–80, 1985.

Rintoul, D. and D. J. Stevenson, The role of large infrequent impacts in the thermal state of the primordial earth (abstract), in *Papers presented to the conference on the origin of the earth*, pp.75–76, Lunar and Planetary Institute, Houston, 1988.

Safronov, V. S., The heating of the earth during its formation, *ICARUS*, *33*, 3–12, 1978.

Sasaki, S. and K. Nakazawa: Metal-silicate fractionation in the growing earth: energy source for the terrestrial magma ocean. *J. Geophys. Res.*, *91*, 9231–9238, 1986.

Tonks, W. B. and H. J. Melosh, The physics of crystal settling and suspension in a turbulent magma ocean, in *Origin of the earth*, 151–174, Oxford, New York, 1990.

Vitense, E., Die Wasserstoff konvektionzone der Sonne, *Z. Astrophys.*, *32*, 135–164, 1953.

Zahnle, K., J. F. Kasting and J. B. Pollack, Evolution of a steam atmosphere during earth's accretion, *ICARUS*, *74*, 62–97, 1988.

Y. Abe, Department of Earth and Planetary Physics, Faculty of Science, University of Tokyo, Bunkyo-ku, Tokyo 113, JAPAN.

Mantle Differentiation Through Continental Crust Growth and Recycling and the Thermal Evolution of the Earth

TILMAN SPOHN AND DORIS BREUER

Institut für Planetologie, Westfälische Wilhelms-Universität, Münster, Germany

Thermal evolution models for the Earth with mantle differentiation through the production of continental crust are presented. The continental crust is enriched in radioactive elements relative to the mantle and the mantle will be depleted in these elements upon the production of crustal rock. The volumetric production rate is taken to be proportional to the mantle convection speed, to the volume of the crustal component in the mantle, and to the square–root of the oceanic surface area. The crust to mantle recycling rate is assumed to be proportional to the convection speed and to the continental surface area. It is assumed that the average concentration of radioactive elements in the crust changes little with time as a consequence of an increasing enrichment of young crustal rock in radioactive elements relative to the mantle. This enrichment may be caused by a decrease with time of the degree of partial melting in the mantle upon cooling. It is further assumed that the continental basal heat flow is also approximately constant in time. The latter two assumptions allow us to satisfy observational constraints on crustal thickness with reasonable crustal temperatures, on the evolution of the subcontinental lithosphere temperature, and on the variation of continental heat production with age. We have systematically varied the factor Σ by which mantle radioactive heat production decreased due to radioactive decay in 4.5 Ga, and the crustal Urey ratio U_c. U_c gives the part of the continental surface heat flow that is due to crustal heat production. The model satisfies observational constraints on continental freeboard and on heat production vs. crustal age distributions for reasonable values of Σ of $4-5$ and $U_c \leq 0.4$. The model satisfies the mean age of continental crustal rock, the present day continental and oceanic surface areas and heat flows, and constraints on mantle temperature, viscosity, and chemistry. Crust production and recycling rates tend towards equilibrium and a decrease of crust volume with time appears to be impossible. The present rate of decrease of mantle heat source density is calculated to be approximately 1.2 times the radioactive decay rate. The thermal effects of differentiation on the mantle are a faster cooling during the Archean due to continental growth followed by a retarded cooling due to thermal blanketing of the mantle by the continents. The thermal blanketing of the mantle also results in a more realistic mantle Urey ratio U_m as compared with most previous thermal evolution models. U_m is the part of the mantle heat flow that is due to mantle heat production. Its present value is calculated to be approximately $0.4-0.5$.

INTRODUCTION

Although the Earth's continental crust is the most accessible part of the planet, its origin and evolution is still the subject of continuing research [see e.g., Kröner, 1985; Taylor and McLennan, 1985; Reymer and Schubert, 1987; for recent reviews]. It is generally accepted that the continental crust is the product of partial melting of the mantle, possibly through a two–stage scenario where the oceanic crust forms by partial melting of the mantle and melt extraction

at mid–oceanic ridges and where the continental crust forms by partial melting of the oceanic crust and melt extraction at subduction zones. Extraction of melt from partially molten subcontinental mantle through plutonism and from partially molten subducted sedimentary rock may also contribute to the formation of young continental crust, however.

Crust formation through mantle differentiation is likely to be of importance for the thermal evolution of the Earth. Although the absolute concentrations of the major heat producing elements ^{238}U, ^{235}U, ^{232}Th, ^{238}U, and ^{40}K are not precisely known, it is well established that radioactive heat production in the crust and mantle accounts for a substantial fraction of the Earth's surface heat flow. The production of crustal rock causes a depletion of mantle heat sources

Evolution of the Earth and Planets
Geophysical Monograph 74, IUGG Volume 14

because radiogenic elements are fractionated into the melt upon partial melting together with other large ion lithophile elements. Therefore, the oceanic crust is enriched in these elements as compared with the mantle and the continental crust is enriched as compared with both the oceanic crust and the mantle. On the other hand, the continents appear to act as thermal shields of the mantle. The average values of continental and oceanic surface heat flow differ by 40 mWm^{-2} [e.g., Sclater et al., 1980; Pollack, 1982], about 40% of the latter. The difference between the average heat flow from the mantle into the continents and the oceanic surface heat flow is still larger and amounts to about 60 mWm^{-2}. Thus, the growth of the continental surface area is likely to have effects on the mantle that go beyond the fractionation of the mantle heat sources.

The bulk of the continental crust volume most likely formed billions of years ago. At present, continental crustal rock is still produced at a rate of $1 - 2$ $km^3\,a^{-1}$ [e.g., Dewey and Windley, 1981; Reymer and Schubert, 1987]. The rate of recycling of crust with the mantle is less well known. Recycling of crust is required to explain the observed isotopic patterns in oceanic basalt [e.g., Allègre, 1982; DePaolo 1983] and may occur through the subduction of sediments [White and Dupré, 1986] and, possibly, through other processes such as crustal delamination [Bird, 1978; Dewey and Windley, 1981]. In older work, present day rates of sediment subduction are given to be about 0.6 to 0.8 km^3a^{-1} [e.g., Reymer and Schubert, 1987] but most recently, von Huene and Scholl [1991] have estimated a contemporary sediment subduction rate of 1.5 $km^3\,a^{-1}$. Geochemical data suggest recycling rates of $1 - 3$ km^3a^{-1} [Armstrong, 1981; DePaolo, 1983]. It is possible, if not likely, that the depletion of mantle heat sources still continues to the present day even if the production rate is approximately balanced by the recycling rate: Young continental rock appears to be enriched in radiogenic elements [e.g., Sclater et al., 1980; Sclater et al., 1981; Pollack, 1982; Morgan, 1985] as compared with, on the average, older crustal rock that is recycled.

Previous models of the thermal evolution of the Earth [e.g., Sharpe and Peltier, 1979; Schubert et al., 1979; Stevenson and Turner, 1979; Spohn and Schubert, 1982; Stevenson et al., 1983; Spohn, 1984; Christensen, 1985] have mostly neglected the effects of the differentiation of the mantle. An exception is the model of Cook and Turcotte [1981]. Assuming that the continental growth rate is proportional to mantle convection speed they concluded that 60 to 80% of the continents formed 3 Ga b.p., that the mean mantle temperature in Archean times was 100 to 150 K larger than it is today, and that the mantle Urey ratio U_m is between 0.70 and 0.75. U_m is the fraction of the mantle heat flow that is due to radiogenic heat production. Cook and Turcotte did not differentiate between continental and oceanic surface heat flow, however. In addition, Cook and Turcotte neglected the possible effects of the recycling of continental crust.

Gurnis and Davies [1985,1986] have presented models of continental growth in which crust production was related to the Earth's heat flux and crust removal was related to the existing crustal volume and the heat flux. For most of their models they have simply assumed that the heat flux declined in proportion to the radioactive heating; for some of their models they have used a more realistic model that also accounted for the cooling of the Earth. However, they have not considered the effects of differentiation on the thermal history of the Earth. Turcotte [1989a] has presented a model of continental growth scaled with the rate of heat loss from the Earth's interior for which he found a present day recycling rate that is consistent with the sediment subduction rate of von Huene and Scholl [1991]. Turcotte [1989b], Spohn [1991], and Schubert et al. [1992] have presented models of mantle differentiation and the thermal evolution of other terrestrial planets.

In the present paper, we present a model of mantle differentiation and thermal evolution that satisfies a large number of geological observational constraints. These constraints are outlined in the next section. We use a parameterization of crust production and recycling rates that does not depend crucially on the tectonic style, in particular, on the operation of plate–tectonics. However, we do not consider specifically changes in tectonic style and the possible effects that these may have on continental crust production and recycling. We assume that the spatially averaged concentration of radioactive elements in the continental crust and the basal heat flow into the crust are approximately constant in time because the observational constraints on the evolution of crust thickness and subcontinental lithosphere temperature and on the variation of crustal heat production with age suggest, as we will discuss below, that the thermal regime of the crust evolved little over the past aeons.

OBSERVATIONAL CONSTRAINTS

Geophysical, geochemical and geological observation provides a number of constraints that the model is required to satisfy. These constraints are:

1. The present day continental surface area of 40% of the Earth's surface area.

2. The present day values of the average continental and oceanic surface heat flows of 58 and 99 mWm^{-2} [e.g., Sclater et al., 1980; Pollack, 1982]

3. The mean age of continental crustal rock of approximately 2 Ga [e.g., Jacobsen and Wasserburg, 1979; Goldstein et al., 1984].

4. The continued (partial) emergence of continents above sea–level at least since the Archean [e.g., Wise, 1974; Hallam, 1984; Schubert and Reymer, 1985; Taylor and McLennan, 1985; Galer, 1991].

5. The formation temperatures of about 900–1200 K of 3.2 Ga old silicate inclusions in diamonds from kim-

berlite pipes located in the Kaapvaal Craton [Richardson et al., 1984; Boyd et al., 1985; Boyd and Gurney, 1986]. These inclusions formed at depths of 150–200 km. The formation temperatures suggest that the subcontinental lithosphere temperature changed little at least since 3.2 Ga ago [Ballard and Pollack, 1988].

6. The trend in continental heat production and heat flow to decrease with the age of the continental heat flow provinces [e.g., Pollack, 1982; Morgan, 1985].

7. The present day value of the upper mantle temperature of approximately 1700 K [Jeanloz and Morris, 1986].

8. The present day value of the mantle viscosity of approximately $10^{17}\, m^2 s^{-1}$ [e.g., Peltier, 1981].

Additional observational constraints on Archean mantle temperatures, on crust and mantle Urey ratios, on crust to mantle radiogenic enrichment factors, on the volume percentage of the depleted mantle, and on the radioactive decay ratio will be used in the text below to discuss the relevance of model results.

MODEL

The model uses the method of parameterized convection and extends the thermal evolution model of Stevenson et al. [1983] by considering mantle differentiation through the production of continental crustal rock. The method of parameterized convection is well established as a means of assessing the thermal evolution of the Earth and the planets [e.g., Sharpe and Peltier, 1979; Schubert et al., 1979; Stevenson and Turner, 1979; Spohn and Schubert, 1982; Stevenson et al., 1983; Spohn, 1984; Schubert et al., 1986; Peltier, 1989; Fischer and Spohn, 1990; Schubert and Spohn, 1990; Spohn, 1991; Spohn and Schubert, 1991]. The model consists of a spherical shell, the Earth's mantle, surrounding a concentric spherical core. Part of the surface of the mantle is covered by continental crust with average thickness D_c. The surface area of the continents is A_c and the volume of the continental crust V_c is given by $A_c D_c$. We are mostly interested in the mantle and pay relatively little attention to the evolution of the core in the present paper. Therefore, we give the equations for calculating the thermal evolution and the differentiation history of the mantle along with the history of the continental crust. Our model includes the evolution of the core, however. The reader is referred to Stevenson et al. [1983] for the equations describing the thermal evolution of the core.

The interaction of the continental crust with the thermal history of the mantle may appear to be hopelessly complex, precluding a simple parameterized model. The thickness of the continental crust, the basal heat flow from the mantle into the crust, and the level of crustal radioactive heat production are quantities that are all required to be calculated consistently as functions of time together with the mantle

parameters. The calculation of the crustal radioactive heat production rate may appear to be particularly difficult requiring a calculation of the age distribution of crustal rock along with the concentrations of radiogenic elements as a function of age. Moreover, there may be strong lateral gradients in radiogenic heat production since, at least at present, new continental crust is added around the older continental cores. Fortunately, some of the observational constraints (#4 – 6) listed above can be combined to argue for a much simpler model in which the average values of the continental crust thickness, heat generation rate, and continental heat flow have remained approximately constant in time.

Schubert and Reymer [1985] have modelled the variation of the continental freeboard, the mean elevation of continents above sea–level, since the Archean and have concluded that an approximately constant D_c is required to explain Precambrian and Phanerozoic emergence of the continents. Our calculations presented below have confirmed their conclusion. A constant D_c is also suggested by seismic data and by a considerable number of other geological data [see Taylor and McLennan, 1985 and Galer, 1991 and references therein].

An approximately constant D_c together with an approximately constant subcontinental lithosphere temperature over Ga is difficult to reconcile with substantial changes with time in crustal heat production and continental basal heat flow. The half–lives of the major radioactive elements suggests that their concentration may have been a factor of about 2 larger in the Archean crust and mantle than at present. Together with temperatures in the present lower crust of about 900 K, a factor of 2 increase in crustal heat production and basal heat flow – due to a factor 2 larger mantle heat production rate – would imply Archean lower crust temperatures of roughly 1500 K. This temperature is larger than the Archean subcontinental lithosphere temperature derived from the kimberlite data and is clearly larger than reasonable solidus temperatures of crustal rock. Ballard and Pollack [1987, 1988] have noted this problem and have argued that the continental basal heat flow may have changed little between the Archean and the present. In their model, the extra mantle heat is diverted away from the continents and removed through the sea–floor by faster mantle convection. The resulting lower crust temperatures are then not excessive.

Reasonable crust and lithosphere temperatures together with a constant D_c are also possible if it is assumed that the average concentration of radiogenics in the Archean continental crust was approximately the same as (or smaller than) it is today. Similar conclusions have been reached by Morgan [1985] and by Ballard and Pollack [1987]. This proposal requires that the Archean crust was less enriched in radioactives relative to the mantle than the present crust. Radioactive decay in the mantle requires that the ratio between crust and mantle heat source density increased by a factor of about 2 under this condition. Since the mantle is depleted in radioactive elements as a consequence of crust production, the ratio must have increased somewhat more rapidly, by about an additional factor of 1.2 as our results suggest. Moreover, young rock added to the crust at any given time or replacing old rock recycled with the mantle must also be enriched relative to the average crust if the average concentration is to be kept constant.

There are at least three possible processes that may have caused an enrichment of young crustal rock with respect to the mantle that increased in time:

1. The mantle was hotter in the past with a larger degree of partial melting and the degree of partial melting decreased as the mantle cooled. The concentration of radiogenic elements in the melt relative to the concentration in the solid should increase with decreasing degree of partial melting. Expanding on work by Shaw [1970], it is easy to show that the concentration of radiogenics c_{me} in the melt as a function of the volume fraction melted S is given by

$$\frac{c_{me}}{c_{sm}} = \frac{1 - (1 - S)^{\frac{1}{\Gamma} - 1}}{S} \qquad (1)$$

where c_{sm} is the concentration of radiogenic elements in the solid and Γ is a distribution coefficient typical values of which are much smaller than 1. c_{me}/c_{sm} increases by about one order of magnitude for melt volume fractions between 10% and 1%. The generation of Archean komatiitic magma requires melt volume fractions of tens of percent [e.g., BVSP, 1981] while modern basalt requires significantly less partial melting. Since the continental crust is derived from the mantle in one way or another continental crustal rock produced in the Archean, accordingly, should have been less enriched in radiogenic elements relative to the mantle as compared with recently formed continental crust.

2. Partial melting of subducted continental sediments in subduction zones may produce enriched magmas that may intrude the continental crust and thereby increase the concentration of crustal radiogenics. The depleted solid rock is recycled with the mantle.

3. Partial melting at the base of the crust and the rise of buoyant, enriched magma may produce enriched crustal rock. The depleted solid rock may be recycled with the mantle through crustal delamination.

The proposal of an approximately constant in time crustal heat source density and the required increase of the enrichment of continental crustal rock with decreasing age can be tested by a comparison between heat production rates of crustal rock of various ages and, although less reliably, by a consideration of the distribution of continental surface heat flow with age. Surface heat flow depends on additional factors such as heat from tectonothermal events and local variations in mantle heat flow. However, if the crustal enrichment factor with respect to the mantle was always constant and if the radioactive decay rates of crust and mantle rock were approximately the same, then present concentrations of radiogenic elements and heat source densities should vary little between rocks of young and old ages. On the contrary, if the average concentration of crustal radioactives was approximately constant then the concentrations of radiogenic elements and the heat production rates in Archean and Phanerozoic rock should differ by roughly a factor of 2.

The observational evidence suggests that the concentrations of radiogenic elements in crustal rock decreases with increasing age. Data compiled by Morgan [1985] (and replotted in Figure 9 of the present paper) show a large scatter measured by their standard deviations and a major exception (central Australia) but the average value of the crustal heat production rate decreases by a factor of 2 to 3 from the Phanerozoic to the early Archean. Similar variations in heat flow have been reported by e.g., Sclater et al. [1980] and Pollack [1982] which have, at least partly, been attributed to variations in crustal heat production.

Consistent with the preceding arguments we have assumed that D_c and the crustal heat production rate Q_c are constant in time. Constant values of D_c and Q_c together with approximately constant values of sublithospheric temperatures imply approximately constant values of mantle heat flow into the base of the continents q_{mc} and approximately constant values of continental surface heat flow q_{cs}. We should note at this point that the arguments presented above are based on the available data which are sparse (Archean subcontinental lithosphere temperatures) and uncertain (heat production values). These data are consistent with a thermal regime of the continents that has not changed throughout the evolution of the Earth but they do not *require* such a conclusion. Taken together our arguments suggest a physically reasonable, simple model worthy of exploration that may be used to illustrate some of the interactions between continental growth and mantle thermal evolution.

Our model allows the calculation of heat production vs. age distributions that can be compared with that of e.g., Morgan [1985] for a more detailed *a posteriori* test of the model. Assuming that the recycled crust has the average crustal concentration of radiogenics $\bar{c}$ we can calculate the concentration $c(t)$ in newly formed crustal units from a mass balance equation for the crust under the assumption that $d\bar{c}/dt$ is zero:

$$0 = \frac{d\bar{c}}{dt}\Big|_{rad} + (c(t) - \bar{c}) \frac{1}{V_c} \frac{dV_{prod}}{dt} \qquad (2)$$

dV_{prod}/dt is the volumetric crust production rate and the first term on the right hand side of (2) is the rate of change of $\bar{c}$ due to radioactive decay. Solving (2) for $c(t)/\bar{c}$ we obtain

$$\frac{c(t)}{\bar{c}} = \frac{\lambda + \frac{1}{V_c} \frac{dV_{prod}}{dt}}{\frac{1}{V_c} \frac{dV_{prod}}{dt}} \qquad (3)$$

where λ is the mean radioactive decay constant. A present day model distribution of crustal radiogenic heat production vs. age of crustal units can be calculated from (3) by transforming $c(t)$ into $c(\tau)$, where τ is age, and by multiplying $c(\tau)$ with $exp(-\lambda\tau)$ to account for radioactive decay and will be discussed below.

The volumetric crustal growth rate dV_c/dt is given by

$$\frac{dV_c}{dt} = \frac{dV_{prod}}{dt} - \frac{dV_{rec}}{dt} \qquad (4)$$

where $\frac{dV_{rec}}{dt}$ is the crust to mantle recycling rate. The crust production and recycling rates are parameterized similar to Cook and Turcotte [1981], Gurnis and Davies [1985, 1986], and Turcotte [1989a]. The volumetric crust production rate is assumed to be given by

$$\frac{dV_{prod}}{dt} = \left(\frac{A_o}{4\pi R_p^2}\right)^{1/2} \frac{(V_{max} - V_c)}{R_p} \chi_p u \qquad (5)$$

In (5), A_o is the surface area of the oceanic crust, R_p is the Earth's radius, V_{max} is the maximum possible volume of the crust for complete differentiation of the mantle, χ_p is a production efficiency factor and u is the average mantle convection speed. The parameterization is quite general and does not depend crucially on tectonic style, in particular, on the operation of plate–tectonics. In fact, a parameterization similar to (5) has been used by Turcotte [1989b], Spohn [1991], and Schubert et al. [1992] to model crust growth on terrestrial one–planet planets. The production efficiency factor χ_p is, essentially, the ratio between the mantle convection turnover time and the characteristic time for crustal fractionation. This factor should actually be a variable depending, among other things, on the degree of partial melting in the mantle and on the tectonic style. For instance, plate–tectonics may be more or less efficient at crust production than e.g., hot–spot tectonics. For simplicity, however, we assume χ_p to be a constant in this paper. We further assume that the mantle is well mixed by convection to keep differentiation from being throttled by compositional layering in the mantle and we assume that the production rate is proportional to the convection speed. The convection speed is the most important and the most variable parameter in (5) because it couples the production rate to the mantle thermal history and because it depends exponentially on mantle temperature as we shall outline below. The dependence of the production rate on u is likely to be independent of tectonic style and the detailed mechanisms of crust production. The first term on the right hand side of (5) takes into account the fact that most of the continental crust is presently produced at active continental margins. We assume that the total length of the subduction zones is proportional to the square root of oceanic surface area. This factor is specific to plate–tectonics but it is actually not a very important parameter since its value only varies roughly between 0.8 and 1. It has been proposed [e.g., Kröner, 1985] that the produc-

tion rate of the Archean crust was proportional to mantle heat flux rather than to convection speed. We have tested this possibility but we found that the results did not depend crucially on the choice between the two parameterizations.

The crust recycling rate is assumed to be given by

$$\frac{dV_{rec}}{dt} = A_c \chi_r u \qquad (6)$$

where χ_r is a recycling efficiency factor. Equation (6) assumes that the recycling rate is proportional to the continental surface area and to the mantle convection speed. If sediment subduction is the primary mode of recycling then the recycling rate should, actually, be proportional to the plate speed [Gurnis and Davies 1985]. Sediment subduction is specific to plate–tectonics. If crustal delamination is the most important recycling mechanism, as it is likely to be with hot–spot tectonics, then the mantle convection speed matters. In any case, we assume that the mantle convection speed and the plate speed are approximately equal. Similar to χ_p, the recycling efficiency factor χ_r is, essentially, the ratio between the mantle convection turnover time and the characteristic time for crustal recycling.

The convection speed u is calculated from

$$u = u_0 Ra^{2\beta} \qquad (7)$$

where u_0 is a speed scale determined from a present-day mantle convection speed of $0.1\ ma^{-1}$, Ra is the mantle Rayleigh number defined as

$$Ra = \frac{\alpha g (\Delta T_s + \Delta T_c)(R_p - R_c)^3}{\kappa\nu}, \qquad (8)$$

and $\beta \approx 0.3$ is a constant. In (8), α is the thermal expansion coefficient, g is gravity, ΔT_s is the temperature difference across the surface thermal boundary layer (Figure 1), and ΔT_c is the temperature difference across the core/mantle thermal boundary layer. R_c is core radius, κ is the mantle thermal diffusivity, and ν is the temperature dependent kinematic mantle viscosity

$$\nu = \nu_o exp(\frac{A}{T_u}) \qquad (9)$$

where ν_o is a constant, A is the activation temperature for viscous flow, and T_u is the temperature of the upper mantle (Figure 1).

The energy balance equation for the mantle is

$$\rho_m C_m V_m \eta_m \frac{dT_u}{dt} = -(A_o q_{os} + A_c q_{mc}) + A_{co} q_{co} + V_m Q_m \qquad (10)$$

where ρ_m is the mantle density, C_m is the mantle specific heat per unit mass, V_m is the volume of the mantle, η_m is the ratio between the mantle temperature representative of the internal energy of the mantle and T_u, q_{os} is the oceanic surface heat flow

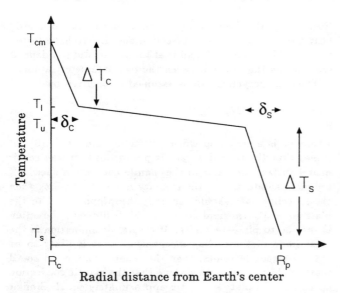

Fig. 1. Temperature as a function of radial distance from the Earth's center. The radius of the core is R_c and R_p is the Earth's radius. Temperature rises by ΔT_s from the surface temperature T_s to the upper mantle temperature T_u across the surface boundary layer of thickness δ_s. It rises by ΔT_c from the lower mantle temperature T_l to the core–mantle boundary temperature T_{cm} across the bottom boundary layer of the mantle which is of thickness δ_c.

$$q_{os} = k\frac{(T_u - T_s)}{\delta_s} \tag{11}$$

with k the thermal conductivity, T_s the surface temperature, and

$$\delta_s = 6.452\frac{(R_p - R_c)}{Ra^\beta} \tag{12}$$

the thickness of the near–surface thermal boundary layer (Figure 1). The numerical constant in (12) was chosen consistent with numerical and laboratory experiments.

$$
\begin{aligned}
q_{mc} &= q_{cs} - Q_c D_c, &(13)\\
&= (1 - U_c)q_{cs}, &(14)
\end{aligned}
$$

where $U_c \equiv Q_c D_c / q_{cs}$. We term U_c the crust Urey ratio in analogy to the mantle Urey ratio. A_{co} is the surface area of the core, and

$$q_{co} = k\frac{(T_{cm} - \epsilon T_u)}{\delta_c} \tag{15}$$

is the heat flow from the core into the mantle, where T_{cm} is the representative core/mantle boundary temperature and ϵ is the ratio between T_l, the temperature at the base of the adiabatic mantle (Figure 1), and T_u. The thickness δ_c of the core/mantle thermal boundary layer is calculated from a local stability criterion [Stevenson et al., 1983]. The local critical Rayleigh number for the breakdown of the boundary layer is

$$Ra_{crb} = \frac{\alpha g \Delta T_c \delta_c^3}{\kappa \nu_c} \tag{16}$$

with

$$\nu_c = \nu_o exp\left(\frac{A}{T_l + \frac{\Delta T_c}{2}}\right) \tag{17}$$

The mantle heat production rate Q_m is given by

$$
\begin{aligned}
Q_m &= Q_o \exp(-\lambda t) - Q_c\frac{V_c}{V_m} &(18)\\
&= Q_o \Sigma^{-\frac{t}{t_p}} - Q_c\frac{V_c}{V_m}, &(19)
\end{aligned}
$$

where Q_o is the initial mantle heat production rate and λ is the mean radioactive decay constant. In (19), Σ is the radioactive decay factor, the factor by which mantle radiogenic heat production decreased in 4.5 Ga due to radioactive decay, and t_p is the present time.

PARAMETER VALUES

In this paper, we mostly vary Σ and U_c. A variation of Σ is equivalent to a variation of the radioactive decay constant λ and equivalent to variations of the representative half live of crustal and mantle heat sources. A variation of U_c is equivalent to a variation of q_{mc} or to a variation of Q_c, both at constant values of q_{cs}. Σ is varied between 7, corresponding to a chondritic value of λ, and 2, a minimum reasonable value for the Earth. The value of Σ calculated from the terrestrial values of the ratios of K/U and Th/U and from the known half–lives of these radiogenic elements as given by e.g., BVSP [1981] is between 4 and 5. U_c is varied between 0.2 and 1.0. Observationally constrained, reasonable present values of U_c are between 0.4 and 0.6, with 0.4 being the preferred value. A value of 0.6 results from assuming a heat flow at the base of the continental crust equal to the average reduced continental heat flow of 27 $mW\,m^{-2}$ [Vitorello and Pollack, 1980; Pollack, 1982]. The former value of U_c of 0.4 was calculated assuming a value of basal heat flow of 35 $mW\,m^{-2}$ corresponding to the sum of the average reduced heat flow and the average contribution to the continental surface heat flow from tectonothermal events [Pollack, 1982]. The value of U_c

TABLE 1. Values of the radioactive decay factor Σ and the crustal Urey ratio U_c that were systematically varied. Also given are the associated values of the initial mantle heat source density Q_o, the present crust over mantle radiogenic enrichment factor Λ, and the production and recycling efficiency factors χ_p and χ_r. Models for which results are shown in Figs. 2 – 9 are indicated by an asterisk.

U_c	Σ	$Q_0(Wm^{-3})$	Λ	χ_p	χ_r	
1	7	1.53×10^{-7}	308	8.3×10^{-3}	2.3×10^{-5}	*
1	2	6.25×10^{-8}	144	5.5×10^{-3}	3.3×10^{-5}	*
0.6	4	1.05×10^{-7}	86	2.7×10^{-3}	2.8×10^{-5}	*
0.6	3	8.50×10^{-8}	75	2.5×10^{-3}	2.8×10^{-5}	
0.4	5	1.21×10^{-7}	53	1.7×10^{-3}	2.7×10^{-5}	*
0.4	4	1.05×10^{-7}	48	1.5×10^{-3}	2.7×10^{-5}	*
0.4	3	8.50×10^{-8}	43	1.5×10^{-3}	2.7×10^{-5}	
0.2	2	6.25×10^{-8}	19	0.7×10^{-3}	3.1×10^{-5}	*

of 0.4 is equivalent to a ratio between the basal heat flow and the continental surface heat flow of 0.6 proposed as a general rule by Vitorello and Pollack [1980]. At any given time, the rate of decrease of mantle heat source density increases with increasing values of both Σ and U_c. Thermal blanketing of the mantle for any given continental surface area is maximum for U_c equal to 1. The effect decreases with decreasing values of U_c.

We require the models to satisfy the observational constraints itemized above. In particular, we choose values of Q_o, χ_p, and χ_r such that a model gives the correct values of the present day surface heat flows, the surface areas of the oceans and the continents, and the mean age of the continental crust. The correct value of the present continental freeboard is then automatically obtained with the additional parameter values of Schubert and Reymer [1985]. Q_o is mostly constrained by the present day value of oceanic surface heat flow and by the chosen values of Σ and U_c. χ_p and χ_r are constrained by the present day values of continental surface area and crustal age. The present enrichment factor in radioactive elements of the crust with respect to the mantle, Λ, and V_{max} can be calculated from Σ, U_c, and Q_o. Values of U_c, Σ, Q_o, Λ, χ_p, and χ_r are given in Table 1. Other parameter values are summarized in Table 2. The latter values can be taken as reasonably well known and/or are not likely to significantly effect the results within their ranges of uncertainty. For the core we have used the parameter values of model E1 of Stevenson et al. [1983].

We assume that crust formation on the Earth started about 4 Ga ago, about 500 Ma after the accretion of the planet. The oldest segments of the present continental crust are approximately 4 Ga old [Taylor and McLennan, 1985; Bowring, 1992]. Geochemical data suggest that any pre-Archean crust was totally recycled back into the mantle by 4 Ga ago [Shirey and Carlson, 1989]. From a consideration of freeboard, Galer [1991] has recently concluded that any very old continental crust could not have been hypsometrically and tectonically distinct from oceanic crust. The lunar crust may provide evidence on the evolution of the early terrestrial crust. The early terrestrial crust most likely was continuously destroyed during the late heavy bombardment just like the early crust on the Moon was continuously pulverized by impacts. The first continuous crust formed only after hundreds of million years on the Moon [e.g., Taylor, 1982].

RESULTS

The results of the model calculations are shown in Figures 2–8. Figure 2 shows the continental surface area and the continental volume as functions of time for six combinations of values of Σ and U_c. With $\Sigma = 7$ and $U_c = 1$, the model with the largest overall rate of decrease (radioactive decay plus fractionation) of mantle heat production, the crust grows rapidly during the first few hundred million years of crustal evolution and remains approximately constant thereafter. For this model, the present production rate of 1.03 $km^3 a^{-1}$ is equal to the recycling rate. The model with the smallest overall rate of decrease with $\Sigma = 2$ and $U_c = 0.2$ has the most sluggish growth of continental crustal volume but the crust grows to about 90% of its present vol-

TABLE 2. Fixed parameter values for thermal evolution models. "CMB" refers to the core-mantle boundary.

Parameter		Value
Mantle thermal conductivity	k $(Wm^{-1}K^{-1})$	4.00
Mantle heat capacity	$\rho_m C_m$ (Jm^{-3})	8×10^6
Viscosity constant	ν_o $(m^2 s^{-1})$	4000.
Activation temperature	A (K^{-1})	5.2×10^4
Thermal expansivity	α (K^{-1})	$2. \times 10^{-5}$
Heat transfer exponent	β	0.3
Planet radius	R_p (km)	6371.
Gravity	g (ms^{-2})	10.
Surface temperature	T_s (K)	293.
Core radius	R_c (km)	3485.
Ratio between representative mantle temperature and upper mantle temperature	η_m	1.3
Ratio between upper mantle temperature and CMB-temperature	ϵ	1.6
Initial upper mantle temperature	T_{uo} (K)	2300.
Initial CMB-temperature	T_{cmo} (K)	4700.
Convection speed scale	$u_0 (ms^{-1})$	3.3×10^{-12}
Volume of oceans	V_0 (km^3)	1.22×10^9
Crustal thickness	D_c (km)	36
Present area of continents	A_c (km^2)	2.17×10^8
Present continental freeboard	h (m)	750.

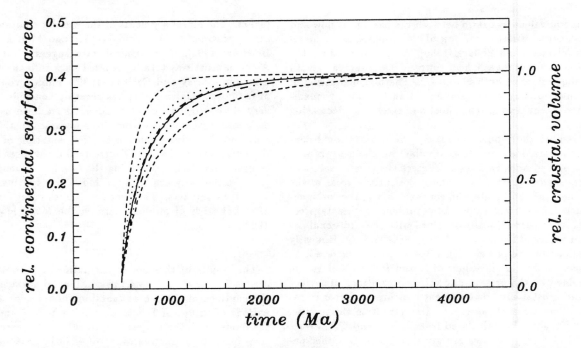

Fig. 2. The continental surface area normalized to the Earth's surface area and the continental volume normalized to the present volume as functions of time for six combinations of values of Σ and U_c. The solid line refers to $\Sigma = 5$ and $U_c = 0.4$, the dash–dotted line refers to $\Sigma = 4$ and $U_c = 0.6$, the dash–double dotted line refers to $\Sigma = 4$ and $U_c = 0.4$, the dotted line refers to $\Sigma = 2$ and $U_c = 1$, the upper dashed curve refers to $\Sigma = 7$ and $U_c = 1$, and the lower dashed curve refers to $\Sigma = 2$ and $U_c = 0.2$.

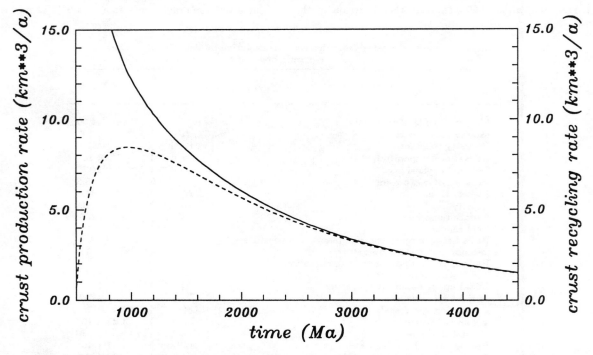

Fig. 3. Crust production (solid line) and recycling (dashed line) rates as a function of time for the model with $\Sigma = 5$ and $U_c = 0.4$.

ume by the end of the Archean even in this model. The present values of the production and recycling rates for this model are $1.78\,km^3a^{-1}$ and $1.70\,km^3a^{-1}$, respectively. In general, the continents grow faster with increasing values of Σ and U_c. For values of Σ between 4 and 5 and U_c between 0.4 and 0.6, present day production rates and recycling rates are approximately equal and amount to between 1.4 and 1.5 km^3a^{-1}. These rates compare well with estimates of present production rates of $1-2\,km^3a^{-1}$ as quoted above and also with the quoted estimates of recycling rates from geochemical and geophysical data. The recycling rates are approximately equal to the most recent estimate of sediment subduction rates von Huene and Scholl [1991] are but larger than previous estimates by about 1 km^3a^{-1}. Figure 3 shows a typical evolution of the crust production and recycling rates with $\Sigma = 5$ and $U_c = 0.4$. The production rate decreases exponentially to its present value. The recycling rate is initially zero and increases to a maximum at 1 Ga taking a value of almost 9 km^3a^{-1}. Thereafter, the recycling rate decreases along with the production rate towards its present value. A computation of the ratio between the time integrals of the two rates for all models showed that the total volume of crust recycled with the mantle is typically approximately half of the total crustal volume produced in 4.0 Ga.

Figure 4 shows the evolution of the ratio between the representative mantle radiogenic element concentration c_m and

the mantle radiogenic element concentration without differentiation

$$c = c_o \exp(-\lambda t) \qquad (20)$$

where c_o is the initial concentration for six combinations of values of Σ and U_c. The ratio c_m/c measures the degree of differentiation of the mantle. The amount of present day depletion increases with increasing values of Σ and U_c. For values of Σ and U_c as large as 7 and 1, respectively, the mantle is depleted of about 60% of its initial radiogenic element inventory. For a model with $\Sigma = 2$ and $U_c = 0.2$ the depletion is by about 10%. For reasonable values of Σ and U_c between 4 and 5 and between 0.4 and 0.6, respectively, the present mantle is depleted by about 20 to 30%. These values compare favorably with geochemically determined values of the volume of the depleted mantle of 25 to 33 vol% [Jacobsen and Wasserburg, 1979; DePaolo, 1981] but other authors have determined significantly larger volumes of the depleted mantle [e.g., O'Nions et al., 1979].

Figure 5 shows the ranges of upper mantle and core/mantle boundary temperatures over time as calculated from the model. Temperatures do not differ between the models by more than 100 K. Core-mantle boundary temperatures decrease by about 400 K during the Archean (2.6 – 4.0 Ga b.p.) and by about 150 K thereafter. The kink in the core/mantle boundary temperature vs. time curves at times of 1.8 to 3 Ga is caused by the onset of the freeze out of an inner core.

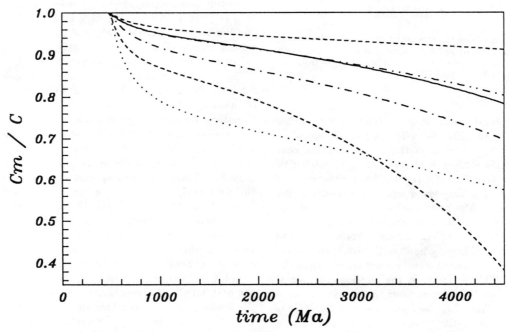

Fig. 4. Evolution of the ratio between the concentration of mantle radiogenic elements c_m and the equivalent concentration without differentiation c for six combinations of values of Σ and U_c. Note that the upper dashed curve refers to $\Sigma = 2$ and $U_c = 0.2$ here and that the lower dashed curve refers to $\Sigma = 7$ and $U_c = 1$. For further explanations see Fig. 2.

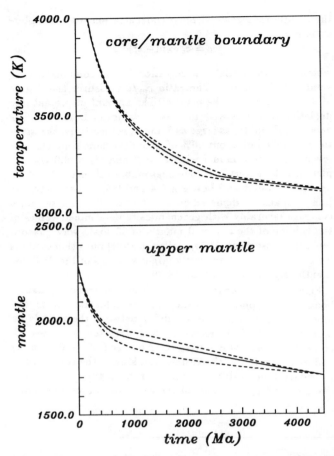

Fig. 5. Ranges of upper mantle and core/mantle boundary temperatures over time. The upper dashed lines refer to $\Sigma = 7$ and $U_c = 1$, the lower dashed lines refer to $\Sigma = 2$ and $U_c = 0.2$, and the solid lines refer to $\Sigma = 5$ and $U_c = 0.4$.

Upper mantle temperatures decrease by 100 to 200 K during the Archean and by roughly 100 K in the post–Archean.

Figure 6 shows the evolution of the average surface heat flow and the oceanic surface heat flow for the model with $\Sigma = 5$ and $U_c = 0.4$. These curves are representative of models with Σ and U_c between 4 and 5 and between 0.4 and 0.6, respectively. Also shown is the constant continental surface heat flow and the evolution of the surface heat flow for a model without differentiation calculated for the same parameter values. During most of the Archean the average heat flow decreases significantly more rapidly than the oceanic heat flow. This is caused by the growing continental surface area. In the Proterozoic, after the continents have formed, the average heat flow decreases less rapidly. This is caused by the constant continental heat flow acounting for an increasing portion of the average heat flow. Approximately half of the decrease of average heat flow is due to radioactive decay in the mantle, the other half is due to secular cooling

and fractionation of heat sources from the mantle into the crust.

A comparison between the average surface heat flows for the differentiation model and the model without differentiation shows, that for the former model heat flow decreases faster during the first two Ga. The more rapid decrease is attributed to the formation of the crust and the associated removal of heat sources from the mantle. Thereafter, the heat flow for the differentiation model decreases at a slower rate and at present the two differ by about 5 mWm^{-2}. This difference is predicted to increase with time, however. The slower decrease in heat flow for the differentiation model is attributed to the thermal blanketing of the mantle by the continents. For the model without differentiation the entire surface area is oceanic while for the differentiation model the oceanic area is reduced by 40%.

Figure 7 shows the evolution of the mantle Urey ratio. U_m increases during the first 1-2 billion years to a maximum value of about 0.65 to 0.7. Thereafter, U_m decreases depending on the values of Σ and U_c. For values of Σ as small as 2, U_m decreases very little from its maximum value. For values of Σ between 4 and 5, the present value of U_m is between 0.45 and 0.5. The present value of U_m becomes as small as 0.25 with Σ equal to 7. Also shown in Figure 7 is the evolution of U_m for a model without differentiation and $\Sigma = 5$. In this model, U_m decreases less rapidly then in the differentiation model with the same value of Σ and attains a present value of about 0.6. The latter evolution of U_m is typical of conventional thermal evolution models neglecting differentiation [e.g., Spohn, 1984]. Geochemical data suggest a value of U_m of 0.4 or may be even smaller [Zindler and Hart, 1986]. Values of U_m between 0.35 and 0.4 require 1 to 2 silicate Earth budgets of K in the core with the present model [Breuer and Spohn, 1992].

Figure 8 shows the freeboard variation calculated after Schubert and Reymer [1985] using their parameter values in addition to ours where required. The variation of freeboard with time in this model is a consequence of the competition between the deepening of the ocean basins as a result of planetary cooling and the increase in sea–level as a result of the shrinking of the ocean surface during continental growth. While a deepening of the ocean basins increases the freeboard, an increase in sea–level decreases the freeboard. Most recently, Galer [1991] has argued that freeboard should depend sensitively on mantle temperature through a dependence of oceanic crust thickness on mantle temperature. Schubert and Reymer's model implicitly assumes a constant oceanic crust thickness. Requiring constant freeboard, Galer has placed a limit on the mantle cooling rate for the past 3.5 Ga of approximately 40 $K\,Ga^{-1}$. In our model, the cooling rates during this time are between 40 and 60 $K\,Ga^{-1}$ depending on the values of Σ and U_c. Considering the similar values of the cooling rates and the uncertainties in Archean freeboard as well as in Galer's model parameters [Schubert, 1991], in particular in the relation between

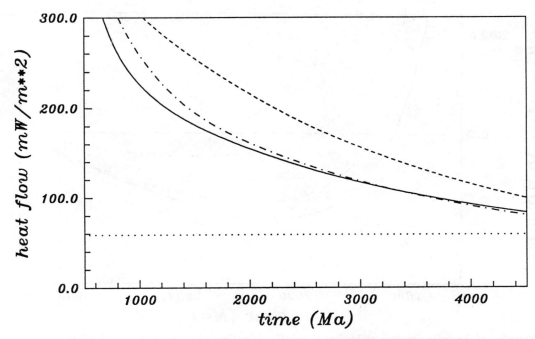

Fig. 6. Evolution of the average surface heat flow (solid line), the oceanic surface heat flow (dashed line), and the continental surface heat flow (dotted line) for a model with $\Sigma = 5$ and $U_c = 0.4$. Also shown for comparison is the surface heat flow for a model without differentiation and $\Sigma = 5$ (dash–dotted line).

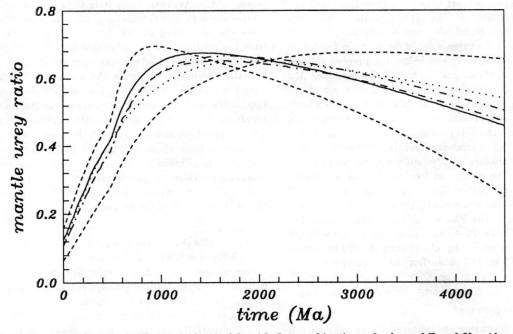

Fig. 7. Evolution of the mantle Urey ratio for models with five combinations of values of Σ and U_c. Also shown for comparison is the mantle Urey ratio for a model without differentiation and with $\Sigma = 5$ (dotted line). The upper dashed curve refers to $\Sigma = 7$ and $U_c = 1$, and the lower dashed curve refers to $\Sigma = 2$ and $U_c = 0.2$. For further explanations see Fig. 2.

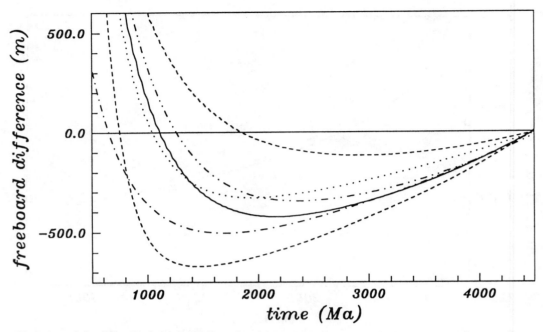

Fig. 8. Variation of the difference between freeboard and its present day value calculated for six combinations of Σ and U_c. Note that the upper dashed line here refers to $\Sigma = 2$ and $U_c = 0.2$ and that the lower dashed line refers to $\Sigma = 7$ and $U_c = 1$. For further explanations see Fig. 2.

oceanic crust thickness and mantle temperature, we have decided not to revise our model for the present purpose.

All models in Figure 8 start with a positive freeboard difference of more than 500 m. The rapid growth of the continental volume for the model with the chondritic value of $\Sigma = 7$ and with $U_c = 1.0$ causes a rapid decrease in freeboard until a minimum of almost 700 m below the present value is reached at a time of about 3.5 Ga b.p., a few hundred Ma after production and recycling reached equilibrium and the continental volume attained its final value (Figure 2). Thereafter, freeboard increases with time as a consequence of the now dominating effect of the continuing deepening of the ocean basins. The other freeboard vs. time curves in Figure 8 show a similar evolution but, because of the slower growth of the continents, the minimum freeboard value is smaller and it is attained at a later time. The model with $\Sigma = 2$ and with $U_c = 0.2$ has a freeboard that may be considered constant during the entire Phanerozoic and Proterozoic. For the models with Σ between 4 and 5 and with U_c between 0.4 and 0.6 freeboard increases by about roughly 200 m during the Proterozoic and by 150 m during the Phanerozoic.

DISCUSSION

Archean Mantle Temperature

The observational constraints itemized above are satisfied by all models. An additional constraint for thermal evolution models that is often quoted is the formation temperature of komatiites [e.g, Christensen, 1985; Richter, 1985; Bickle,

1986]. The komatiite formation temperatures have been used to suggest that the mantle temperature was 200 to 300 K larger in the Archean than it is today. The mean mantle temperature in our calculations decreases by 200 to 300 K since the beginning of the Archean (Figure 5), consistent with the evidence. It is likely, however, that the variation of the core/mantle boundary temperature is more relevant to komatiite formation because the komatiites are believed to have been formed in upwelling plumes. Plumes that rise from the core/mantle boundary have temperatures that decrease adiabatically from the core/mantle boundary temperature. The variation of the core/mantle boundary temperature in our models is by about 400 to 550 K since the beginning of the Archean (Figure 5). Considering the uncertainty in the temperature estimate of komatiite formation [Richter, 1985; Galer, 1991] we conclude that our models are consistent with the evidence from Archean komatiites.

Freeboard

Our results show a considerable variation of freeboard although there is continental emergence during the entire evolution of the crust. The results are more consistent with the conclusion of Hallam [1984] that there was a significant secular fall in sea–level during the Phanerozoic then with the claim of Wise [1974] and Schubert and Reymer [1985] of an essentially constant post–Archean freeboard. Wise's conclusion of approximately constant Phanerozoic freeboard for North America has been critizized by Hallam to be based on inadequate data. Hallam finds that seal–level fell by 600 m since the late Ordovician. Part of the sea–level fall is attributed to a secular change while the remainder is probably mostly due to tectonoeustasy. Galer [1991] has concluded that there was secular increase in freeboard during the Phanerozoic by about 200 m. These results compare well with the approximately 150 m increase in freeboard that we have calculated.

The simplest explanation for a secular increase in freeboard is a deepening of the ocean basins due to secular cooling at approximately constant continental volume. This has previously been pointed out by Turcotte and Burke [1978] and Hallam [1984]. A secular decrease in oceanic crust thickness as a consequence of mantle cooling may also cause an increase in freeboard [Galer, 1991]. Extrapolating this trend back into the Proterozoic and Archean suggests a further decrease in freeboard unless there is significant continental growth during that time. Schubert and Reymer found that the crustal volume must grow by 25% during the post–Archean to keep the freeboard constant. Since reliable eustatic data are available only for the Phanerozoic, extrapolations are necessarily speculative. However, there are several ways of reducing the freeboard variation in our model. For instance, the assumption of a constant volume of ocean water may be relaxed. It is possible that there was a net degassing of the mantle during the Proterozoic [McGovern and Schubert, 1989]. A net increase of ocean water volume by a few tens of percent would cause the freeboard to be constant during the Proterozoic in our model.

A probably more appealing way of reducing the freeboard variation may be possible through the assumption that the recycling rate, or a significant part thereof, is proportional to the area of the continents above sea–level. The present model simply assumes that the recycling rate is proportional to the entire continental surface area and neglects the effects of flooding on erosion. Flooding should reduce erosion because the continental volume above sea level is eroded much more effectively [Wise, 1974] as compared with the volume under water. Since the area of flooding increases and decreases with freeboard, by taking inundation into account the evolution of the recycling rate and the continental growth rate will be coupled to the evolution of the freeboard. This coupling may set up a regulating system that may reduce the variation in freeboard. Preliminary results of calculations to be presented in detail elsewhere suggests that the freeboard may be at a minimum near the Proterozoic/Phanerozoic boundary. The continental crust grows by about 15 vol. % during the Proterozoic in this model.

To obtain a rough estimate of how a reduction of the recycling rate may influence the evolution of freeboard we have calculated a few models neglecting recycling altogether. For these models with reasonable values of Σ and U_c we found large positive freeboard differences for the past aeons in contrast with the negative differences for models with recycling. The models without recycling could satisfy our remaining constraints except for the mean crustal age, however. Neglecting crustal recycling typically results in mean crustal ages of 500 Ma greater than the accepted value of 2 Ga.

Parameterization of Production and Recycling Rates

We have assumed that the crust production and recycling rates are proportional to the mantle convection speed. It has been proposed that the continental crust production rate may be proportional to the heat flow instead of being proportional to the convection speed [Kröner, 1985]. In additional calculations we have tested this proposal and we have found that the differences between such models and the present models are relatively minor and that our general conclusions are not affected. In models of continental growth, Gurnis and Davies [1985, 1986] have tried several parameterizations of the relation between mantle convection speed and the crustal growth rate and have also found that results depended little on the details of the parameterization as long as the heat flux declined in proportion to radioactive decay. Moreover, they have emphasized the importance of recycling in pointing out that peaks of the continental crustal age distribution do not require episodic growth of the continents arising from an episodically evolving mantle. Rather, these peaks can be explained as arising from a smoothly evolving mantle with crustal recycling.

The volumetric crust growth curves in Figure 2 are similar to those of Cook and Turcotte [1981] and of Gurnis and Davies [1985]. They are mostly determined by the dependence of the production and recycling rates on the convection speed u and do not depend critically on other details of the parameterization, e.g., the factor $(A_o/(4\pi R_p^2))^{1/2}$ in (5), as additional calculations have shown. The rate of change of u with time is proportional to the rate of change of mantle heat source density with time and the absolute value of du/dt increases with increasing values of Σ and U_c. The convection speed through its dependence on the viscosity, which itself depends exponentially on temperature, is subject to the same self–regulating mechanism described first by Tozer [1965] for the convective heat flux. It is interesting to note that production and recycling always tend towards an equilibrium in our model. Even additional calculations with χ_r increased by two orders of magnitude from the values given in Table 1 have not resulted in a recycling rate larger than the production rate. A decrease of crustal volume with time as was proposed by Fyfe [1978] appears to be impossible with our model.

Crust Thickness, Heat Source Density, and Heat Flow

In further additional calculations we have relaxed our assumption of a constant crust thickness and have allowed D_c to grow under the condition that the temperature at the base of the crust was always smaller than or equal to 1025 K, a reasonable upper bound on crustal melting temperatures. Moreover, the crustal heat production rate was assumed to decrease with time according to radioactive decay and the crust over mantle radiogenic enrichment factor was assumed a constant in these calculations. The freeboard for this model was found to increase from zero to the present value in about 1.5 Ga. Accordingly, the continents must have been flooded during the Archean and early Proterozoic. Submerged continents during the Archean have indeed been proposed by

Hargraves [1976] but this model is clearly inconsistent with the evidence for Archean and post–Archean land emergence. Consistent with Schubert and Reymer [1985] we require a constant thickness crust to satisfy this constraint.

To allow a constant thickness continental crust with reasonable crust and lithosphere temperatures and to satisfy the observed trend of continental heat production to decrease with the age of the heat flow province we have assumed that the spatially averaged continental heat production rate is approximately constant with time. The decrease of the concentration of radiogenic elements in the crust due to radioactive decay must then be balanced by the addition of radiogenic elements to the crust through relatively enriched young crustal rock that replaces old crustal units recycled with the mantle or that is added to the crustal volume. For our models, the ratio between the concentration in young rock, $c(t)$, and the average concentration, $\bar{c}$, increases from an initial value of 1 to about 3 at present. The accompanying decrease of mantle heat source density can be further illustrated by writing (19) in its differential form:

$$\frac{1}{Q_m}\frac{dQ_m}{dt} = -\lambda - \frac{V_c Q_c}{V_m Q_m}\left(\lambda + \frac{1}{V_c}\frac{dV_c}{dt}\right) \quad (21)$$

$$= -\lambda \frac{Q}{Q_m} - \frac{V_c Q_c}{V_m Q_m}\frac{1}{V_c}\frac{dV_c}{dt} \quad (22)$$

where $Q = Q_o exp(-\lambda t)$. The first term on the right hand side of (21) is the rate of decrease of mantle heat source density due to radioactive decay. The second term gives the rate of decrease of mantle heat source density due to replacement of recycled crust with young crust. The third term gives the rate of decrease of mantle heat source density due to net growth of the continental crust. At present, assuming that dV_c/dt is approximately zero and $Q_m/Q = c_m/c \approx 0.8$ (Figure 4),

$$\frac{1}{Q_m}\frac{dQ_m}{dt} \approx 1.2\lambda \quad (23)$$

a reasonably small deviation from λ.

In Table 2 the present crust over mantle radiogenic enrichment factor Λ is given. Λ is defined here as the ratio between the crust and mantle radiogenic heat production rates per unit mass or, equivalently, as the ratio between the representative concentrations of radiogenic elements in the crust and mantle. For reasonable values of Σ and U_c, Λ is between 50 and 60. This compares well with geochemical estimates of the ratio between the concentrations of heat production elements in the crust and mantle.

In Figure 9 we compare model heat production vs. age distributions, calculated using (3), with the empirical distribution compiled by Morgan [1985]. The model distributions were converted into surface heat flow - reduced heat flow vs. age distributions for consistency with the empirical data. It was assumed that the reduced heat flow equals the basal continental heat flow.

Although the data scatter is large, there appears to be a clear trend of enrichment of young crust as compared with older crust. As we have already pointed out above, if the crust over mantle enrichment factor was always constant and if the radioactive decay rates of crust and mantle rock were approximately the same, then present concentrations of radiogenic elements and heat source densities should, contrary to the observational evidence, vary little between rock of young and old ages. The model distributions depend mostly on the assumed value of U_c. Model distributions with $\Sigma \le 5$ and $U_c \le 0.4$ compare quite well with the empirical distributions. The model distributions are also in nice agreement with data compiled by Pollack [1982] and with a value of up to 0.35 $\mu W m^{-3}$ difference in crustal heat production density between Archean and post–Archean continental rock quoted by Ballard and Pollack [1987].

It may be argued that the average concentration of radiogenic elements in recycled crust is larger than $\bar{c}$ and that the quality of the fit of the model distributions in Figure 9 is dependent on an invalid assumption. An larger than average concentration of radiogenics in the recycled crust would increase the variation of the model heat production rate with age and, at least for some models, decrease the fit. After all, it may be argued, a significant part of the recycled crustal volume is made of subducted sediments. These sediments come preferentially from mountain belts which tend to be younger and more enriched than the average crust. It is also widely believed that the upper crust is enriched with respect to the average crust [e.g., Sclater et al., 1981]. However, it is still possible that sediments account only for a fraction

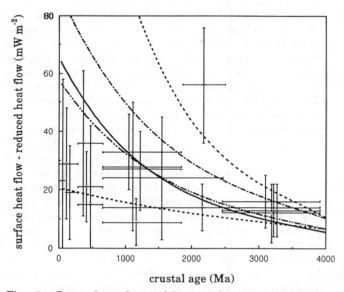

Fig. 9. Comparison of crustal heat production vs. age distributions calculated for five combinations of Σ and U_c with data compiled by Morgan (1985). For consistency with the empirical data, the model distributions are given as surface heat flow - reduced heat flow vs. age distributions. For further explanations see Fig. 2.

of the present recycling rate and crustal delamination may account for the rest of the present recycling rate. Crust recycled through delamination is likely to have a concentration that is less than $\bar{c}$ because crustal delamination preferentially recycles lower crust which is widely believed to be depleted in radioactives relative to the average crust. Recycling of sediments and delamination combined may very well recycle crust of approximately average concentration.

Other explanations for the continental heat production vs. age relation that emphasize the effects of erosion have been suggested. For instance, Morgan [1985] proposes that the present data on Archean heat source density may be biased and that Archean crustal rock with larger heat source density may have been preferentially recycled with the mantle through crustal erosion. In effect, this proposal is similar to earlier proposals that attribute the decrease in continental surface heat flow with time to erosion [e.g., England and Richardson, 1980]. While these explanations are reasonable, they are not quantitative and they do not consider the constraints on crustal thickness and subcontinental mantle temperature and their consequences for the evolution of the thermal regime of the crust that motivated our model. Our assumption of a constant with time crustal heat source density together with the assumption of a constant continental basal heat flow is certainly likely to be an oversimplification. The assumption must eventually break down at some point in the albeit distant future when the production and recycling rates will tend towards zero together with the convection speed. Some increase (or decrease) over time of average crustal heat production rate is certainly consistent with the data and would allow deviations of the concentration of radiogenic elements in recycled rock from $\bar{c}$. But the assumption of an approximately constant in time thermal regime of the continental crust is consistent with the observational evidence and has reasonable implications for the evolution of the crust and the mantle. It appears to be more realistic than the assumption of a constant radiogenic crust over mantle enrichment factor.

Comparison With Previous Models

There are some important differences between thermal evolution models such as those of e.g., Stevenson et al. [1983] and Spohn [1984] that neglect mantle differentiation and crust formation and our present models. Among these differences are a smaller rate of decrease of the mantle temperature in the present models. Both models remove about the same amount of heat from the Earth's interior over the age of the planet, however, for reasonable values of Σ and U_c. A comparison of heat removal for our models shows that the latter quantity increases with U_c: For instance, a model with $U_c = 1$ removes 25% more heat from the interior as compared with a model with $U_c = 0.2$. As the comparison between the average surface heat flow for our model with differentiation and the model without differentiation in Figure 6 showed, the evolution of the heat flow is also different due to the combined effects of mantle cooling upon heat source removal and thermal blanketing by the continental crust.

A most important difference between models with and without differentiation is the evolution of the mantle Urey ratio as was shown in Figure 7. Conventional thermal history models have been criticized by Christensen [1985] to give too large values of U_m of 0.7 to 0.8. The comparatively small Urey ratios calculated from the present models are partly a consequence of the larger and more realistic value of the heat capacity (calculated after Stacey [1981]) that we use and partly a consequence of the thermal blanketing of the mantle by the continents. Christensen has proposed that the heat transfer exponent β should be as low as 0.05 to account for the temperature dependence of the mantle rheology and to arrive at mantle Urey ratios of about 0.4. In fact, Christensen claims that the Urey ratio provides the strongest argument for very small values of β. Our calculations show that such values of U_m are obtained for models with β equal to 0.3 if mantle differentiation is taken into account.

Models with small values of β are difficult to reconcile with the evidence for a secular fall in sea–level during the Phanerozoic. These models lack the required deepening of the ocean basins. The deepening of the ocean basins is proportional to the decrease in surface heat flow. Since the heat flow decreases little in models with values of β as small as 0.05 or even smaller as proposed by Christensen [1985], freeboard must be constant if crustal thickness and volume are constant.

CONCLUSIONS

Thermal evolution models with mantle differentiation parameterized by continental production and recycling rates proportional to the mantle convection speed and assuming a thermal regime of the continental crust that evolved little over the past aeons can satisfy important geological constraints on crust and mantle evolution. An important constraint is Archean and post–Archean land emergence which requires constant crustal thickness. Constant crustal thickness together with reasonable crustal temperatures and constant subcontinental mantle temperatures suggests approximately constant in time continental basal heat flow. Heat generation vs. age distributions calculated from the model compare favorably with empirical distributions for radioactive decay ratios of mantle heat source density less or equal to 5 and crust Urey ratios less or equal to 0.4. The thermal effects of mantle differentiation on the mantle are a faster cooling during the Archean during continental growth and fractionation of mantle heat sources followed by a retarded cooling due to the thermal blanketing of the mantle by the continents. At present, the rate of change of mantle heat source density is approximately 1.2 times the radioactive decay rate. The thermal blanketing also results in a realistic present mantle Urey ratio of 0.4 – 0.5. There is no need for small values of β to obtain reasonable values of the mantle Urey ratio. Crust production and recycling rate tend towards an equilibrium and a decrease of crustal volume with time appears to be impossible.

Acknowledgments. We thank two anonymous reviewers for valuable suggestions. This work was supported by the Deutsche Forschungsgemeinschaft.

REFERENCES

Armstrong, R. L., Radiogenic isotopes: The case for crustal recycling on a near–steady–state no–continental–growth earth, *R. Soc. Lond. Phil. Trans.*, *A301*, 433–471, 1981.

Allègre, C. J., Chemical geodynamics, *Tectonophysics*, *81*, 109–132, 1982.

Ballard, S. and H. N. Pollack, Diversion of heat by Archean cratons: a model for southern Africa, *Earth Planet. Sci. Lett.*, *85*, 253–264, 1987.

Ballard, S. and H. N. Pollack, Modern and ancient geotherms beneath southern Africa, *Earth Planet. Sci. Lett.*, *88*, 132–142, 1988.

Basaltic Volcanism Study Project, *Basaltic Volcanism on the Terrestrial Planets*, 1286pp., Pergamon, New York, 1981.

Bickle, M. J., Implications of melting for stabilisation of the lithosphere and heat loss in the Archean, *Earth Planet. Sci. Lett.*, *80*, 314–324, 1986.

Bird, P., Initiation of intracontinental subduction in the Himalaya, *J. Geophys. Res.*, *83*, 4975–4987, 1978.

Boyd, F. R., J. J. Gurney, and S. H. Richardson, Evidence for a 150–200 km thick Archean lithosphere from diamond inclusion thermobarometry, *Nature*, *315*, 387–389, 1985.

Boyd, F. R. and J. J. Gurney, Diamonds and the African lithosphere, *Science*, *232*, 472–477, 1986.

Bowring S. A., Earth's early crust, *EOS Trans. AGU*, *73*, 33, 1992.

Breuer, D. and T. Spohn, Urey ratios and the distribution of heat producing elements between crust, mantle, and core reservoirs (Abstract), *Annal. Geophys.*, *10 Suppl.*, in press, 1992.

Christensen, U., Thermal evolution models for the Earth, *J. Geophys. Res.*, *90*, 2995–3007, 1985.

Cook, F. A. and D. L. Turcotte, Parameterized convection and the thermal evolution of the Earth, *Tectonophysics*, *75*, 1–17, 1981.

DePaolo, D. J., Nd isotopic studies: Some new perspectives on Earth structure and evolution, *EOS Trans. AGU*, *62*, 137–140, 1981.

DePaolo, D. J., The mean life of continents: Estimates of continental recycling from Nd and Hf isotopic data and and implications for mantle structure, *Geophys. Res. Lett.*, *10*, 705–708, 1983.

Dewey, J. F. and B. F. Windley, Growth and differentiation of the continental crust, *R. Soc. Lond. Phil. Trans.*, *A301*, 189–206, 1981.

England, P. C. and S. W. Richardson, Erosion and the age dependence of continental heat flow, *Geophys. J. R. Astr. Soc.*, *62*, 421–437, 1980.

Fischer, H.J. and T. Spohn, Thermal-orbital histories of viscoelastic models of Io (J1), *Icarus*, *83*, 39-65, 1990.

Fyfe, W. S., The evolution of the earth's crust: Modern plate tectonics to ancient hot spot tectonics?, *Chem. Geol.*, *23*, 89–114, 1978.

Galer, S. J. G., Interrelationships between continental freeboard, tectonics and mantle temperature, *Earth Planet. Sci. Lett.*, *105*, 214–228, 1991.

Goldstein, S. L., R. K. O'Nions, P. J. and Hamilton, A Sm-Nd isotopic study of atmospheric dusts and particulates from major river systems, *Earth Planet. Sci. Lett.*, *70*, 221–236, 1984.

Gurnis, M. and G. F. Davies, Simple parametric models of crustal growth, *J. Geodynamics*, *3*, 105-135, 1985.

Gurnis, M. and G. F. Davies, Aparent episodic crustal growth arising from a smoothly evolving mantle, *Geology*, *14*, 396–399, 1986.

Hallam, A., Pre–quaternary sea–level changes, *Annu. Rev. Earth Planet. Sci.*, *12*, 205–243, 1984.

Hargraves, R. B., Precambrian geologic history, *Science*, *193*, 363–371, 1976.

Jacobsen, S. B. and G. J. Wasserburg, The mean age of mantle and crustal reservoirs, *J. Geophys. Res.*, *84*, 7411–7427, 1979.

Jeanloz, R. and S. Morris, Temperature distribution in the crust and mantle, *Annu. Rev. Earth Planet. Sci.*, *14*, 377–415, 1986.

Kröner, A., Evolution of the Archean continental crust, *Annu. Rev. Earth Planet. Sci.*, *13*, 49–74, 1985.

McGovern, P. J. and G. Schubert, Thermal evolution of the Earth: effects of volatile exchange between atmosphere and interior, *Earth Planet. Sci. Lett.*, *96*, 27-37, 1989.

Morgan, P., Crustal radiogenic heat production and the selective survival of ancient continental crust, *J. Geophys. Res.*, *90*, Suppl. C561–570, 1985.

O'Nions, R.K., N. M. Evenson, and P. J. Hamilton, Geochemical modelling of mantle differentiation and crustal growth, *J. Geophys. Res.*, *84*, 6091–6101, 1979.

Peltier, W. R., Ice age geodynamics, *Annu. Rev. Earth Planet. Sci.*, *9*, 199-225, 1981.

Peltier, W. R., Models of the thermal history of the Earth, *EOS Trans. Am. Geophs. U.*, *70*, 1000, 1989.

Pollack, H. N., The heat flow from the continents, *Annu. Rev. Earth Planet. Sci.*, *10*, 459–481, 1982.

Reymer, A. P. S. and G. Schubert, Phanerozoic and Precambrian crustal growth, in *Proterozoic Lithospheric Evolution*, edited by A. Kröner, pp. 1–9, AGU Geodynamics Series Vol. 17, Am. Geophys. U., Washington, 1987.

Richardson S. H., J. J. Gurney, A. J. Erlank, and J. W. Harris, Origin of diamonds in old enriched mantle, *Nature*, *310*, 198–202, 1984.

Richter, F. M., Models for the Archean thermal regime, *Earth Planet. Sci. Lett.*, *73*, 350–360, 1985.

Schubert, G., The lost continents, *Nature*, *354*, 358, 1991.

Schubert, G., P. Cassen, R. E. and Young, Subsolidus convective cooling histories of the terrestrial planets, *Icarus*, *38*, 192–211, 1979.

Schubert, G. and A. P. S. Reymer, Continental volume and free-board through geological time, *Nature*, *316*, 336–339, 1985.

Schubert, G., T. Spohn, and R. T. Reynolds, Thermal histories, compositions, and internal structures of the moons of the solar system, in *Satellites*, edited by J. A. Burns and M. S. Matthews, pp. 224–292, Univ. of Arizona Press, Tucson, 1986.

Schubert, G. and T. Spohn, Thermal history of Mars and the sulfur content of its core, *J. Geophys. Res.*, *95*, 14095–14104, 1990.

Schubert, G., S. C. Solomon, D. L. Turcotte, M. J. Drake, and N. H. Sleep, Origin and thermal evolution of Mars, in *Mars*, edited by H. Kieffer, B. Jakowsky, C. Snyder, and M. S. Matthews, Univ. Arizona Press, Tucson, in press, 1992.

Sclater, P., C. Jaupart, and D. Galson, The heat flow through oceanic and continental crust and the heat loss of the Earth, *Rev. Geophys. Space Phys.*, *18*, 269–311, 1980.

Sclater, P., B. Parsons, and C. Jaupart, Oceans and continents: Similarities and differences in the mechanisms of heat loss, *J. Geophys. Res.*, *86*, 11535–11552, 1981.

Sharpe, H. N. and W. R. Peltier, A thermal history for the Earth with parameterized convection, *Geophys. J. R. Astr. Soc.*, *59*, 171–203, 1979.

Shaw, D. M., Trace element fractionation during anatexis, *Geochim. Cosmochim. Acta*, *34*, 237–243, 1970.

Shirey D. M. and R. W. Carlson, The Pb and Nd isotopic evolution of the Archean mantle, in *Workshop on The Archean Mantle*, edited by L. D. Ashwal, pp. 82 – 84, LPI Tech Rpt. 89–05, Lunar and Planetary Institute, Houston, 1989.

Spohn, T., Die thermische Evolution der Erde, *J. Geophys.*, *54*, 77–96, 1984.

Spohn, T., Mantle differentiation and thermal evolution of Mars, Mercury, and Venus, *Icarus*, *90*, 222–236, 1991.

Spohn, T. and G. Schubert, Modes of mantle convection and the removal of heat from the Earth's interior, *J. Geophys. Res.*, *87*, 4682–4696, 1982.

Spohn, T. and G. Schubert, Thermal equilibration of the Earth following a giant impact, *Geophys. J. Int.*, *107*, 163–170, 1991.

Stacey, F. D., Cooling of the Earth – A constraint on paleotectonic hypotheses, in *Evolution of the Earth*, edited by R. J. O'Connell and W. S. Fyfe, pp. 272–276, AGU Geodynamics Series Vol. 5, Am. Geophys. U., Washington, 1981.

Stevenson, D. J. and S. J. Turner, Fluid models of mantle convection, in *The Earth, Its Origin, Structure, and Evolution*, edited by M. McElhinny, pp. 227–263, Academic Press, New York, 1979.

Stevenson, D.J., T. Spohn, and G. Schubert, Magnetism and thermal evolution of the terrestrial planets, *Icarus*, *54*, 466–489, 1983.

Taylor, S. R., *Planetary Science: A Lunar Perspective*, 481pp., Lunar Planet. Inst., Houston. 1982

Taylor, S. R. and S. M. McLennan, *The continental crust: its composition and evolution*, 312pp., Blackwell Scientific Publications, Oxford. 1985

Tozer, D. C., Heat transfer and convection currents, *R. Soc. Lond. Phil. Trans.*, *A258*, 252-271, 1965.

Turcotte, D. L. and K. Burke, Global sea–level changes and the thermal structure of the earth, *Earth Planet. Sci. Lett.*, *41*, 341–346, 1978.

Turcotte, D. L., Dynamics of recycling, in *Crust/Mantle Recycling at Convergence Zones*, edited by S. R. Hart and L. Gülen, pp. 245–257, Kluwer Academic Publishers, Dordrecht, 1989a.

Turcotte, D.L., Thermal evolution of Mars and Venus including irreversible fractionation, *20th Lunar Planet. Sci. Conf., Abstr. Vol.*, 1138-1139, 1989b.

Vitorello, I. and H. N. Pollack, On the variation of continental heat flow with age and the thermal evolution of continents, *J. Geophys. Res.*, *85*, 983–995, 1980.

von Huene, R. and D. W. Scholl, Observations at convergent margins concerning sediment subduction, subduction erosion, and the growth of continental crust, *Rev. Geophys.*, *29*, 279–316, 1991.

White, W. M. and B. Dupré, Sediment subduction and magma genesis in the Lesser Antilles: isotopic and trace element constraints, *J. Geophys. Res.*, *91*, 5927–5941, 1986.

Wise, D. U., Continental margins, freeboard and the volumes of continents and oceans through time, in *The Geology of Continental Margins*, edited by C. A. Burk and C. L. Drake, pp. 45–58, Springer–Verlag, New York, 1974.

Zindler, A. and S. Hart, Chemical geodynamics, *Annu. Rev. Earth Planet. Sci.*, *14*, 493–571, 1986.

D. Breuer and T. Spohn, Institut für Planetologie, Westfälische Wilhelms-Universität, W. Klemmstraße 10, D-4400 Münster, Federal Republic of Germany.

Ferric Iron in the Upper Mantle and in Transition Zone Assemblages: Implications for Relative Oxygen Fugacities in the Mantle

H. ST.C. O'NEILL, D. C. RUBIE, D. CANIL, C. A. GEIGER[1], C. R. ROSS II,

F. SEIFERT AND A. B. WOODLAND

Bayerisches Geoinstitut, Universität Bayreuth, W-8580 Bayreuth, Federal Republic of Germany
[1] *Present address: Mineralogisch-Petrographisches Institut, Universität Kiel, Olshausenstr. 40, W-2300 Kiel 1, Federal Republic of Germany*

The $Fe^{3+}/\Sigma Fe$ ratio in least-depleted upper mantle spinel lherzolites is estimated to be 2.3 ± 1%, from Mössbauer determinations of $Fe^{3+}/\Sigma Fe$ in each phase, multiplied by the total Fe determined by electron microprobe analysis, and the phase's modal abundance. This ratio is lower than previous direct determinations of $Fe^{3+}/\Sigma Fe$ in mantle samples by wet chemistry, but is compatible with $Fe^{3+}/\Sigma Fe$ measurements in primitive MORB glasses. This low $Fe^{3+}/\Sigma Fe$ leads to a relatively high oxygen fugacity in the spinel lherzolite upper mantle (i.e. ΔQFM = -1 to +0.5 log-bar units) because the modally dominant phase in the upper mantle, olivine, almost completely excludes Fe^{3+}, and the next most abundant phase, orthopyroxene, accepts only limited amounts of Fe^{3+} into its structure. This concentrates Fe^{3+} in the modally minor phases, clinopyroxene and spinel. In contrast, experimental data show that the major phases in the mantle's transition zone, majorite garnet plus β-phase or silicate spinel, can all accommodate substantial amounts of Fe^{3+}, thus lowering the concentrations, and hence activities, of the Fe^{3+} components (i.e. Fe_3O_4 in spinel, $Fe^{2+}_3Fe^{3+}_2Si_3O_{12}$ in garnet). This will tend to result in lower relative fO_2's.

The minimum $Fe^{3+}/\Sigma Fe$ in either majorite garnet or silicate spinel occurs at any given T, P and Mg/(Mg+Fe) ratio at its low fO_2 stability limit, when it is in equilibrium with excess SiO_2 and Fe. Mössbauer spectroscopy shows that majoritic garnet of composition $(Mg_{0.85}Fe_{0.15})SiO_3$ synthesized from Fe^{3+}-free orthopyroxene, in equilibrium with excess SiO_2 and Fe metal, at 18 GPa and 1900°C, contains 10 ± 2% $Fe^{3+}/\Sigma Fe$, while $(Mg_{0.85}Fe_{0.15})_2SiO_4$ spinel synthesised from the same starting material at 18 GPa and 1700°C, also with excess SiO_2 and Fe metal, contains approximately 3% $Fe^{3+}/\Sigma Fe$. This leads to a minimum whole rock $Fe^{3+}/\Sigma Fe$ for a pyrolite-like composition of 5%. Any material with upper mantle $Fe^{3+}/\Sigma Fe$ in the transition zone must produce the extra Fe^{3+} by disproportionation of Fe^{2+} to Fe^{3+} plus Fe^o, or by reduction of oxidized volatile components such as CO_2 and H_2O to CH_4; upper mantle material would thus probably be at an fO_2 close to metal saturation in the transition zone.

If in lower mantle assemblages $Fe^{3+}/\Sigma Fe$ is similar in coexisting perovskite and magnesiowüstite such that the magnesiowüstite has the same $Fe^{3+}/\Sigma Fe$ as that estimated for the upper mantle, the relative fO_2 in the lower mantle will be approximately similar to that of the upper mantle, well above metal saturation. In a mantle without major chemical stratification (including oxygen content), the transition zone will thus form a shell of reducing conditions between relatively oxidizing upper and lower mantle.

INTRODUCTION

The application of phase equilibrium methods to measure the oxygen fugacity at which upper mantle peridotites equilibrated (principally using the olivine-orthopyroxene-spinel assemblage, but also making use of other equilibria [Eggler, 1983; Haggerty and Tompkins, 1983; O'Neill and Wall, 1987; Wood and Virgo, 1989; Luth et al., 1990]) has led to a consensus that the upper mantle is nearly everywhere relatively oxidized, the majority of samples falling within -1.5 to +0.5 log-bar units of the quartz-fayalite-magnetite

Evolution of the Earth and Planets
Geophysical Monograph 74, IUGG Volume 14
Copyright 1993 by the International Union of Geodesy and Geophysics and the American Geophysical Union.

oxygen buffer (ΔQFM = -1.5 to +0.5). So far, the most reduced samples reliably documented are a few abyssal peridotites and some specimens from orogenic lherzolite massifs at ΔQFM = -2.5 to -3 [Bryndzia and Wood, 1990; Woodland et al., 1992]. One consequence of this observation is that even the parts of the upper mantle at the lower end of the recorded range have oxygen fugacities >3 log-bar units more oxidized than that appropriate for chemical equilibrium with the Fe-rich metal now in the Earth's core. The oxidized nature of the upper mantle thus supports the evidence from the mantle's apparently anomalous overabundance of Ni and other siderophile elements [Ringwood, 1966] that the fundamental differentiation of the Earth into its metal core and silicate mantle could not have occurred by a simple one-stage metal segregation process, at least under upper mantle pressures. It has therefore often been argued that the composition and oxidation state of the mantle is the net result of heterogenous accretion and more than one core-forming event [e.g. O'Neill, 1991a,b].

A major precept in this type of argument is that the metal-silicate equilibrium occurred at comparatively low pressures, within the upper mantle regime. Too little is known about how planets accrete to say if this precept is the only one likely. For instance, it has been argued that oxygen is the best candidate for the light element in the Earth's core [e.g. Ringwood, 1984]; if so, then this would require metal/silicate interactions at substantially higher pressures, as >16 GPa is needed to dissolve sufficient O in liquid Fe [Ringwood and Hibberson, 1990]. If metal segregation did occur at very high pressures, then the chemical constraints on core formation will have to take this into account. Several authors have pointed out that metal/silicate partition coefficients at higher pressures and temperatures tend to approach unity, thus to some extent diminishing the siderophile element anomaly [e.g. Urakawa et al., 1989; Ringwood et al., 1990; Ohtani et al., 1991; Rama Murthy, 1991; Urakawa, 1991]. This raises the question, what happens to the constraint on metal/silicate equilibria from the mantle's fO_2 at very high pressures? The purpose of this paper is to attempt to answer this question, by estimating the oxygen content of the upper mantle, and then, assuming constant chemical composition, using this to estimate the relative oxygen fugacity in the mantle's transition zone.

Knowledge of fO_2 in the deep mantle is also needed to predict fluid speciation in C-O-H volatiles, and its potential role in "redox melting" [e.g. Taylor and Green, 1988], which may occur where volatiles are locally concentrated. Briefly, the argument runs thus: below the topmost part of the mantle (i.e., below about 60 km), only fluids rich in CH_4 can exist, as CO_2-rich compositions would result in carbonation of mantle peridotite, while H_2O-rich compositions would result either in partial melting, or in phlogopite/amphibole production in the upper mantle [e.g. Eggler, 1983], or would be held in potentially hydrous phases such as the β-phase (and structurally related phases) in the transition zone [Smyth, 1987]. Large scale movement of volatiles in the mantle should thus be restricted to CH_4-rich fluids, but CH_4-rich fluids imply quite reducing conditions (roughly, below IW+1 in peridotite-C-O-H systems [Woermann and Rosenauer, 1985; Taylor and Foley, 1989]), such as are rarely if ever found in upper mantle samples. With a lower fO_2 in the transition zone, the existence of CH_4-rich fluid becomes a possibility. On passing to the higher fO_2 environment of the upper mantle, such fluid will be oxidized, triggering partial melting (e.g. $CH_{4(fluid)} + 4 Fe_2O_3 = CO_{2(melt)} + H_2O_{(melt)} + 8 FeO$). This kind of mechanism has been suggested for the origin of kimberlites.

The treatment of oxygen in the upper mantle has differed from that of most other elements in that it has largely been concerned with chemical potential (i.e. the oxygen fugacity), rather than actual amounts or concentrations of O. The amount of O in the mantle is most easily handled as $Fe^{3+}/\Sigma Fe$ ratios, as Fe is far more abundant than any other element with differing oxidation states, with the possible exceptions of the volatiles C, H, and S [Canil and O'Neill, in preparation]. Oxygen fugacity and $Fe^{3+}/\Sigma Fe$ are of course related, but in a complicated fashion, involving bulk composition, phase assemblage, temperature and pressure. The focus on fO_2 may be due to it being easier to measure with useful accuracy than $Fe^{3+}/\Sigma Fe$ ratios. Oxygen fugacity is also the useful quantity for constraining the nature of any inaccessible part of the original system, i.e. a C-O-H fluid phase, or for testing if the system could have been in equilibrium

with metal, and in situations when fO_2 behaves as an extrinsic variable, i.e. if oxidation state is controlled by oxygen exchange with a fluid reservoir (which may be the case for some types of mantle melting). In contrast, it seems likely that material circulating through the mantle will do so essentially isochemically, and in such situations oxygen fugacity will behave as an intrinsic variable, and it is oxygen content (as reflected in $Fe^{3+}/\Sigma Fe$, for example) which is the controlling quantity. The approach taken in this paper will be to assess the oxygen fugacity of the transition zone under the assumption that the entire mantle is essentially isochemical, i. e., that the transition zone has the same chemical composition as typical, least-depleted or "primitive" upper mantle [e.g. Jagoutz et al., 1979], including $Fe^{3+}/\Sigma Fe$. Note that for the purposes of this paper, we consider the mantle to be divided into three parts, namely the "upper mantle" down to 400 km depth with olivine as the dominant component, the "transition zone" from 400 to 670 km consisting mainly of β- or γ- phase plus majoritic garnet, and the "lower mantle" below the 670 km seismic discontinuity, consisting mainly of silicate perovskite plus magnesiowüstite. Since upper mantle $Fe^{3+}/\Sigma Fe$ is not at present well known, the first step in our argument is to provide an estimate for this ratio.

FE^{3+}/ΣFE (OXYGEN CONTENT) OF THE UPPER MANTLE

For reasons that are discussed in Appendix 1, it seems that the determination of $Fe^{3+}/\Sigma Fe$ in upper mantle samples by traditional wet-chemical methods gives unreliable results, and that better estimates of $Fe^{3+}/\Sigma Fe$ in upper mantle samples may be made by using Mössbauer spectroscopy to determine $Fe^{3+}/\Sigma Fe$ in each phase individually, and then multiplying this ratio by total Fe, determined by electron microprobe analysis, and the modal abundance of the phase. Unfortunately, this restricts our data base to five spinel lherzolite xenoliths (four from the continental USA) analyzed by Dyar et al. [1989], for which approximate modes are available, eight spinel lherzolite xenoliths from British Columbia analyzed by Luth and Canil [1992], for which modes have to be estimated, and four high temperature garnet lherzolites hosted in Southern African kimberlites, for which reliable modes are available, but for which the $Fe^{3+}/\Sigma Fe$ Mössbauer determinations on the individual phases are incomplete.

The mineral chemistry of the spinel lherzolite xenoliths indicate that they are typical representatives of the continental spinel lherzolite suite, and nearly all of them appear to lie at the least depleted end of this suite. Importantly for our purpose, they have also all equilibrated at typical upper mantle oxygen fugacities (i.e. ~1 log unit below QFM [O'Neill, and Wall, 1987; Wood and Virgo, 1989]). We will first discuss the abundances of Fe^{3+} in the four phases of the upper mantle's spinel lherzolite facies (i.e. olivine, orthopyroxene, clinopyroxene and spinel), and then use this information to estimate $Fe^{3+}/\Sigma Fe$ for the primitive upper mantle.

Spinel

The need to determine $Fe^{3+}/\Sigma Fe$ in mantle spinels accurately, in order to calculate fO_2's from the olivine-orthopyroxene-spinel equilibrium, has led to a number of detailed Mössbauer studies [e.g. Wood and Virgo, 1989; Canil et al., 1990], which may be used to supplement the information from Dyar et al. [1989] and Luth and Canil [1992]. Taken together, these studies indicate that spinels from continental spinel lherzolite xenoliths have $Fe^{3+}/\Sigma Fe$ ranging from 0.15 to 0.35, varying mainly with fO_2. The mean of 70 samples is 0.24, and there is no obvious correlation of $Fe^{3+}/\Sigma Fe$ with the major variable in mantle spinel chemistry, $Cr/(Al+Cr)$. All samples studied by Dyar et al. [1989] and Luth and Canil [1992] fall within this range and thus appear typical. The Mössbauer determinations have been checked against synthetic spinels with known $Fe^{3+}/\Sigma Fe$ [Wood and Virgo, 1989], and have been shown to be free of systematic error.

Mantle spinel has the highest total iron ($FeO* = 10\text{-}12\%$) of the four phases, so that with its high $Fe^{3+}/\Sigma Fe$, it contributes significantly to whole rock $Fe^{3+}/\Sigma Fe$, despite its low modal abundance of 1 to 2%. The corollary of this is that we need to know its mode quite precisely, and this is difficult to do by direct methods such as point counting [see, e.g. Dick et al., 1984]. Indeed, many xenoliths are simply too small for a reliable estimate to be made.

We avoid this problem by making use of the systematics of Cr_2O_3 in mantle peridotites. Cr is a compatible element, which during partial melting seems to partition about equally between melt and residue [e.g. Liang and Elthon, 1990a], and is not expected to be altered subsequently by metasomatism. Consequently, Cr shows an almost constant abundance in mantle peridotites, irrespective of the degree of depletion [e.g. Maaløe and Aoki, 1977; Frey et al, 1985]. From Maaløe and Aoki [1977] we take $Cr_2O_3 = 0.42$ wt% at 40% MgO as the mantle average, which agrees well with the estimate for the primitive mantle abundance for Cr of Jagoutz et al. [1979] (that is, 2870 vs. 3010 ppm). Since Cr is very heavily concentrated into spinel, it typically contains about 50% of the whole rock Cr content, despite its low modal abundance. The remaining Cr is held in orthopyroxene and clinopyroxene. The olivine of spinel peridotites usually contains <0.03 wt% Cr_2O_3 [e.g. Archbald, 1979; Stosch, 1981; Liang and Elthon, 1990b], low enough to be ignored. We then calculate the mode of spinel from a simple mass balance, as:

$$Z_{sp} = [0.42 - (Cr_2O_{3(opx)} \times Z_{opx} + Cr_2O_{3(cpx)} \times Z_{cpx})]/Cr_2O_{3(sp)} \quad (1)$$

For typical peridotites, the uncertainty in the calculated modal percentage of spinel from this method is about ± 0.5% (absolute). If modes were to be estimated from whole rock analyses (information not available with the present data), it might also be preferable to use an idealized rather than the actually analyzed whole rock Cr_2O_3, because of the danger of the latter also suffering from spinel sampling errors. Probably some of the reported variation in Cr_2O_3 in mantle samples [e.g. as shown in Maaløe and Aoki, 1977] is caused by this [Liang and Elthon, 1990a]. In the present case, we note, for example, that the mode given by Dyar et al. [1989] for their otherwise fairly primitive-looking xenolith Ki-5-31 (10% clinopyroxene, 25% orthopyroxene, 5% spinel) implies whole rock Cr_2O_3 of 0.9 wt%, which is over twice

the estimated primitive upper mantle value, and larger than in any of the 302 whole rock analyses of continental spinel lherzolites treated by Maaløe and Aoki [1977], 300 of which have $Cr_2O_3 < 0.65$ wt%.

Clinopyroxene

Dyar et al. [1989] and Luth and Canil [1992] found that clinopyroxene has nearly as high $Fe^{3+}/\Sigma Fe$ as coexisting spinel, i.e. 0.12 to 0.23 (Dyar et al.) and 0.06 to 0.24 (Luth and Canil). Luth and Canil [1992] showed that such values are consistent with thermodynamic calculations (i.e. the activities of $CaFe^{3+}AlSiO_6$ and $NaFe^{3+}Si_2O_6$ components) at the T, P and fO_2 at which their samples equilibrated. Clinopyroxene is much more abundant than spinel in the least depleted lherzolites, but contains less total Fe ($FeO*$ is typically only ~ 3%), so its net contribution to whole rock Fe_2O_3 is about the same (see Table 1).

Orthopyroxene

Both Dyar et al. [1989] and Luth and Canil [1992] found that their orthopyroxenes have $Fe^{3+}/\Sigma Fe = 0.03$ to 0.06. Although this ratio is substantially less than in spinel or clinopyroxene, orthopyroxene typically has 6-7 wt% $FeO*$, and is present at about 25% in the mode; it thus holds a comparable portion of the whole rock's Fe_2O_3 to spinel or clinopyroxene (see Table 1).

Olivine

Olivine is the most abundant phase in the upper mantle (i.e. 55-60% in least depleted spinel lherzolites), and also contains the most $FeO*$ (Fo_{89} has 10.7 wt % FeO , about equal to co-existing spinel). The combination of high $FeO*$ and high abundance means that even quite small fractions of Fe^{3+} in olivine might contribute non-negligible amounts to whole rock $Fe^{3+}/\Sigma Fe$. For example, the limit of detection for $Fe^{3+}/\Sigma Fe$ by Mössbauer is commonly quoted as 0.02; if mantle olivine actually had 0.02 $Fe^{3+}/\Sigma Fe$, this would contribute 0.12 wt % Fe_2O_3 to the bulk rock, i.e. at about the same level as the contributions from the three other phases.

There are two lines of evidence which indicate that Fe^{3+} in mantle olivines is well below such a level. Firstly, the detailed thermogravimetric study of Nakamura and Schmalzried [1983] on synthetic Fe-Mg olivines shows that $Fe^{3+}/\Sigma Fe$ in fayalite (Fe_2SiO_4) remains very small (< 0.001) even at the maximum fO_2 for fayalite stability (i.e. QFM), and that substitution of Mg for Fe^{2+} causes $Fe^{3+}/\Sigma Fe$ to decrease exponentially. Since mantle olivines have generally equilibrated at $fO_2 < $ QFM, and have $Fe^{2+}/(Fe^{2+}+Mg) = 0.1$, this kind of intrinsic Fe^{3+} should be negligible.

TABLE 1. The contribution of each phase to whole rock Fe_2O_3 in spinel peridotite xenoliths.

Sample	Ba-2-3	Ep-1-13	Ki-5-31	Sc-1-1	H30-b2	SL32	SL47	SL125	LBR1	JL8	JL1	TKN15	RR222
Mode sp (%)	1.8	2.9	1.6	3.0	0.9	1.8	1.3	0.7	1.1	1.2	2.6	2.0	2.4
Fe_2O_3 fom sp (wt%)	0.05	0.08	0.06	0.08	0.04	0.07	0.06	0.02	0.07	0.03	0.06	0.04	0.05
Fe_2O_3 from cpx (wt%)	0.07	0.12	0.08	0.11	0.04	0.10	0.10	0.13	0.12	0.05	0.08	0.06	0.11
Fe_2O_3 from opx (wt%)	0.10	0.11	0.10	0.11	0.06	0.07	0.05	0.04	0.05	0.03	0.03	0.03	0.03
Whole rock Fe_2O_3 (wt%)	0.22	0.31	0.24	0.30	0.14	0.24	0.21	0.19	0.24	0.11	0.17	0.13	0.19
Whole rock $Fe^{3+}/\Sigma Fe$ (%)	2.5	3.4	3.0	3.4	3.0	2.0	1.9	2.0	2.2	1.4	2.1	1.5	2.3

All data from Dyar et al. [1989] and Luth and Canil [1992]. The amount of Fe_2O_3 contributed by each phase is given by $Fe^{3+}/\Sigma Fe$ determined from Mössbauer spectroscopy, multiplied by the total Fe from electron microprobe analysis and the phase's modal abundance. Idealized modes for orthopyroxene and clinopyroxene in least depleted spinel lherzolites of 25% and 15% are assumed, and the mode of spinel is calculated assuming Cr_2O_3 of 0.42% in the whole rock.

However, in chemically more complex environments heterogenous substitutions such as $Fe^{3+}_{(VI)} + Al_{(IV)} = Mg + Si$ (e.g. as in pyroxenes) become possible, and may enhance the solubility of Fe^{3+} in olivine. The maximum additional substitution by this mechanism (i.e. octahedral Fe^{3+} charge-balanced by tetrahedral Al), over that found in the Al-free system, should be given by $N_{Fe3+} = N_{Al} - N_{Cr}$, where the N´s are the number of atoms per formula unit. Good quality microprobe analyses of olivine from typical spinel lherzolites show Al_2O_3 <0.05 wt%, which, allowing for 0.02wt% Cr_2O_3, suggests a maximum solubility for Fe_2O_3 of ~0.06 wt%. With an olivine mode of 60%, this sums to a contribution of < 0.04 wt % to the whole rock Fe_2O_3, which is small enough to be considered negligible.

Primitive upper mantle $Fe^{3+}/\Sigma Fe$

Since our goal is to calculate $Fe^{3+}/\Sigma Fe$ for primitive upper mantle (i.e. least depleted by extraction of partial melt; equivalent to pyrolite [Ringwood, 1975], or the model composition of Jagoutz et al. [1979]) we use idealized figures for pyroxene modal abundances, namely 25% for orthopyroxene and 15% for clinopyroxene. These abundances are similar to those reported by Takahashi [1986] for KLB1, and also to those from several other xenoliths with primitive major element chemistry, and which have equilibrated at similar P, T and fO_2 conditions to the samples in our data base [e.g. Stosch, 1981; Preß et al., 1986; Xue et al., 1990]. For spinel, we use the nominal mode computed by assuming whole rock Cr_2O_3 of 0.42 wt%, as explained above.

Our justification for this procedure is that, for the pyroxenes, we observe no correlation of $Fe^{3+}/\Sigma Fe$ with depletion, within the limited range of depletion covered by the available data. For spinel, while we expect Fe_2O_3 to increase with increasing depletion, as Cr/(Cr+Al) increases [Reid and Woods, 1978], this tends to be cancelled by the concomitant decrease in calculated modal abundance.

The results are summarized in Table 1, and plotted in Figure 1. We find that calculated $Fe^{3+}/\Sigma Fe$ in the 13 spinel lherzolite samples of our data base is 0.023, with a statistical uncertainty of ±0.006. Additional systematic errors might accrue from systematic errors in the Mössbauer determination of $Fe^{3+}/\Sigma Fe$, or errors in our assumed modes. As for the former, we note that the Mössbauer determinations of $Fe^{3+}/\Sigma Fe$ in spinel have been checked against synthetic standards [Wood and Virgo, 1989], and have been shown to be accurate, while the $Fe^{3+}/\Sigma Fe$ in clinopyroxene accords well with that expected from thermodynamic calculation [Luth and Canil, 1992]. However, in orthopyroxene $Fe^{3+}/\Sigma Fe$ is near the Mössbauer limit of detection, and at these low levels peak areas often tend to be overestimated [Hawthorne and Waychunas, 1988]. For the modal abundances, any error for the assumed modal abundance of orthopyroxene is likely to be mostly compensated for by a change in the modal abundance of clinopyroxene, and vice versa: i.e. our estimate of 55 to 60% modal olivine in least-depleted spinel lherzolite is fairly robust. Taking all this into account, we estimate that primitive upper mantle $Fe^{3+}/\Sigma Fe$ must lie in the range 0.015 to 0.04. Taking the mean value of 0.023 $Fe^{3+}/\Sigma Fe$, and assuming a primitive mantle abundance for FeO* of 8 wt% [Maaløe and Aoki, 1977; Jagoutz et al., 1979], we obtain whole rock Fe_2O_3 of 0.20 wt%. This value is much lower than estimates from whole-rock wet chemical analyses of mantle peridotites, for reasons discussed in Appendix 1.

The relatively constant calculated $Fe^{3+}/\Sigma Fe$ in Figure 1 is of course partly an artefact of our assumption of particular pyroxene modal abundances, but it also reflects the fact that whole rock $Fe^{3+}/\Sigma Fe$ is a surprisingly insensitive function of fO_2 for peridotites with typical

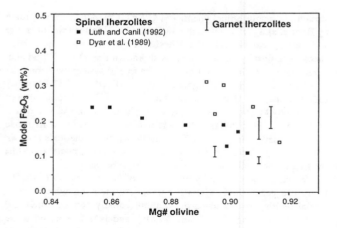

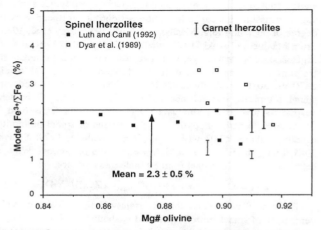

Fig. 1. Fe^{3+} in the primitive (i.e. least depleted) upper mantle, estimated from Mössbauer determinations of $Fe^{3+}/\Sigma Fe$ in orthopyroxene, clinopyroxene and spinel, and ideal primitive mantle modes of 25% orthopyroxene, 15% clinopyroxene, and spinel as calculated from a mass balance for Cr_2O_3. a) Model Fe_2O_3 vs. Mg/(Mg+Fe) in coexisting olivine, and b) Model $Fe^{3+}/\Sigma Fe$ vs. Mg/(Mg+Fe) in coexisting olivine. The mean value ($Fe^{3+}/\Sigma Fe$ = 2.3%) is from the spinel lherzolites only. Primitive upper mantle olivine has Mg/(Mg+Fe) = 0.89.

upper mantle mineralogy. Conversely, fO_2 will vary greatly with small changes in $Fe^{3+}/\Sigma Fe$ - for example, metal saturation in the upper mantle occurs at fO_2's near the IW buffer, about 3 to 4 log-bar units below typical upper mantle values [O'Neill and Wall, 1987], yet this large change in fO_2 is only accompanied by a change in $Fe^{3+}/\Sigma Fe$ of ~0.023, if $Fe^{3+}/\Sigma Fe \cong 0$ at metal saturation.

We expect that increasing depletion will result in lower whole rock $Fe^{3+}/\Sigma Fe$ simply because of the large decrease in modal abundance of the pyroxenes with increasing depletion, and the consequent increase of modal olivine [e.g Dick et al., 1984]. This holds regardless of whether the partial melting event which caused the depletion occurred with fO_2 extrinsically buffered (perhaps implying vapour saturation), or whether fO_2 varied intrinsically, or whether the depleted peridotite was then subsequently metasomatized - for example, if our argument that olivine contains virtually no Fe^{3+} is correct, then a residual dunite will have almost zero $Fe^{3+}/\Sigma Fe$. In other words, Fe_2O_3 behaves as an incompatible component during partial melting, whereas FeO is compatible.

If fO_2 is not controlled extrinsically, $Fe^{3+}/\Sigma Fe$ in primary magma will reflect the degree of partial melting, and high degree partial melts such as MORB will have relatively low $Fe^{3+}/\Sigma Fe$. For simple equilibrium batch melting of component M:

$$\frac{c_M^{liq}}{c_M^o} = \frac{1}{D_M^{sol/liq} + f(1 - D_M^{sol/liq})} \qquad (2)$$

where c denotes concentration, the superscript ^{o} the initial state, f the melt fraction, and D the average solid silicate/liquid silicate partition coefficient. We take $D_{FeO}^{sol/liq} = 1$, from the empirical observation that FeO* ($\approx$FeO) remains nearly constant with increasing depletion [e.g. Maaløe and Aoki, 1977; Frey et al., 1985], hence for small $Fe^{3+}/\Sigma Fe$ ratios:

$$\left(\frac{Fe^{3+}}{\Sigma Fe}\right)_{liq} \cong \left(\frac{Fe^{3+}}{\Sigma Fe}\right)_o \left[\frac{1}{D_{Fe_2O_3}^{sol/liq} + f(1 - D_{Fe_2O_3}^{sol/liq})}\right] \qquad (3)$$

For high MgO MORB we take $f \approx 0.15$ [e.g. Hofmann, 1981]. We estimate $D_{Fe_2O_3}^{sol/liq} \cong 0.2$, suitable for a moderately incompatible component, and similar to the empirical partition coefficient for Ga [Goodman, 1972; Frey et al., 1985; O'Neill, 1991a], which is geochemically a close analogue for Fe^{3+}, and shows a similar distribution amongst the solid phases of mantle peridotite [McKay and Mitchell, 1988]. Hence $Fe^{3+}/\Sigma Fe$ for primary MORB should be about 0.07, in good agreement with the value measured by Christie et al. [1986] on MORB quench glass rims of 0.07 ± 0.03. The oxidation state of MORB is thus consistent with our estimate of low upper mantle $Fe^{3+}/\Sigma Fe$. At very low degrees of partial melting Equation (3) reduces to:

$$\left(\frac{Fe^{3+}}{\Sigma Fe}\right)_{liq} \cong \left(\frac{Fe^{3+}}{\Sigma Fe}\right)_o / D_{Fe_2O_3}^{sol/liq} \qquad (4)$$

that is, about 0.012 - so that using the relationship between fO_2 and $Fe^{3+}/\Sigma Fe$ of Kilinc et al. [1983], we estimate that, if fO_2 is an intrinsic variable during melting, primary low degree partial melts (alkali basalts?) should be ~1 log-bar unit more oxidized than MORBs - c.f. Carmichael and Ghiorso [1986]. Any primary melt with higher $Fe^{3+}/\Sigma Fe$ must reflect oxidizing metasomatism (e.g. many island arc basalts).

$Fe^{3+}/\Sigma Fe$ in garnet lherzolites

Modal data plus some $Fe^{3+}/\Sigma Fe$ analyses of individual phases exist for four high temperature garnet lherzolites hosted in Southern African kimberlites. These lherzolites are thought to represent samples of fertile asthenospheric mantle, originating from depths of ~200 km beneath the Kaapvaal craton [Boyd and Mertzman, 1987; Boyd, 1987]. Such "fertile" peridotites have similar bulk chemistry to least-depleted spinel lherzolite xenoliths [Boyd and Mertzman, 1987].

We present a preliminary estimate of $Fe^{3+}/\Sigma Fe$ in these fertile garnet lherzolites from the available data, summarized in Table 2. Luth et al. [1990] found that the garnets typically contain substantial amounts of ferric iron ($Fe^{3+}/\Sigma Fe = 0.1 - 0.13$). Of clinopyroxene from this facies only that from FRB1033 has been examined by Mössbauer spectroscopy: Luth and Canil [1992] found $Fe^{3+}/\Sigma Fe = 0.28$, which is at the high end of the range reported for clinopyroxenes from spinel lherzolites (see above). In the absence of further data, we use this value as typical for all garnet lherzolite facies clinopyroxene. There are no Mössbauer data for $Fe^{3+}/\Sigma Fe$ in orthopyroxene, so we assume that the

TABLE 2. Estimates of $Fe^{3+}/\Sigma Fe$ for four high temperature garnet lherzolite xenoliths

	FRB1033	BD2501	FRB76	PHN5267
Mg# olivine	0.895	0.910	0.910	0.914
Modes (%)				
garnet	4.9	5.1	9.0	7.7
clinopyroxene	3.5	2.6	4.0	7.6
orthopyroxene	7.5	6.0	15.0	20.0
wt% Fe_2O_3 in gt	0.94	0.86	0.89	0.83
wt% Fe_2O_3 in cpx[a]	1.03	(0.9)	(1.1)	(1.0)
wt% Fe_2O_3 in opx[b]	(0.2-0.6)	(0.2-0.5)	(0.2-0.6)	(0.2-0.5)
Whole Rock				
$Fe^{3+}/\Sigma Fe$ (%)	1.1-1.5	1.0-1.2	1.7-2.3	1.8-2.4
Fe_2O_3 (wt%)	0.10-0.13	0.08-0.10	0.15-0.21	0.18-0.24

The amount of Fe_2O_3 in each phase is given by $Fe^{3+}/\Sigma Fe$ determined from Mössbauer spectroscopy, multiplied by the total Fe from electron microprobe analysis. Modal data for BD2501 from Cox et al. [1987], others from F. R. Boyd (personal communication to D. Canil). $Fe^{3+}/\Sigma Fe$ in garnet from Luth et al. [1991] and in cpx from Luth and Canil [1992].
[a] Values in parentheses estimated using $Fe^{3+}/\Sigma Fe = 0.28$ from FRB1033.
[b] Estimated using $Fe^{3+}/\Sigma Fe = 0.03$ to 0.10, which is the range found for opx in spinel lherzolites (samples listed in Table 1).

range found for spinel lherzolite orthopyroxenes also holds (i.e. $Fe^{3+}/\Sigma Fe = 0.03$ to 0.10). Using these approximations, calculated whole rock $Fe^{3+}/\Sigma Fe$ for the four garnet lherzolites is given in Table 2, and plotted in Figure 1. These preliminary data suggest that there is no significant difference in $Fe^{3+}/\Sigma Fe$ between these garnet lherzolites, and the spinel lherzolites previously discussed.

THE REDOX STATE OF THE TRANSITION ZONE

The fundamental reason that the fO_2 of the upper mantle is quite high (i.e. near QFM), whereas its oxygen content is quite low ($Fe^{3+}/\Sigma Fe \cong 0.023$), is that Fe^{3+} is excluded for crystal-chemical reasons from entering the most abundant phase in the upper mantle assemblage, olivine, thereby forcing up its concentration in the minor phase, spinel. If, for the sake of argument, $Fe^{3+}/\Sigma Fe$ were the same in each phase of the upper mantle, so that $Fe^{3+}/\Sigma Fe$ in spinel were also 0.023 rather than ~0.24 (the average observed value), fO_2, as given by the olivine-orthopyroxene-spinel equilibrium:

$$3\,Fe_2SiO_4 + O_2 = 2\,Fe_3O_4 + 3\,SiO_2 \qquad (5)$$
olivine   spinel   ol/opx

would decrease by approximately $(0.023/0.24)^4$, i.e. nearly 4 log units, since $a_{Fe_3O_4}^{sp} \propto (X_{Fe^{3+}}^{sp})^2$, and the small amount of Fe^{3+} means that its hypothetical redistribution will not affect the activities of the other participants in reaction (5). For most of the upper mantle, 4 log units more reducing would put the fO_2 near the level of metal saturation [e.g. O'Neill and Wall, 1987], and would also be in the regime where CH_4 would be the dominant gas species in C-O-H fluids at upper mantle pressures [e.g. Woermann and Rosenhauer, 1985].

Now, consider the distribution of Fe^{3+} in the next layer below the olivine rich upper mantle, the transition zone.

The transition zone in the mantle

Below 400 km in the mantle $(Mg,Fe)_2SiO_4$ olivine (with $Mg/(Mg+Fe)=0.9$) transforms over a narrow interval to a phase with a spinel-like structure, known as the β-phase, or wadsleyite. The β-phase in turn transforms to $(Mg,Fe)_2SiO_4$ with the spinel structure proper (the γ-phase) at 500-550 km, and this silicate spinel in turn disproportionates to $(Mg,Fe)SiO_3$ perovskite plus $(Mg,Fe)O$ magnesiowüstite at 670 km. The olivine/β-phase transformation coincides with the 400 km seismic discontinuity in the mantle, and the spinel/perovskite reaction with the 670 km discontinuity, leading to the view now held widely, albeit not unanimously, that the seismic structure of the mantle can be explained solely by these phase changes without recourse to major large-scale chemical layering [e.g. Weidner and Ito, 1987; Ringwood and Irifune, 1988; Bukowinski and Wolf, 1990]. The region in between the olivine/β-phase and spinel/perovskite phase changes is known as the transition zone (i.e. transition between the upper and lower mantles), and constitutes about 13% of the mantle by volume.

In a system with typical upper mantle bulk chemistry (i.e. similar to the pyrolite of Ringwood [1975]), as pressure increases above ~6 GPa, $(Ca,Mg,Fe)SiO_3$ pyroxenes become increasingly soluble in the garnet phase, e.g. as the majorite component $Mg_3(MgSi)Si_3O_{12}$. By pressures corresponding to the 400 km discontinuity, all or nearly all the pyroxene is reacted out [Irifune, 1987], so that the transition zone has an essentially bi-mineralic assemblage of β-phase plus majoritic garnet in its top half, and spinel plus majoritic garnet in the bottom half. Pyrolite composition consists of 57% β-phase or spinel, and 43% majoritic garnet [Ringwood, 1975]. In detail, a very small amount of a Ca-rich phase, pyroxene or Ca-silicate perovskite, may also be present [e.g. Ito and Takahashi, 1987], and, at slightly lower Mg# than pyrolite, perhaps also some stishovite.

If we assume that the transition zone has the same chemical composition as the upper mantle, including $Fe^{3+}/\Sigma Fe$, then the oxygen fugacity of the transition zone will depend on how well these phases can accommodate the small amounts of Fe^{3+}. We will now discuss the compatibility of Fe^{3+} in each of the major transition zone phases.

The crystal chemistry of Fe^{3+} in transition zone phases

γ-phase (silicate spinel). Fe_3O_4 of course also has the spinel structure, and there is complete solid solution of Fe_3O_4 with Fe_2SiO_4 at 9 GPa and 1100°C. At lower pressures (e.g. 7 GPa) the completeness of the solid solution is interrupted by the appearance of a spinelloid phase [Canil et. al., 1991; Ross et al., 1992], but there is no reason to suppose that the solution becomes intrinsically unstable. A summary of experimental results on the synthesis of spinel solid solutions along the Fe_3O_4 - Fe_2SiO_4 binary is given in Table 3, and the molar volumes are plotted as a function of composition in Figure 2. The linear dependence of the molar volume on composition reflects the fact that the Fe_3O_4 component retains the inverse cation arrangement across the solid solution [O'Neill and Navrotsky, 1984]. Karpinskaya et al. [1982] have also reported a spinel nearly halfway between Fe_3O_4 and Fe_2SiO_4 in composition at a temperature as low as 800°C at 10 GPa. The implication from the apparent lack of a solvus at such temperatures is that activity coefficients in Fe_3O_4-Fe_2SiO_4 spinel cannot be very large. Activity-composition relations have yet to be determined in any binary spinel system between a 4-2 normal spinel (like Fe_2SiO_4) and a 3-2 inverse spinel [O'Neill and Navrotsky, 1984], and therefore it is not yet possible to predict activity-composition relations in the Fe_3O_4 - Fe_2SiO_4 spinel system. However, the activity of magnetite ($a^{sp}_{Fe_3O_4}$) in silicate spinel in equilibrium with Fe metal at the fO_2 of the quartz-fayalite-iron buffer (the low fO_2 breakdown of Fe_2SiO_4 olivine at low pressure), is 0.007-0.015 at 1100-1500°C, implying Fe_3O_4 contents of around 1% if $a^{sp}_{Fe_3O_4} \approx X^{sp}_{Fe_3O_4}$, or 10% if $a^{sp}_{Fe_3O_4} \approx (X^{sp}_{Fe_3O_4})^2$. Thus even at the lowest oxygen fugacities at which "Fe_2SiO_4" spinel is stable, it is likely to contain Fe^{3+} at levels which are significant when compared to the low $Fe^{3+}/\Sigma Fe$ of the upper mantle. These levels also mostly fall within the range detectable by Mössbauer spectroscopy.

Figure 3 shows a Mössbauer spectrum of Fe_2SiO_4 spinel synthesized at 1000°C and 8 GPa in an Fe capsule. There was no detectable Fe^{3+} in the olivine starting material, but there is 0.07 ± 0.02 $Fe^{3+}/\Sigma Fe$ in the run product. The charge contained excess $FeSiO_3$ (mainly around the edge of the capsule), suggesting that fO_2 is buffered at the Fe_2SiO_4-$FeSiO_3$-Fe equilibrium.

TABLE 3. Synthesis data for Fe_2SiO_4 - Fe_3O_4 spinels*

Run#	Starting Material	P (GPa)	T (°C)	Run time (hours)	Products	Spinel lattice constant (Å)	mol% Fe_3O_4§
196	$Fa_{85}Mt_{15}$	7.0	950	15	sp,spd	8.2541(3)	16
195	$Fa_{85}Mt_{15}$	7.0	1075	12	sp,spd	8.2512(2)	18
195	$Fa_{40}Mt_{60}$	7.0	1075	12	sp,spd	8.3453(2)	74
205	$Fa_{85}Mt_{15}$	7.0	1150	12	sp,spd	8.2511(4)	18
205	$Fa_{40}Mt_{60}$	7.0	1150	12	sp,spd	8.3532(3)	72
269	$Fa_{90}Mt_{10}$	7.0	1175	11	sp,spd,cfs	8.2462(1)	10
219	Mt_{100}	7.0	1200	5	sp	8.3962(3)	100
155	$Fa_{85}Mt_{15}$	7.0	1200	5	sp,spd,cfs	8.2463(1)	10
321	$Fa_{25}Mt_{75}$	7.0	1200	5	sp	8.3541(2)	75
170**	Fa_{100}	8.0	1000	9	sp,cfs	n.d.	14
524	$Fa_{50}Mt_{50}$ + SiO_2	9.0	1100	12	sp,coes	8.315(2)	52

* lattice constants and phase relations for spinelloids will be presented elsewhere
** synthesized in Fe capsule, all other runs used Au capsules except #524 which used Ag
§ from electron microprobe analysis
Fa = fayalite, Mt = magnetite, coes = coesite, sp = spinel, spd = spinelloid, cfs = clinoferrosilite, n.d. not determined

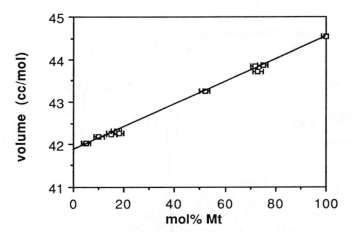

Fig. 2. Volume (in cm³/mol) vs. composition for spinels synthesized at 7 to 9 GPa along the join Fe_2SiO_4 - Fe_3O_4. The linear trend extrapolates to 42.86 cm³/mol at Fe_2SiO_4, which corresponds to a cell edge of 8.224 Å.

The effect of adding Mg to the spinel system will be to raise the activity coefficient of Fe_3O_4, because the free energy of the reciprocal reaction:

$$1/2\ Fe_2SiO_4 + MgFe_2O_4 = 1/2\ Mg_2SiO_4 + Fe_3O_4 \qquad (6)$$

spinel   spinel   spinel   spinel

is strongly negative (~10 kJ/mol at 1000 K - thermo-dynamic data from Bina and Wood [1987] for the silicate spinels, and Robie et al. [1978] for the ferrite spinels). Adding Mg to the system also enlarges the stability field of spinel to lower fO_2's. Nevertheless, even Mg-rich silicate spinels contain sufficient Fe^{3+} to be detectable by the Mössbauer method (see below).

β–phase. The β–phase is a spinelloid polytype, and like spinel proper consists of an approximately cubic-close-packed oxygen lattice with 1/8 of the tetrahedral and 1/2 of the octahedral interstices occupied by cations, leading to an octahedral to tetrahedral cation site ratio (M:T) of 2:1. However, the spinelloids differ as to the degree of polymerization of the various sites, most importantly the polymerization of the T sites. In particular, β–phase has a single T site [Horiuchi and Sawamoto, 1981], polymerized to form T_2O_7 groups. The bridging oxygen, O2, is also bonded to M2, one of the three crystallographically distinct octahedral sites. Based on simple Pauling bond-strength sums, O3 and O4 (each coordinating one T and three M sites) are exactly charge balanced, whereas O1 (coordinating 5 M sites) is severely underbonded (1.67 v.u.) and O2 (the bridging oxygen) is severely overbonded (2.33 v.u.). Distortion of the structure, in particular the T site, causes some compensation, and based upon the bond-strength bond-length formulae of Brown and Altermatt [1985] the bonding to O1, O2, O3 and O4 is 2.07, 1.96, 1.91, and 2.01 v.u. respectively.

Although no systematic experiments investigating the compatibility of Fe^{3+} with the β-phase structure have been carried out, there are several lines of evidence that suggest that significant quantities of Fe^{3+} might easily be incorporated into this structure.

Hazen et al. [1990] have synthesized single crystals of β-phase with nominal compositions on the join Mg_2SiO_4 - Fe_2SiO_4 at 1800°C and 16 GPa. Mössbauer analysis of these samples revealed up to 0.08

$Fe^{3+}/\Sigma Fe$, indicating that at least under the conditions of synthesis (possible fO_2, in particular, is not reported) substantial amounts of ferric iron may be incorporated without difficulty. This substitution of Fe^{3+} had no clear effect on bulk modulus, linear compressibility, or unit cell dimensions.

Ohtani et al. [1986], Ito and Takahashi [1987] and Ohtani and Sawamoto [1987] have shown that β-phase in equilibrium with melt at 20 GPa contains up to 1.5 % Al_2O_3 and Cr_2O_3; thus there is no intrinsic difficulty in including trivalent cations larger than Si into the β-phase structure, although trivalent iron nominally requires larger M-O distances than either Al^{3+} or Cr^{3+} (1.865 Å vs. 1.757 Å or 1.803 Å for tetrahedral coordination, based on Brown and Altermatt [1985]).

A spinelloid phase has been recently described in the system Fe_3O_4 - Fe_2SiO_4 at 1200°C and 7 GPa [Canil et al., 1991; Ross et al, 1992], with $Fe^{3+}/\Sigma Fe \approx 0.46$. Although not isostructural with the β-phase, both are spinelloids and therefore have many structural similarities. In particular, the bridging oxygen (O2) in β-phase is topologically identical with O5 in this phase, but Pauling bond-sum charge-balance is improved by partial substitution of Fe^{3+} for Si in the tetrahedral sites; the reduced valence of the tetrahedrally-coordinated cation allows relaxation of the structure and regularization of the tetrahedral sites. Similarly, O1 in the β-phase shows severe underbonding based on Pauling bond-sums, although distortions in the adjoining octahedrally-coordinated cation sites result in good charge balance (based on Brown and Altermatt [1985]); this oxygen is topologically equivalent to O6

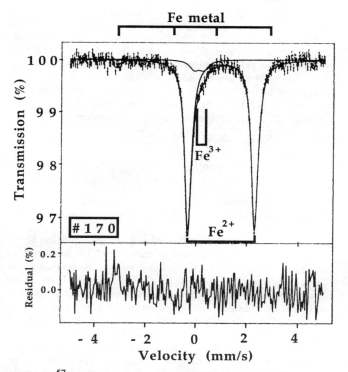

Fig. 3. ^{57}Fe Mössbauer spectrum (298 K) of "Fe_2SiO_4" spinel synthesized at 1000°C, 8GPa in an Fe capsule (run 170 in Table 3), fitted to two symmetrical Lorentzian doublets. A small amount of metallic iron (not fitted) is also visible. The residual gives the deviation of the observed spectrum from the calculated envelope, divided by the square-root of the background. Velocity scale relative to ^{57}Fe in Rh; for conversion relative to metallic iron, add 0.114 mm/s.

in the new phase, which also has a good Brown and Altermatt charge balance, but with considerably less distortion of the cation sites. As in spinel, half the Fe^{3+} must substitute on the T site (thus, incidentally, improving charge balance around O2). The octahedral substitution is less clear, but the smaller M-O distance for Fe^{3+} as compared to Mg or Fe^{2+} suggests that the M2 site might be preferred.

Smyth [1987] suggested that the β-phase might be a host for H_2O; the bonding pattern around O1 is such that occupation by hydroxyl might be energetically favorable. It was suggested that charge balance would be maintained by vacancy on M1. Another possible mechanism for maintaining charge balance in such a case would be substitution of Fe^{3+} for Si.

Majorite garnet. Many types of Fe^{3+}-bearing garnets are known, for example the magnetic REE ferrite garnets ($R_3Fe^{3+}_5O_{12}$), in which Fe^{3+} fills both the octahedral and the tetrahedral sites. Among silicate garnets Fe^{3+} is generally restricted to the octahedral site, although tetrahedral Fe^{3+} can occur under exceptional circumstances (e.g. Amthauer et al. [1976]). At low pressures, substantial fractions of Fe^{3+} only occur in Ca-garnets, i.e. as $Ca_3Fe^{3+}_2Si_3O_{12}$ (andradite). In the $MgO-SiO_2-Fe-O$ system, Fe^{3+} can be considered in terms of a skiagite ($Fe^{2+}_3Fe^{3+}_2Si_3O_{12}$) or a khoharite ($Mg_3Fe^{3+}_2Si_3O_{12}$) component. These garnet end-members, which have not been found in nature, are expected to become more stable with increasing pressure [Schreyer and Baller, 1981]. This expectation has been borne out for skiagite in a series of synthesis experiments along the almandine ($Fe^{2+}_3Al_2Si_3O_{12}$) - skiagite join [Woodland and O'Neill, 1992]. The solubility of skiagite in almandine increases steadily with increasing pressure. At 1100°C, the skiagite component saturates at 12, 35, and 90 mole % at 2.7, 6.0, and 9.0 GPa respectively. The skiagite end-member itself becomes stable above ~9.5 GPa at 1100°C. Mössbauer spectra of these garnets indicate that all Fe^{3+} resides on the octahedral sites. The molar volume of end-member skiagite is 121.4 ± 0.1 cm^3, in agreement with the estimate from the systematics of garnet crystal chemistry of 121.2 cm^3 [Novak and Gibbs, 1971]. This implies that the molar $\Delta V_{(1,298)}$ for the reciprocal exchange reaction:

$$Fe^{2+}_3Fe^{3+}_2Si_3O_{12} + Ca_3Al_2Si_3O_{12} =$$
garnet   garnet

$$Fe^{2+}_3Al_2Si_3O_{12} + Ca_3Fe^{3+}_2Si_3O_{12} \qquad (7)$$
garnet   garnet

is positive, and is relatively large for a reciprocal reaction (~0.1 J/bar). This acts to lower the activity coefficient of $Fe^{2+}_3Fe^{3+}_2Si_3O_{12}$ at high pressures.

We have synthesized a number of majorite garnets in the $MgO-SiO_2-Fe-O$ system, from starting material which contained no detectable Fe^{3+} ($Fe^{3+}/\Sigma Fe < 0.02$). Run conditions are summarized in Table 4, and experimental details are given in Appendix 2. Mössbauer spectroscopy shows that all these garnets contain substantial amounts of Fe^{3+}. A representative spectrum at 80 K for one sample, synthesized in a Re capsule, is shown in Figure 4. Visually, the spectum consists of a very intense outer doublet and two well-resolved, weaker inner peaks. The low-velocity peak of the latter two is much more intense than the high velocity peak, and therefore these two cannot belong to a single doublet. Rather, an additional doublet has to be fitted under the low-velocity peak and the minimum number of doublets in a fit must be three. Although structural considerations for these tetragonal garnets lead to a large number of possible doublets, in the absence of fine structure in the Mössbauer spectra the number of doublets was kept to the minimum of three. Based on the

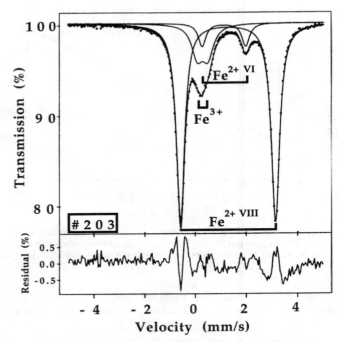

Fig. 4. ^{57}Fe Mössbauer spectrum (80 K) of majorite garnet (Mg/(Mg+Fe) = 0.9) synthesized at 1800°C, 18 GPa in a Re capsule (run 203 in Table 4), fitted to three symmetrical Lorentzian doublets that correspond to Fe^{2+} in the octahedral site, Fe^{2+} in the dodecahedral site, and Fe^{3+}. Residuals and velocity scale as in Figure 3.

resulting hyperfine parameters [Geiger et al., 1991], the intense outer doublet is due to Fe^{2+} in the dodecahedral site(s). Of the weaker inner doublets the one with the larger quadrupole splitting and isomer shift has to be assigned to Fe^{2+} in octahedral coordination, and the doublet with the smaller isomer shift and quadrupole splitting has to be assigned to Fe^{3+}. Although there is no certainty as to the site occupancy of Fe^{3+} in these garnets, the fractional area of the Fe^{3+} doublet and therefore the $Fe^{3+}/\Sigma Fe$ is well constrained, because it is determined by the intensity difference of the low-velocity and high velocity envelopes.

$Fe^{3+}/\Sigma Fe$ ratios for three garnet samples have been calculated from the Mössbauer data (collected both at 80 and 298 K) and structural formulae assigned under the assumption that Fe^{3+} occupies the octahedral site [Geiger et al., 1991]. $Fe^{3+}/\Sigma Fe$ ratios lie between 0.12 and 0.16 (Table 4), implying 3 to 5% of a $(Mg,Fe^{2+})_3Fe^{3+}_2Si_3O_{12}$ component. The $Fe^{3+}/\Sigma Fe$ ratios are derived from the 80 K spectra, to account for possibly different recoiless fractions of Fe^{3+} and Fe^{2+} [Amtauer et al., 1976]; however, the differences in $Fe^{3+}/\Sigma Fe$ between the 80 K and 298 K spectra were found to be less than 10% (relative), in agreement with results from synthetic $Fe^{2+}_3Al_2Si_3O_{12}$ - $Fe^{2+}_3Fe^{3+}_2Si_3O_{12}$ garnets [A. B. Woodland, unpublished data]. The two runs using Mo as the capsule material give slightly lower $Fe^{3+}/\Sigma Fe$ ratios than the run in the Re capsule. This may be significant if the Mo capsule exerts some buffering influence.

In summary: spinel, β-phase and high pressure majoritic garnet are all capable of containing substantial amounts of Fe^{3+} under transition zone conditions of temperature, pressure and likely Mg/(Mg+Fe) ratio. In fact, all syntheses of these phases seem to contain enough Fe^{3+} to

TABLE 4. $(Mg,Fe)SiO_3$ garnet and $(Mg,Fe)_2SiO_4$ spinel syntheses

Run#	Starting Material	Capsule	P(GPa)	T (°C)	Time (min)	Products	Lattice Constants (Å) §		$Fe^{3+}/\Sigma Fe$
							a	c	
(Mg,Fe)SiO₃ garnet									
203	Fs_{10}	Re	18.0	1800	30	gt (st)	11.5328(3)	11.4444(3)	0.16
307	Fs_{20}	Mo	18.0	1800	15	gt (st)	11.5291(10)	11.4458(10)	0.14
319	Fs_{10}	Mo	18.0	1800	15	gt (st)	11.5307(5)	11.4412(5)	0.12
526	Fs_{15} + Fe + SiO_2	Fe	18.0	1900*	5	gt+st+Fe (sp)	11.5282(9)	11.4659(12)	0.10
(Mg,Fe)₂SiO₄ spinel									
519	Fs_{15} + SiO_2	Fe	18.0	1700*	15	sp+st+Fe (gt)	8.101(2)		0.06**

Fs = ferrosilite, i.e. Fe/(Fe + Mg)
§ using stishovite as an internal standard, with a = 4.1790 Å, c = 2.6649 Å.
* Temperature estimated on power (± 100°C)
** probably a maximum value, from the fitting procedure - see discussion in text
gt = garnet, sp = spinel, st = stishovite, Fe = Fe metal, phases listed in brackets are in trace amounts

be detected by Mössbauer spectroscopy, even when these syntheses are accomplished using Fe^{3+}-free starting materials, and (supposedly) reducing Mo capsules. The importance of this observation for the transition zone fO_2 is that Fe^{3+} will not be concentrated into modally minor phases as in the upper mantle, but will be distributed throughout the major phases, lowering its chemical potential. This implies lower fO_2 (see Equations 10 and 11 below). In order to quantify this, we now report some experiments at the low fO_2 stability limits of spinel and majoritic garnet, under buffered conditions, to determine the minimum Fe^{3+} content of these phases.

Minimum $Fe^{3+}/\Sigma Fe$ in silicate spinel and majorite garnet in the system MgO-SiO_2-Fe-O

The low fO_2 stability limit of $(Mg,Fe)_2SiO_4$ spinel is given by the reaction:

$$Fe_2SiO_4 = 2 Fe + SiO_2 + O_2 \qquad (8)$$
$$\text{spinel} \quad \text{metal} \quad \text{stish}$$

and that of $(Mg,Fe)_4Si_4O_{12}$ majorite garnet by:

$$Fe_4Si_4O_{12} = 4 Fe + 4 SiO_2 + 2 O_2 \qquad (9)$$
$$\text{garnet} \quad \text{metal} \quad \text{stish}$$

Both reactions show that in the presence of excess SiO_2 and Fe metal, fO_2 depends on the activities of Fe_2SiO_4 and $Fe_4Si_4O_{12}$, i.e. approximately on the $Fe^{2+}/(Mg+Fe)$ ratio in the simple MgO - SiO_2 - Fe - O system.

The amount of Fe^{3+} in each phase is controlled by the reactions:

$$3 Fe_2SiO_4 + O_2 = 2 Fe_3O_4 + 3 SiO_2 \qquad (10)$$
$$\text{spinel} \qquad \text{spinel} \quad \text{stish}$$

and:

$$5 Fe_4Si_4O_{12} + 2 O_2 = 4 Fe_3Fe_2Si_3O_{12} + 8 SiO_2 \qquad (11)$$
$$\text{garnet} \qquad \text{garnet} \qquad \text{stish}$$

The minimum amount of Fe^{3+} in either spinel or garnet at constant $Mg/(Mg+Fe)$ occurs at the low fO_2 stability limits as given by reactions (8) and (9) - that is, buffered by excess Fe metal and SiO_2.

In order to determine these minimum Fe^{3+} contents, we synthesized spinel+stishovite+Fe metal and garnet+stishovite+Fe metal, both from a composition of $(Mg_{0.85}Fe_{0.15})SiO_3$ plus excess Fe and SiO_2. The spinel was synthesized at 18 GPa and 1700°C, the garnet at 18 GPa and 1900°C, the higher temperature being to the right hand side of the divariant reaction:

$$2 (Mg_{0.85}Fe_{0.15})_2SiO_4 + 2 SiO_2 = (Mg_{0.85}Fe_{0.15})_4Si_4O_{12} \qquad (12)$$
$$\text{spinel} \qquad\qquad \text{stish} \qquad\qquad \text{garnet}$$

c.f. Ohtani and Kagawa [1989]. Both experiments were done using Fe capsules. Experimental details are described in Appendix 2, and the results summarized in Table 4.

The Mössbauer spectrum for the garnet is shown in Figure 5. The spectrum is similar to that found for other majorite garnets (e.g. Figure 4), but with one extra peak between the high velocity $Fe^{2+}(VIII)$ and $Fe^{2+}(VI)$ peaks. Fitting an extra doublet to this peak results in hyperfine parameters essentially identical to those found for spinel (Figure 3), and this interpretation is consistent with the small amount of spinel identified in the run product by powder XRD. The $Fe^{3+}/\Sigma Fe$ ratio is found to be 0.10 ± 0.02, which is only slightly less than that found for the garnets synthesized in Mo or Re, implying that the fO_2 in these other runs was not anomalously high. This is obviously an amount of some consequence when compared to the estimated upper mantle whole rock $Fe^{3+}/\Sigma Fe$ of 0.023.

The Mössbauer spectrum from the spinel run (Figure 6) is similar to that described earlier for Fe_2SiO_4 spinel (Figure 3), but contains a small high-velocity peak consistent with <5% majorite, a small amount of which was also identified by XRD. There is a small but distinct shoulder at 0.3 mm/s, which we assign to Fe^{3+}, as in other silicate spinel spectra (e.g. Figure 3). The amount of Fe^{3+} is difficult to estimate because the shoulder is not resolved, and fitted areas in these regions tend to overestimate the actual area [Hawthorne and Waychunas, 1988]. The fit gives 0.06 $Fe^{3+}/\Sigma Fe$, but the uncertainty is such that ratios as low as 0.03 are not excluded.

DISCUSSION

We apply these experimental results to the problem of the oxidation state of the transition zone bearing in mind differences in

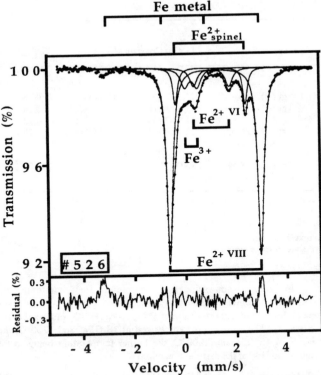

Fig. 5. ^{57}Fe Mössbauer spectrum (298 K) of majorite garnet (Mg/(Mg+Fe) = 0.85) synthesized at 1900°C, 18 GPa in an iron capsule, coexisting with metallic iron, stishovite and a trace of silicate spinel (run 526 in Table 4). The Mössbauer spectra of the contaminant spinel (cf. Figure 3) and the Fe metal (not fitted) are displayed, in the form of bar diagrams, at the top of the figure. Residuals and velocity scale as in Figure 3.

temperature (the mantle probably has a temperature of about 1600°C at our experimental pressure of 18 GPa - e.g. Ito and Katsura [1989]), and in composition, for example the presence of substantial Al_2O_3 in garnet (~10wt% in the garnet phase for a primitive upper mantle composition, e.g. Takahashi and Ito [1987]), and a slight difference in Mg# (0.88 versus 0.85). The pyrolite composition in the transition zone is also undersaturated in SiO_2 (no stishovite present), allowing higher fO_2's and Fe^{3+} contents in equilibrium with Fe metal (see Equations 8 to 11 above). Nevertheless, all these differences are of a second order nature, and should have a relatively minor influence on our main point, which is that Fe^{3+} is so compatible in the major transition zone phases that the oxygen fugacity of a system with primitive upper mantle composition will be very low, probably near the level of metal saturation. This conclusion and some of its implications will now be discussed in more detail.

The exact pressure-temperature co-ordinates of the olivine/β-phase transition, and the spinel-to-perovskite+magnesiowüstite reaction, and the widths of the associated divariant fields, are important clues to the temperature distribution in the mantle, and as to whether the mantle might be compositionally layered [e.g. Bina and Wood, 1987; Ito and Takahashi, 1989; Wood, 1990]. A major aim of high pressure experimental petrology is to match the results from the experimentally determined phase equilibria to the seismic observations. The strong preference of Fe^{3+} for the transition zone phases may influence the

experimentally determined phase boundaries at a level which is significant compared to the desired precision. For example, in experiments on the olivine/β-phase transition in which fO_2 is not buffered, and Fe^{3+} not determined, the β-phase will be additionally stabilized to an unknown extent, perhaps resulting in an erroneous impression of the width of the olivine/β-phase (pseudo-) divariant field. There are also obvious implications for interpreting physical property measurements, e.g. the jump in electrical conductivity at the olivine/spinel transition [Omura, 1989].

In order to estimate the minimum possible bulk $Fe^{3+}/\Sigma Fe$ for a primitive upper mantle composition in the transition zone, we take the compositions of co-existing spinel and garnet experimentally determined by Takahashi and Ito [1987] at 1600°C and 20 GPa, assume that each phase has $Fe^{3+}/\Sigma Fe$ as in our experiments (0.10 for majoritic garnet, and a conservatively estimated 0.03 for silicate spinel), and assume 60% spinel, 40% garnet (ignoring the small amount of Ca-perovskite phase in their experiment). We find $Fe^{3+}/\Sigma Fe = 0.05$, with uncertainties suggesting a possible range of 0.03 to 0.08. At 8.0 wt% FeO*, this implies 0.44 wt % Fe_2O_3. This minimum $Fe^{3+}/\Sigma Fe$ occurs at saturation with Fe metal, and is calculated for equilibrium in the simple $MgO-SiO_2-Fe-O$ system with excess SiO_2; silica undersaturated conditions will raise $Fe^{3+}/\Sigma Fe$ (Equations 8 to 11 above).

We have estimated that primitive upper mantle has 0.2 wt % Fe_2O_3, i.e. lower than this estimate for the transition zone minimum. In the absence of the oxidized volatile components H_2O and CO_2,

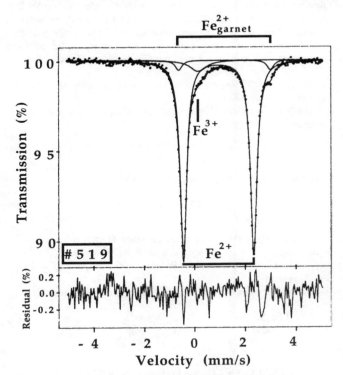

Fig. 6. ^{57}Fe Mössbauer spectrum (298 K) of $(Mg_{0.85}Fe_{0.15})_2SiO_4$ spinel (run 519 in Table 4) synthesized at 1700°C, 18 GPa in coexistence with metallic iron, stishovite and a trace of majorite garnet (indicated by the bar diagram on top of the figure). For Fe^{3+} in spinel, only a broadened singlet instead of the required doublet could be fitted. Residuals and velocity scale as in Figure 3.

adjustment of the $Fe^{3+}/\Sigma Fe$ ratio might procede through partitioning of Fe (actually, mostly Ni at first) into the mantle sulfide phase, i.e. $3\ Fe^{2+}_{(silicate)} = Fe^0_{(sulfide)} + 2\ Fe^{3+}_{(silicate)}$ (there is ~200 ppm S in the primitive upper mantle, e.g. O'Neill [1991b, Appendix 1]. The implication is that in the absence of any buffering effect from volatiles, the present $Fe^{3+}/\Sigma Fe$ of the upper mantle is more or less consistent with metal saturation under transition zone conditions. The metal would be an Fe-Ni-S alloy.

Extrapolation of fluid equilibria in the H-O and C-H-O systems to transition zone pressures is uncertain because of lack of high pressure data, but the trends calculated for upper mantle pressures indicate that at the low fO_2 of metal saturation, CH_4 is the dominant species in a fluid phase [e.g. Woermann and Rosenhauer, 1985]. Therefore, provided that the amounts of C and H components do not exceed the oxygen buffering capacity of the Fe^{3+}/Fe^{2+} equilibrium, CH_4 may be produced in transition zone assemblages by reduction of the hydroxyl component in the silicates in the presence of carbon, together with any carbonate, i.e., according to the reactions:

$$2\ H_2O\ +\ C\ +\ 4\ FeO = 2\ Fe_2O_3\ +\ CH_4 \qquad (13)$$
as hydroxyl   diamond   fluid

$$2\ H_2O\ +\ CO_2\ +\ 8\ FeO = 4\ Fe_2O_3\ +\ CH_4 \qquad (14)$$
as hydroxyl   as carbonate   fluid

Increasing Fe_2O_3 from 0.2 wt % to 0.44 wt % could, according to reaction (13), reduce 270 ppm H_2O if sufficient elemental C is present; alternatively, the required increase in Fe_2O_3 according to reaction (14) could reduce 130 ppm H_2O plus 330 ppm CO_2. These amounts are comparable to the estimated primitive mantle abundances of H_2O and CO_2.

So far as is presently known, CH_4, unlike H_2O and CO_2, does not dissolve into, or react with, mantle silicates at deep mantle temperatures and pressures, and may therefore co-exist with mantle peridotite as a discrete fluid phase. If this low density, low viscosity fluid phase can separate from its silicate matrix and migrate upwards, then the transition zone would form a barrier against re-circulation of C-H volatiles from the upper mantle (and crust) into the lower mantle. This effect may also apply to subducted oceanic crust, the basaltic part of which transforms to >90% majoritic garnet (plus stishovite) at transition zone pressures [Irifune and Ringwood, 1987]. Transfer of material between upper and lower mantles (the extent of which is, at present, under debate) would be a one way process for C, H and any other component which partitions into a CH_4-rich fluid, concentrating them relative to silicate-compatible components in the upper mantle over geological time.

CH_4-rich fluids upwelling from the transition zone, perhaps from subducted basaltic material, are expected to undergo the reverse of reactions (13) and (14) on encountering the higher fO_2's of the olivine upper mantle, triggering redox melting [e.g. Taylor and Green, 1987].

The lower mantle

A lower mantle of pyrolite composition would consist of 65 to 75% silicate perovskite with Mg/(Mg+Fe) = 0.96, 16 to 18% magnesiowüstite with Mg/(Mg+Fe) = 0.8, plus some Ca-silicate perovskite and perhaps an Al-rich phase [Ringwood, 1989; Ito et al., 1984; Ito and Takahashi, 1989]. Since Mg-Fe silicate perovskite is manifestly very poor in total Fe, it may be assumed that perovskite holds only a fairly small proportion of the whole rock's total Fe^{3+}, the major part of which would therefore be concentrated into the coexisting magnesiowüstite. From the estimate of the lower mantle mineralogy given above, magnesiowüstite would contain about 75% of the whole rock FeO* in a pyrolitic mantle. Assuming equal $Fe^{3+}/\Sigma Fe$ in perovskite and magnesiowüstite, the latter would have the estimated mantle $Fe^{3+}/\Sigma Fe$ of about 0.023, i.e. a composition near $(Mg_{0.8}Fe^{2+}_{0.195}Fe^{3+}_{0.005})_{0.998}O$. At atmospheric pressure and 1500°C, such a magnesiowüstite would exist at log fO_2 = -7.4 [Valet et al., 1975, their eqns. 21 and 27], which is about half way between the (extrapolated) QFM and IW buffers - in other words, unless pressure has a large effect on the thermochemistry of Fe^{3+} in magnesiowüstite, and preliminary data indicate that it does not [Kato et al., 1989], the lower mantle has a roughly similar relative oxygen fugacity to that of the upper mantle, and, like the upper mantle, is far removed in fO_2 from metal saturation (by ~2.5 log-bar units according to the above calculation). It seems that at least this postulated level of $Fe^{3+}/\Sigma Fe$ in magnesiowüstite is needed to explain the high electrical conductivity of the lower mantle [Peyronneau and Poirier, 1989; Wood and Nell, 1991]. It would also be difficult to obtain the necessary high electrical conductivity if the lower mantle had a non-pyrolitic, more silica-rich bulk composition, and consequently less magnesiowüstite [Wood and Nell, 1991].

The transition zone therefore forms a shell of reduced conditions sandwiched between relatively oxidized upper mantle and lower mantle. The further implications of this for core formation in the early Earth must depend largely on the ability of metal to separate from a silicate matrix. It has been argued that metal cannot segregate unless it is molten, and unless either the silicate is also partially molten, or the high surface potential of the metal is somehow lowered sufficiently for it to wet the silicate [Stevenson, 1990]. Certainly, the empirical evidence is that small amounts of sulfide have existed in the upper mantle since core formation was completed [O'Neill, 1991b], which, from Pb isotopic evidence, is constrained to be shortly after accretion [Patterson, 1956]. We do not think, therefore, that the putative presence of Fe-Ni-S metal in the transition zone in the present Earth need imply continuing loss of metal to the core.

Metal saturation in the transition zone might seem to resurrect the possibility that core formation took place in the early Earth by a one step segregation of metal from this region, for example, from the bottom of a terrestrial magma ocean. Whether or not this did happen needs to be tested by comparing the mantle abundances of the siderophile elements with their appropriate high pressure metal/silicate partition coefficients. Estimates from mineral/melt partitioning at high pressure coupled with the existing low pressure melt/metal data suggest that this scenario cannot satisfactorily explain either the overabundance of the moderately siderophile elements, e.g. Ni and Co, or the depletion relative to chondritic of the least siderophile of the potential siderophiles, Cr and V [Ohtani et al., 1991]: a two-stage model [e.g. O'Neill, 1991b] is still required.

APPENDIX 1

Comparison of $Fe^{3+}/\Sigma Fe$ from Mössbauer spectroscopy with wet chemical analyses

Our estimate of $Fe^{3+}/\Sigma Fe$ is so much lower than published wet chemical analyses of mantle peridotites as to require further comment. We will first compare the Mössbauer determinations of $Fe^{3+}/\Sigma Fe$ for the individual phases with the available wet chemical data.

For Cr-spinels of mantle-like compositions, Osborne et al. [1981] found wet chemical analyses to be in good agreement with Mössbauer determinations of $Fe^{3+}/\Sigma Fe$, whereas Lucas et al. [1989] found that wet chemistry gave inflated Fe^{3+} values. This illustrates a well-known problem with the wet-chemical data, that results differ markedly from one laboratory to another, so that it is always difficult to be sure of the reliability of the data. For pyroxenes, we are aware of only one set of $Fe^{3+}/\Sigma Fe$ wet chemical analyses from mantle peridotites which is accompanied by full major element analyses of co-existing phases, namely that published by Frey and Green [1974] on six spinel lherzolite xenoliths from Victoria, Australia (analyst E. Kiss). The chemistry of these samples indicates that they are more depleted than the ones in the Mössbauer studies of Dyar et al. [1989] and Luth and Canil [1992], but have equilibrated at comparable P, T, and fO_2; the two analytical methods should give similar results. By and large, this they do, but the wet chemical analyses show greater dispersion than the Mössbauer determinations. For the wet chemical analyses, $Fe^{3+}/\Sigma Fe$ in orthopyroxene ranges from 0.03 to 0.10 (cf. 0.03 to 0.06 by Mössbauer), and in clinopyroxene from 0.18 to 0.36 (cf. 0.06 to 0.24). There is no correlation of $Fe^{3+}/\Sigma Fe$ in orthopyroxene with that in clinopyroxene, nor any correlation of either orthopyroxene or clinopyroxene with $Fe^{3+}/\Sigma Fe$ in spinel, or with calculated fO_2. We believe that the most straightforward interpretation of these data is that they support the Mössbauer results, but are less precise, and tend to be skewed towards higher Fe_2O_3, by oxidation during the analytical process. This latter point probably accounts for the generally higher $Fe^{3+}/\Sigma Fe$ in comparable pyroxenes from various laboratories listed by Deer et al. [1978].

There are many more whole rock determinations of $Fe^{3+}/\Sigma Fe$ in mantle samples - a compilation is given in Table 5. We have not included samples with obvious alteration (e.g. high H_2O or CO_2 contents, which includes most garnet peridotites).

We believe that certainly the higher of these $Fe^{3+}/\Sigma Fe$ estimates, and probably even the lower ones, are an inflated measure of the true oxidation state of the upper mantle, as an inversion of the modal abundance/Fe_2O_3-content argument will show. Consider a typical mantle peridotite with $FeO* = 8.0$ wt%, and consisting of 60% modal

olivine, composition $Mg/(Mg+\Sigma Fe) = 0.90$. For reasons given above, we assume that olivine contains no Fe_2O_3, so that all the nominal 9.8% FeO in olivine of this composition is indeed FeO; the olivine therefore contributes 5.9% FeO to the whole rock total. If all the remaining Fe in the pyroxenes, spinel or garnet were Fe^{3+}, the net $Fe^{3+}/\Sigma Fe$ would be 0.26. Clearly, not all the Fe in these phases can possibly be Fe^{3+}, particularly in orthopyroxene, the next most abundant phase after olivine. Fe-Mg partitioning experiments between olivine and Fe^{3+}-free orthopyroxene in the simple system $MgO-FeO-SiO_2$ show that $KD_{ol-opx}^{R^{2+}-Mg}$ is ~1 at mantle temperatures for $Mg/(Mg+Fe)$ ~0.90 [e.g. von Seckendorff and O'Neill, 1992]. Since in mantle peridotites $KD_{ol-opx}^{\Sigma Fe-Mg}$ is invariably near unity (actually, usually slightly greater than unity due to the effects of other components), very little of the Fe in orthopyroxene can be Fe^{3+}, in agreement with the Mössbauer data. Orthopyroxene in mantle peridotite contains ~6% "FeO"; if we take $Fe^{3+}/\Sigma Fe = 0.20$ in orthopyroxene to be an upper limit, with 25% orthopyroxene in the mode, this contributes a further 1.2 wt% FeO to the whole rock, leaving a maximum whole rock $Fe^{3+}/\Sigma Fe$ of 0.12. Furthermore, if half the remaining Fe in clinopyroxene, spinel and garnet is Fe^{2+} (again an obvious underestimate, as may similarly be deduced from olivine-clinopyroxene or olivine-garnet Fe-Mg partitioning relations), the maximum possible $Fe^{3+}/\Sigma Fe$ is reduced to ~0.06. This we regard to be a maximum realistic limit for $Fe^{3+}/\Sigma Fe$ in any peridotite of upper mantle mineralogy, even for the most oxidized samples (i.e. QFM+1); higher $Fe^{3+}/\Sigma Fe$ would have to be reflected in lower olivine modal abundance and high modal abundance of Fe_3O_4-rich spinel. Only the analyses reported by Frey and Green [1974] and Preß et al. [1986] fall near this maximum limit. Even in these cases, we argue that our lower, Mössbauer-based estimate is a more realistic determination of upper mantle $Fe^{3+}/\Sigma Fe$, since these wet-chemical whole rock Fe_2O_3 analyses seem to vary erratically with either fO_2, as calculated from the olivine-orthopyroxene-spinel equilibrium, or with degree of depletion, as indicated by wt% MgO. In the case of the data of Frey

TABLE 5. A compilation of Fe_2O_3 and $Fe^{3+}/\Sigma Fe$ determinations on mantle peridotites by wet chemistry.

Locality	No. of samples	Fe_2O_3 range (wt%)	$Fe^{3+}/\Sigma Fe$ range (%)	$Fe^{3+}/\Sigma Fe$ mean (%)	Reference
Spinel lherzolites and harzburgites					
Itinome-gata, Japan	11	0.83 - 2.20	10 - 21	17	Kuno and Aoki [1970]
Dreiser Weiher, Germany	12	1.06 - 2.07	13 - 21	17	"
Victoria, Australia	6	0.38 - 1.50	4.6 - 18	6.3[a]	Frey and Green [1974]
Puy Beaunit, France	27	0.56 - 2.09	-	15	Hutchison et al. ([1975]
Volcan de Zanière, FR	18	1.08 - 2.13		17	"
Montboissier, FR	41	1.33 - 2.60	-	23	"
Monistrol d'Allier, FR	16	0.63 - 1.92	-	15	"
Tarreyres, FR	40	0.80 - 2.29	-	16	"
Mongolia	11	0.17 - 0.75	2.0 - 8.3	5.5	Preß et al. [1986]
Victoria, Australia[b]	12	0.42 - 0.99	4.5 - 11	7.7	Stolz and Davies [1988]
Garnet lherzolites and associated harzburgites					
Lashaine, Tanzania[c]	8	0.71 - 1.30	9.2 - 19	12	Rhodes and Dawson [1975]
"	8	0.72 - 1.39	8.7 - 21	15	"

Only samples with 6 to 9.5% FeO* and MgO > 36% are included.

[a] not including an outlier with 18% $Fe^{3+}/\Sigma Fe$.

[b] 8 of these contain amphibole (mode 3 to 17%); there is no correlation of $Fe^{3+}/\Sigma Fe$ with modal amphibole.

[c] described as "remarkably fresh", but analyses report ~0.3% H_2O^+. The first 8 xenoliths contain 4 - 8% modal garnet and 63 - 79% olivine; the second 8 are more depleted and contain no garnet but >80% ol.

and Green [1974], the separate determinations of Fe_2O_3 in the spinels and pyroxenes, when multiplied by the modes for these phases, are also inconsistent with whole rock Fe_2O_3. The Mongolian peridotite suite analyzed by Preß et al. [1986] shows particularly good trends for all other moderately incompatible elements versus MgO, but almost random scatter for Fe_2O_3.

In some cases, some part of the high $Fe^{3+}/\Sigma Fe$ may be due to alteration, e.g. serpentinisation or in the kelyphitic rims of garnet. An example of this is presented in Figure 7, which shows a plot of whole rock Fe_2O_3/FeO versus H_2O^+, obtained by Frey et al. [1985] on a suite of variably serpentinized high temperature peridotites from the Ronda Massif. This suite shows very well defined major element and trace element depletion trends vs. MgO, indicating that the serpentinisation has not significantly affected most of the whole rock chemistry. This does not apply to the oxidation state of the samples. There is a good correlation between Fe_2O_3/FeO and H_2O^+ (R=0.98), and the effect is clearly going to be strong enough to destroy more subtle effects, such as a correlation with MgO (i.e. degree of depletion). The trend extrapolates to $Fe_2O_3/FeO = 0.14$ (equivalent to $Fe^{3+}/\Sigma Fe = 0.11$) at the $H_2O^+ = 0$ intercept, which is still somewhat high.

The data summarized in Table 5, though, are for spinel peridotites xenoliths which are often very fresh, containing virtually no low temperature, hydrous alteration, and yet are still reported to have unrealistically high $Fe^{3+}/\Sigma Fe$. It is not possible to state categorically that this is always due to analytical error, but we suspect that this may be the case. An initial difficulty is caused simply by the low $Fe^{3+}/\Sigma Fe$ of mantle samples, so that any oxidation during the analytical process is magnified in its effects. Thus an error of, say, ±0.05 in $Fe^{3+}/\Sigma Fe$ which would have little consequence at $Fe^{3+}/\Sigma Fe = 0.5$, has larger implications at $Fe^{3+}/\Sigma Fe = 0.05$. Secondly, about one third of the Fe^{3+} in a mantle peridotite is held in spinel, which is notoriously difficult to dissolve prior to wet chemical analysis. It is conceivable

that the severe level of acid attack needed to dissolve spinel may result in extra oxidation of easily soluble olivine in the whole rock powder.

APPENDIX 2

Experimental details

The Mg-Fe majorite garnets and silicate spinels synthesized in this study were made from three separate starting compositions. Two compositions on the $MgSiO_3-FeSiO_3$ join at $Mg/(Mg+Fe) = 0.8$ and 0.9 were prepared from mixtures of MgO, SiO_2 and ^{57}Fe-enriched Fe_2O_3. The mixtures were heated at 900°C for several hours in a CO-CO_2 atmosphere at $fO_2 <$ QFM to reduce Fe_2O_3 to FeO. The mixes were then sealed in iron capsules with a small amount of H_2O and run in a conventional piston cylinder apparatus at 900°C and 1.5 GPa for 24 hours. From optical, X-ray and Mössbauer investigations, the products consisted of >99% orthopyroxene containing no detectable Fe^{3+} ($Fe^{3+}/\Sigma Fe < 0.02$).

A third composition corresponding to $(Mg_{0.85}Fe_{0.15})SiO_3$ plus 20% excess SiO_2 and 20% excess Fe metal was prepared from MgO, SiO_2, ^{57}Fe-enriched Fe_2O_3 and isotopically normal Fe metal. This mixture was first crystallized to orthopyroxene + quartz + Fe metal in a piston-cylinder apparatus at 1000°C and 1.9 GPa for 6.5 hours, using a silver capsule with a tight fitting lid. The run product was examined optically, and by Mössbauer spectroscopy, which revealed a spectrum characteristic of (Mg,Fe)SiO_3 orthopyroxene, with no discernible trace of Fe^{3+}. There is little if any isotopic equilibration between the ^{57}Fe-enriched Fe_2O_3 and the isotopically normal Fe metal, which experimentally has the very convenient consequence that the excess metal in this material (and in the high pressure syntheses) is barely visible in the Mössbauer spectra, and does not interfere with the silicate spectra.

High pressure syntheses were done in a 1200 ton uniaxial split-sphere multianvil apparatus. Toshiba F grade tungsten carbide anvils with a 5 mm truncation edge length were used with 10 mm MgO octahedral sample assemblies containing a $LaCrO_3$ heater. Pressure was calibrated at room temperature using transitions in Bi, ZnS and GaAs, and at 1200°C and 1600°C using the olivine/β-phase and β-phase/spinel transitions in Mg_2SiO_4 [Akaogi et al., 1989]. Temperatures were monitored using a W3%Re/W25%Re thermocouple. In each experiment, ~3 mg of sample was contained in a capsule made from pure Re, Mo or Fe foil. Ceramic components of the sample assembly were fired at 1000°C and prior to each high-pressure experiment the complete sample assembly was dried at 230°C in a vacuum for at least 12 hours. Rubie et al. [1992] have demonstrated that this procedure results in effectively anhydrous conditions during high temperature, high pressure multianvil experiments. A summary of the experimental conditions for each run is given in Table 4.

The experiments with excess Fe metal were performed in Fe capsules, and in addition contained finely divided Fe metal powder in the starting mixture. This Fe metal was found to have remained evenly dispersed throughout the capsule at the end of the run, suggesting that all parts of the experimental charge were in equilibrium with Fe metal. Clinopyroxene synthesized from the $(Mg_{0.85}Fe_{0.15})SiO_3 + SiO_2 + Fe$ starting material at 15 GPa and a nominal temperature of 1600°C, using the same experimental procedures and sample assembly as for the majorite and silicate spinel syntheses, was found to contain no discernible Fe^{3+} by Mössbauer spectroscopy.

Silicate spinels along the join $Fe_2SiO_4 - Fe_3O_4$ were synthesized at 7 to 9 GPa also using a multianvil apparatus (Table 3). Pressure

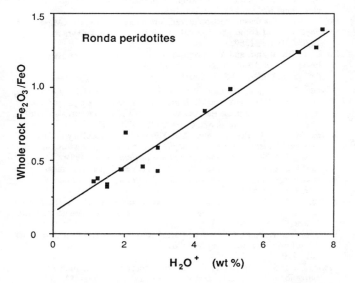

Fig. 7. Possible influence of hydrous low temperature alteration of mantle peridotites on whole rock Fe_2O_3/FeO, analyzed by wet chemistry. Data from Frey et al. [1985].

assemblies and calibrations employed in these experiments are identical to those described in Canil [1991] except that Pt/Pt10%Rh instead of W3%Re/W25%Re thermocouples were used to measure temperature. Starting materials were mixtures of pure fayalite synthesized in an evacuated silica tube from a stoichiometric mix of Fe, Fe_2O_3 and SiO_2, and puratronic grade magnetite (Johnson Matthey).

Mössbauer spectra were collected using a variable temperature Mössbauer spectrometer with a ^{57}Co in Rh source of 50 mCi nominal activity in the constant acceleration mode. Left-hand and right-hand sides of the spectra were recorded independently. Fe metal foil served for calibration of the velocity scale, and isomer shifts were determined relative to metallic Fe at room temperature. Symmetrical doublets with Lorentzian lineshapes have been fitted to the folded data.

REFERENCES

Akaogi, M., E. Ito, and A. Navrotsky, Olivine-modified spinel-spinel transitions in the system Mg_2SiO_4-Fe_2SiO_4: Calorimetric measurements, thermo-chemical calculation , and geophysical application, *J. Geophys. Res., 94*, 15,671-15,685, 1989.

Amthauer, G., H. Annersten, and S. S. Hafner, The Mössbauer spectrum of ^{57}Fe in silicate garnets, *Zeit. Kristal., 143*, 14-55, 1976.

Archbald, P. N., Abundances and dispersions of some compatible volatile and siderophile elements in the mantle, *Unpublished M. Sc. Thesis*, Australian National University, 1979.

Bina, C.R., and B. J. Wood, Olivine-spinel transitions: Experimental and thermodynamic constraints and implications for the nature of the 400-km seismic discontinuity, *J. Geophys. Res., 92*, 4853-4866, 1987.

Boyd, F.R., High - and low-tmperature garnet peridotite xenoliths and their possible relation to the asthenosphere-lithosphere boundary beneath Souther Africa, in *Mantle Xenoliths*, edited by P. H. Nixon, 403-412, John Wiley, New York, 1987.

Boyd, F. R. and S. A. Mertzman, Composition and structure of the Kaapvaal lithosphere, Southern Africa, in *Magmatic Processes: Physiochemical Principles*, edited by B. O. Mysen, The Geochemical Society Spec. Publ. No.1, Pennsylvania, U.S.A., 13-24, 1987.

Brown, I. D., and D. Altermatt, Bond-valence parameters obtained from a systematic analysis of the Inorganic Crystal Structure Database, *Acta Cryst., B41*, 244-247, 1985.

Bryndzia, L. T., and B. J. Wood, Oxygen thermobarometry of abyssal spinel peridotites: The redox state and C-O-H volatile composition of the Earth's sub-oceanic mantle, *Amer. J. Sci., 290*, 1093-1116, 1990.

Bukowinski, M. S., and G. H. Wolf, Thermodynamically consistent decompression: Implications for lower mantle composition, *J. Geophys. Res., 95*, 12,583-12,593, 1990.

Canil, D., Experimental evidence for the exsolution of cratonic peridotite from high-temperature harzburgite, *Earth Planet. Sci. Lett., 106*, 64-72, 1991.

Canil, D., H. O'Neill, and C. R. Ross II, A preliminary look at phase relations in the system Fe_3O_4 - γ-Fe_2SiO_4 at 7 GPa, *Terra Abstr., 3*, 65, 1991.

Canil, D., D. Virgo, and C. M. Scarfe, Oxidation state of mantle xenoliths from British Columbia, Canada, *Contrib. Mineral. Petrol., 104*, 453-462, 1990.

Carmichael, I. S. E., and M. S. Ghiorso, Oxidation-reduction relations in basic magma: A case for homogeneous equilibria, *Earth Planet. Sci. Lett., 78*, 200-210, 1986.

Christie, D. M., I. S. E. Carmichael, and C. H. Langmuir, Oxidation states of mid-ocean ridge basalt glasses, *Earth Planet. Sci. Lett., 79*, 397-411, 1986.

Cox, K. G., M. R. Smith, and S. Beswetherick, Textural studies of garnet lherzolites: Evidence of probable exsolution origin from high temperature harzburgite, in *Mantle Xenoliths*, edited by P. H. Nixon, 537-550 John Wiley, New York, 1987.

Deer, W. A., R. A. Howie, and J. Zussman, *Rock Forming Minerals, Volume 2A: Single Chain Silicates*, 668pp., Longman, London, second edition, 1978.

Dick, H. J. B., R. L. Fisher, and W. B. Bryan, Mineralogic variability of the uppermost mantle along mid-ocean ridges, *Earth Planet. Sci. Lett., 69*, 88-106, 1984.

Dyar, M. D., A. V. McGuire, and R. D. Ziegler, Redox equilibria and crystal

chemistry of coexisting minerals from spinel lherzolite mantle xenoliths, *Amer. Mineral., 74.*, 969-980, 1989.

Eggler, D. H., Upper mantle oxidation state: Evidence from olivine-orthopyroxene-ilmenite assemblages, *Geophys. Res. Lett., 10*, 365-368, 1983.

Frey, F. A., and D. H. Green, The mineralogy, geochemistry and origin of lherzolite inclusions in Victorian basanites, *Geochim. Cosmochim. Acta, 38*, 1023-1059, 1974.

Frey, F. A., C. J. Suen, and H. W. Stockman, The Ronda high temperature peridotite: Geochemistry and petrogenesis, *Geochim. Cosmochim. Acta, 49*, 2469-2491, 1985.

Geiger, C., D. C. Rubie, C. R. Ross II, and F. Seifert, Synthesis and ^{57}Fe Mössbauer study of (Mg,Fe)SiO_3 garnet, *EOS, Trans. Amer. Geophys. Union, 72*, 564, 1991.

Goodman, R. J., The distribution of Ga and Rb in coexisting groundmass and phenocryst phases of some basic volcanic rocks, *Geochim. Cosmochim. Acta, 36*, 303-317, 1972.

Haggerty, S. E., and L. A. Tompkins, Redox state of the Earth's upper mantle from kimberlitic xenoliths, *Nature, 303*, 295-300, 1983.

Hawthorne, F. C., and G. A. Waychunas, Spectrum-fitting methods, in *Reviews in Mineralogy Volume 8*, edited by F. C. Hawthorne, Mineralogical Society of America, Washington D.C., 63-98, 1988.

Hazen, R. M., J. Zhang, and J. Ko, Effects of Fe/Mg on the compressibility of synthetic wadsleyite: β-(Mg$_{1-x}$Fe$_x$)$_2SiO_4$ (x≤0.25), *Phys. Chem. Min., 17*, 416-419, 1990.

Hofmann, A. W., Chemical differentiation of the Earth: the relationship between mantle, continental crust, and oceanic crust, *Earth Planet. Sci. Lett., 90*, 297-314, 1988.

Horiuchi, H. and H. Sawamoto, β-Mg_2SiO_4: Single-crystal X-ray diffraction study, *Amer. Mineral., 66*, 568-575, 1981.

Hutchison, R., A. L. Chambers, D.K. Paul, and P. Harris, Chemical variation among French ultramafic xenoliths - evidence for a heterogeneous mantle, *Mineralog. Mag., 40*, 153-170, 1975.

Irifune, T., An experimental investigation of the pyroxene-garnet transformation in a pyrolite composition and its bearing on the constitution of the mantle, *Phys. Earth Planet. Interiors, 45*, 324-336, 1987.

Irifune, T. and Ringwood, A. E., Phase Transformations in primitive MORB and pyrolite compositions to 25 GPa and some geophysical implications, in *High Pressure Research in Mineral Physics*, edited by M. H. Manghnani and Y. Syono, Terra Pub/Amer. Geophys. Union, Tokyo/Washington D. C., 231-242, 1987.

Ito, E., and T. Katsura, A temperature profile of the mantle transition zone, *Geophys. Res. Lett., 16*, 425-428, 1989.

Ito, E., and E. Takahashi, Melting of peridotite at uppermost mantle conditions, *Nature, 328*, 514-517, 1987.

Ito, E., and E. Takahashi, Postspinel transformations in the system Mg_2SiO_4-Fe_2SiO_4 and some geophysical implications, *J. Geophys. Res., 94*, 10,637-10,646, 1989.

Ito, E., E. Takahashi, and Y. Matsui, The mineralogy and chemistry of the lower mantle: An implication of the ultrahigh-pressure phase relations in the system MgO-FeO-SiO_2, *Earth Planet. Sci. Lett., 67*, 238-248, 1984.

Jagoutz, E., H. Palme, H. Baddenhausen, K. Blum, M. Cendales, G. Dreibus, B. Spettel, V. Lorenz, and H. Wänke, The abundances of major, minor, and trace elements in the Earth's mantle as derived from ultramafic nodules, *Proc. 10th Lunar Planet. Sci. Conf.*, 2031-2050, 1979.

Kato, M., S. Urakawa, and M. Kumazawa, Stability fields of nonstoichiometric wüstite and magnesiowüstite at high pressure, *Abstract of DELP 1989 Misasa International Symposium*, 80-82, 1989.

Karpinskaya, T. B., I. A. Ostrovsky, and T. L. Yevstigneeva, Synthetic pure iron skiagite garnet, *Izvestia AN SSSR (Moscow), ser. geol., no. 9*, 128-129, 1982, [in Russian].

Kilinc, A., I. S. E. Carmichael, M. L. Rivers, and R. O. Sack, The ferric-ferrous ratio of natural silicate liquids equilibrated in air, *Contrib. Mineral. Petrol., 83*, 136-140, 1983.

Kuno, H., and K. Aoki, Chemistry of ultramafic nodules and their bearing on the origin of basaltic magmas, *Phys. Earth Planet. Interiors, 3*, 273-301,1970.

Liang, Y., and D. Elthon, Evidence from chromium in mantle rocks for extraction of picrite and komatiite melts, *Nature, 343*, 551-553, 1990a.

Liang, Y., and D. Elthon, Geochemistry and petrology of spinel lherzolite xenoliths from Xalapasco de La Joya, San Luis Potosi, Mexico: Partial melting and mantle metasomatism, *J. Geophys Res., 95*, B10, 15,859-15,877, 1990b.

Lucas, H., M. T. Muggeridge, and D. M. McConchie, Iron in kimberlitic ilmenites and chromian spinels: A survey of analytical techniques, in *Kimberlites and Related Rocks volume 2*, edited by J. Ross, Geol. Soc. Australia Spec. Publ. 14, Blackwell, Australia, 311-319, 1989.

Luth, R. W., and D. Canil, Ferric iron in mantle-derived pyroxenes and a new oxybarometer for the mantle, *Contrib. Mineral. Petrol.*, in press.

Luth, R. W., D. Virgo, F. R. Boyd, and B. J. Wood, Ferric iron in mantle-derived garnets, implications for thermobarometry and for the oxidation state of the mantle, *Contrib. Mineral. Petrol.*, 104, 56-72, 1990.

Maaløe, S., and K. Aoki, The major element composition of the upper mantle estimated from the composition of lherzolites, *Contrib. Mineral. Petrol.*, 63, 161-173, 1977.

McKay, D. B., and R. H. Mitchell, Abundance and distribution of gallium in some spinel and garnet lherzolites, *Geochim. Cosmochim. Acta*, 52, 2867-2870, 1988.

Nakamura, A., and H. Schmalzried, On the stoichiometry and point defects of olivine, *Phys. Chem. Mineral.*, 10, 27-37, 1983.

Novak, G. A., and G. V. Gibbs, The crystal chemistry of the silicate garnets, *Amer. Mineral.*, 56, 791-825, 1971.

Ohtani, E., and N. Kagawa, Stability of tetragonal garnet $(Mg,Fe)SiO_3$ and hydrous mineral phase B at high pressure and temperature, *Abstract of DELP 1989 Misasa International Symposium*, 87-89, 1989.

Ohtani, E., T. Kato, and E. Ito, Transition metal partitioning between lower mantle and core materials at 27 GPa, *Geophys. Res. Lett.*, 18, 85-88, 1991.

Ohtani, E., T. Kato, and H. Sawamoto, Melting of a model chondritic mantle to 20 GPa, *Nature*, 322, 352-353, 1986.

Ohtani, E., and H. Sawamoto, Melting experiment on a model chondritic mantle at 25 GPa, *Geophys. Res. Lett.*, 14, 733-736, 1987.

Omura, K., Change of electrical conductivity of olivine associated with the olivine-spinel transition, *Abstract of DELP 1989 Misasa International Symposium*, 23-24, 1989.

O'Neill, H. St.C., The origin of the Moon and the early history of the Earth - a chemical model. part 1: The Moon, *Geochim. Cosmochim. Acta*, 55, 1135-1157, 1991a.

O'Neill, H. St.C., The origin of the Moon and the early history of the Earth - a chemical model. part 2: The Earth, *Geochim. Cosmochim. Acta*, 55, 1159-1172, 1991b.

O'Neill, H. St.C., and A. Navrotsky, Cation distributions and thermodynamic properties of binary spinel solid solutions, *Amer. Mineral.*, 69, 733-753, 1984.

O'Neill, H. St.C., and V. J. Wall. The olivine-orthopyroxene-spinel oxygen geobarometer, the nickel precipitation curve, and the oxygen fugacity of the Earth's upper mantle, *J. Petrol.*, 28, 1169-1191, 1987.

Osborne, M. D., M. E. Fleet, and G. M. Bancroft, Fe^{2+}-Fe^{3+} ordering in chromite and Cr-bearing spinels, *Contrib. Mineral. Petrol.*, 77, 251-255, 1981.

Patterson, C. C., Age of meteorites and the Earth, *Geochim. Cosmochim. Acta*, 10, 230, 1956.

Peyronneau, J., and J. P. Poirier, Electrical conductivity of the Earth's lower mantle, *Nature*, 342, 537-539, 1989.

Preß, S., G. Witt, H. A. Seck, D. Eonov, and V. I. Kovalenko, Spinel peridotite xenoliths from the Tariat Depression, Mongolia II: Geochemistry and Nd and Sr isotopic composition and their implications for the evolution of the subcontinental lithosphere, *Geochim. Cosmochim. Acta*, 50, 2601-2614, 1986.

Rama Murthy, V., Early differentiation of the Earth and the problem of siderophile elements: A new approach, *Science*, 253, 303-306, 1991.

Reid, J. B., and G. A. Woods, Oceanic mantle beneath the Rio Grande Rift, *Earth Planet. Sci. Lett.*, 41, 303-316, 1978.

Rhodes, J. M., and J. B. Dawson, Major and trace element chemistry of peridotite inclusions from the Lashaine volcano, Tanzania, *Phys. Chem. Earth*, 9, 545-557, 1975.

Ringwood, A. E., Chemical evolution of the terrestrial planets, *Geochim. Cosmochim. Acta*, 30, 41-104, 1966.

Ringwood, A. E., *Composition and Petrology of the Earth's Mantle*, McGraw-Hill, 1975.

Ringwood, A. E., The Earth's core: Its composition, formation, and bearing upon the origin of the Earth, *Proc. Royal Soc. London*, A395, 1-46, 1984.

Ringwood, A. E., Constitution and evolution of the mantle, in *Kimberlites and Related Rocks volume 2*, edited by J. Ross, Geol. Soc. Australia Spec. Publ. 14, Blackwell, Australia, 457-485, 1989.

Ringwood, A. E., and W. Hibberson, The system Fe-FeO revisited, *Phys. Chem. Mineral.*, 17, 313-319, 1990.

Ringwood, A. E., and T. Irifune, Nature of the 650-km seismic discontinuity: implications for mantle dynamics and differentiation, *Nature*, 331, 131-136, 1988.

Ringwood, A. E., T. Kato, W. Hibberson, and N. Ware, High pressure geochemistry of Cr, V, and Mn and implications for the origin of the Moon, *Nature*, 347, 174-176, 1990.

Robie, R. A., B. S. Hemingway, and J. R. Fisher, Thermodynamic properties of minerals and related substances at 298.15K and 1 bar (10^5 Pa) pressure and at higher temperature, *U. S. Geol. Surv. Bull. 1452*, Washington D.C., 456pp, 1978.

Ross II, C. R., T. Armbruster, and D. Canil, Crystal structure refinement of a spinelloid in the system Fe_3O_4 - Fe_2SiO_4, *Amer. Mineral.*, in press, 1992.

Rubie, D. C., S. Karato, H. Yan, and H. St. C. O'Neill, Low differential stress and controlled chemical environment in multi-anvil high-pressure experiments, *J. Geophys. Res.*, submitted.

Schreyer, W., and Th. Baller, Calderite, $Mn^{+2}_3Fe^{+3}_2Si_3O_{12}$, a high-pressure garnet, *Proc. XII M. A. Meeting Novosibirsk 1978* (publ. Nauka), *Experimental Mineralogy*, 68-77, 1981.

Seckendorff, V. von, and H. St. C. O'Neill, An experimental study of Fe-Mg partitioning between olivine and orthopyroxene at 1173, 1273, and 1423 K and 1.6 GPa, *Contrib. Mineral.Petrol.*, in press.

Smyth, J. R., b-Mg_2SiO_4: A potential host for water in the mantle?, *Amer. Mineral.*, 72, 1051-1055, 1987.

Stevenson, D. J., Fluid dynamics and core formation, in *Origin of the Earth*, edited by H. E. Newsom and J. H. Jones, Oxford University Press, New York, 231, 1990.

Stolz, A. J., and G. R. Davies, Chemical and isotopic evidence from spinel lherzolite xenoliths for episodic metasomatism of the upper mantle beneath southeatern Australia, in *Oceanic and Continental Lithosphere: Similarities and Differences*, edited by M. A. Menzies and K. G. Kox, 303-330, Oxford University Press, Oxford, 1988.

Stosch, H.-G., Sc, Cr, Co, and Ni partitioning between minerals from spinel peridotite xenoliths, *Contrib. Mineral. Petrol.*, 78, 166-174, 1981.

Takahashi, E., Melting of a dry peridotite KLB-1 up to 14 GPa: Implications on the origin of peridotitic upper mantle, *J. Geophys Res.*, 91, 9367-9382, 1986.

Takahashi, E., and E. Ito, Mineralogy of mantle peridotite along a model geotherm up to 700 km depth, in *High Pressure Research in Mineral Physics*, edited by M. H. Manghnani and Y. Syono, Terra Pub/Amer. Geophys. Union, Tokyo/Washington D. C., 427-438, 1987.

Taylor, W. R. and S. F. Foley, Improved oxygen buffering techniques for C-O-H fluid-saturated experiments at high pressure, *J. Geophys. Res.*, 94, 4146-4158, 1989.

Taylor, W. R., and D. H. Green, Measurement of reduced peridotite-C-O-H solidus and implications for redox melting of the mantle, *Nature*, 332, 349-52, 1988.

Urakawa, S., Partitioning of Ni between magnesiowüstite and metal at high pressure: Implications for core-mantle equilibrium, *Earth Planet. Sci. Lett.*, 105, 293-313, 1991.

Urakawa, S., M. Kato, and M. Kumazawa, An experimental approach to core formation process of the earth, *Abstract of DELP 1989 Misasa International Symposium*, 114-115, 1989.

Vallet, P.-M., W. Pluschkell, and H-J. Engell, Gleichgewichte von MgO-FeO-Fe_2O_3 Mischkristallen mit Sauerstoff, *Arch. Eisenhüttenwes.*, 46, 383-388, 1975.

Weidner, D. J., and E. Ito, Mineral physics constraints on a uniform mantle composition, in *High Pressure Research in Mineral Physics*, edited by M. H. Manghnani and Y. Syono, Terra Pub/Amer. Geophys. Union, Tokyo/Washington D. C., 439-446, 1987.

Woermann, E., and M. Rosenhauer, Fluid phases and the redox state of the Earth's mantle, *Fortschr. Mineral.*, 83, 263-349, 1985.

Wood, B. J., Postspinel transformations and the width of the 670-km discontinuity: A comment on "Postspinel transformations in the system Mg_2SiO_4-Fe_2SiO_4 and some geophysical implications", *J. Geophys. Res.*, 95, 12,681-12,685, 1990.

Wood, B. J., and J. Nell, High-temperature electrical conductivity of the lower mantle phase (Mg,Fe)O, *Nature*, 351, 309-311, 1991.

Wood, B. J., and D. Virgo, Upper mantle oxidation state: Ferric iron contents of lherzolite spinels by ^{57}Fe Mössbauer spectroscopy and resultant oxygen fugacities, *Geochim. Cosmochim. Acta*, 53, 1277-1291, 1989.

Woodland, A. B., J. Kornprobst, and B. J. Wood, Oxygen thermobarometry of orogenic lherzolite massifs, *J. Petrol.*, 33, 203-230, 1992.

Woodland, A. B., and H. St. C. O'Neill, Synthesis of skiagite ($Fe^{+2}_3Fe^{+3}_2Si_3O_{12}$) and phase relations with almandine ($Fe^{+2}_3Al_2Si_3O_{12}$) - skiagite solid solutions (abstract), *Terra Abstr.*, 4, 47, 1992.

Xue, X., H. Baadsgaard, A. J. Irving, and C. M. Scarfe, Geochemical and isotopic characteristics of lithospheric mantle beneath West Kettle River, British Columbia: Evidence from ultramafic xenoliths, *J. Geophys. Res.*, 95, B10, 15,879-15,891, 1990.

H. St. C. O'Neill, D. C. Rubie, D. Canil, C. R. Ross II, F. Seifert and A. B. Woodland, Bayerisches Geoinstitut, Universität Bayreuth, W-8580 Bayreuth, Federal Republic of Germany.

C. A. Geiger, Mineralogisch-Petrographisches Institut, Universität Kiel, Olshausenstr. 40, W-2300 Kiel 1, Federal Republic of Germany.

Constraints on the Large-Scale Structure of the Earth's Mantle

Robert L. Woodward, Alessandro M. Forte, Wei-Jia Su,
and Adam M. Dziewonski

*Department of Earth and Planetary Sciences, Harvard University,
Cambridge, MA 02138*

We describe the results of several seismological experiments which are aimed at obtaining a better understanding of heterogeneity in the Earth's mantle.

The use of $PP-P$, $SS-S$, and $ScS-S$ differential travel time observations to map heterogeneity in the Earth's upper and lower mantle is reviewed. These experiments indicate the presence of clear signal in the upper mantle which is related to tectonic structures (shields, spreading centers, etc.). The overall pattern of heterogeneity in the upper mantle is dominated by long-wavelength features, although there can be relatively sharp changes between features. The $ScS-S$ observations clearly reveal the presence of significant heterogeneity in the lowermost mantle, and demonstrate that the amplitude of this heterogeneity is greater than at middle mantle depths.

Observation of SS absolute travel time residuals allow one to obtain very complete coverage of the upper mantle. These residuals indicate that the spectrum of heterogeneity in the mantle is dominated by the longest wavelengths (harmonic degrees 1 to 8). Such observations are critical to determining the dominant scale of convection in the mantle. In addition, these results allow us to safely consider only the longest wavelengths when using long-period SS observations and similar data to model large-scale mantle structure.

We report the results of using a combined data set of seismic waveforms and differential travel times to obtain a 3-D model of shear velocities in the mantle. We find that the boundary between the upper and lower mantle seems to be a barrier to material flux in some regions, but heterogeneities appear to be continuous across it in other areas, particularly those near spreading centers. By testing different model parameterizations we find that the structure in the vicinity of the upper–lower mantle transition is well constrained by the data.

By scaling the shear velocity models to density perturbations and considering the convective flow which results from these seismically-inferred buoyancy forces, we are able to calculate predictions for several geodynamic observables. We find that the new models of mantle shear velocity do a good job of predicting the nonhydrostatic geoid, the tectonic plate motions, and the dynamic topography of the core-mantle boundary. We find that such good predictions may only be obtained with viscous flow models which assume a whole-mantle style of flow. In the course of fitting the geoid and plate motion data we infer a new radial viscosity profile for the mantle and depth-dependent $\delta ln\rho/\delta lnv_s$ proportionality factors.

1. Introduction

Accurate characterization and mapping of the three-dimensional (3-D) variations in shear and compressional velocity in the mantle have important bearing on a variety of fields. In particular, characterizing the overall pattern of mantle flow and how the upper and lower mantles are related has remained a difficult and unresolved issue [e.g. Olson et al., 1990; Tanimoto, 1991a; Romanowicz, 1991]. Determining if mass flux between the upper and lower mantle is occurring has proven especially difficult. Knowledge of the power spectrum of heterogeneity, and

Evolution of the Earth and Planets
Geophysical Monograph 74, IUGG Volume 14

its depth variation is also critical as this information is necessary to understand the mode of convection governing different regions of the mantle and the relative importance of heating from below and from within [Jarvis and Peltier, 1986].

In the past, various geophysical approaches, including seismic tomography, have yielded conflicting answers to these questions. However, new techniques can help resolve these issues. In this paper we discuss the results of several such recent studies which place new constraints on the nature of large-scale mantle heterogeneity and allow a much more precise characterization of its structure.

Recently, graphics-oriented workstation computers have been used to make large numbers of highly accurate observations of seismic travel times [e.g. Woodward and Masters, 1991a,b; Su and Dziewonski, 1991; Sheehan and

Solomon, 1991]. In the first section which follows, we review the results of using large numbers of differential travel time observations to make simple and direct observations of shear and compressional velocity heterogeneity in the mantle.

Next, we review the results of using a large data set of absolute travel times to accurately characterize the spectrum of heterogeneity in the mantle. This information is crucial to understanding the importance of small scale structure in convection, as well as assessing the adequacy of low order basis function expansions for global tomographic modeling.

In section 4 we present the results of new modeling experiments. These experiments make use of two important developments. The first is the use of large numbers of highly accurate differential travel time observations, as described above. The low noise level of such data sets allow them to surpass the much larger ISC database for utility in constraining large-scale 3-D Earth structure. The second development is the demonstration of a reasonable consistency between models obtained from different data sets [Giardini et al., 1987] and the utility of combining them to obtain improved coverage and resolution [Woodward and Masters, 1992]. We have built on this work by performing a joint inversion of differential travel times and complete seismic waveforms to obtain improved images of the Earth's interior. These two data sets are very complementary. The waveforms provide a significant amount of information on heterogeneity, but are difficult to quantify in terms of errors and fit. The differential travel times have the advantage of being very accurate and highly quantifiable. The travel times thus provide an extremely sensitive indicator of the overall accuracy of a model.

Finally, we can perform important tests on these new models by testing their ability to predict independent observations. Scaling the perturbations in shear velocity given by the model to perturbations in density allows the calculation of mantle flow. Using viscous flow models of a compressible mantle [Forte and Peltier, 1991a], we calculate predictions for the nonhydrostatic geoid, CMB topography, and tectonic plate motions. Section five of this paper describes briefly the results of such calculations.

2. OBSERVATIONS OF MANTLE HETEROGENEITY

The goal of differential travel times studies is to obtain measurements which are sensitive to a particular portion of the Earth, yet have reduced sensitivity to source mislocation and source/receiver region structure. In addition, they provide the possibility for obtaining measurements in regions which are not otherwise sampled by seismic sources or receivers. Figure 1 illustrates the basic ray geometry involved. One sees that, for example, the S and SS ray paths are very similar near the source and receiver. Thus the measurement of an $SS - S$ differential travel time is dominantly sensitive to structure near the SS surface

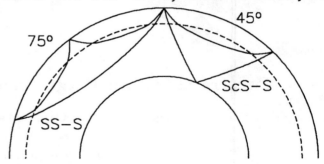

Fig. 1. Diagram of the ray paths of S, SS, and ScS phases. The geometry of the $SS - S$ observations is shown on the left. These observations are made in the epicentral distance range of roughly 50° to 100°. The S and SS phases have similar raypaths near the source and receiver and both bottom in the lower mantle, thus an $SS - S$ measurement is dominantly sensitive to structure in the upper mantle beneath the SS surface bounce point. The geometry of the $ScS - S$ observations is shown on the right. These measurements can be made in the distance range of 45° to 75°. The ScS and S raypaths are extremely similar in the upper mantle, so these observations are effectively only sensitive to lower mantle structure.

bounce point. We describe here some recent results from systematic studies of $PP - P$ and $SS - S$ differential travel time residuals for constraining upper mantle structure [Woodward and Masters, 1991a], and $ScS - S$ residuals for constraining lower mantle structure [Woodward and Masters, 1991b].

A similar large data set of $SS - S$ observations for the North Atlantic region have been used by Sheehan and Solomon [1991] in inversions for lateral variations in upper mantle shear velocity, temperature and composition. Other studies, which have used much smaller numbers of direct $SS - S$ observations to investigate various regions, include those of Butler [1979], Stark and Forsyth [1983], Grand and Helmberger [1985], and Kuo et al. [1987], as well as a study using $PP - P$ observations by Girardin [1980].

Figure 2 shows smoothed $PP - P$ and $SS - S$ residuals. The observations are made by a waveform cross-correlation procedure. The observed residuals are then mapped to the reflected phases' bounce point locations. A running mean smoothing procedure is then used to produce the maps shown in Figure 2. The signal in the raw, unsmoothed, observations is very clear but the smoothing procedure makes the large-scale patterns in the data even more apparent (especially on the small maps shown in Figure 2). The smoothing procedure consists of: (1) mapping the individual observations to their bounce point locations; (2) averaging all measurements which fall within a 5° radius circular cap; (3) plotting the cap average at the location of the cap center; (4) moving the cap in latitude or longitude such that there is always an overlap of one radius with the previous cap; (5) repeating the averaging

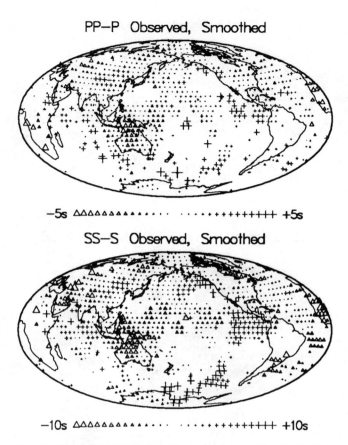

Fig. 2.   Smoothed $PP - P$ residuals (top) and $SS - S$ residuals (bottom) from the study of Woodward and Masters [1991a]. The individual measurements have been mapped to the surface bounce point of the reflected phase (PP or SS) and then smoothed by a running mean smoothing procedure (by averaging in 5° spherical caps; see text). The negative residuals (triangles) indicate faster than average upper mantle and the positive residuals (pluses) indicate slower than average upper mantle velocities.

procedure. Additional details of the data collection and interpretation can be found in Woodward and Masters [1991a]. What we would like to draw attention to here is how these observations clearly reveal the patterns of large-scale heterogeneity in the upper mantle (Figure 2). We would also like to emphasize that no modeling or assumptions are involved at this stage of the interpretation. The raw observations have only been smoothed.

Woodward and Masters [1991a] have shown the existence of a clear correlation between $PP - P$ and $SS - S$ residuals and the tectonic province of the PP or SS bounce point locations, as well as a correlation with lithospheric age for those observations in oceanic areas. In addition, they found that the ocean-age related signal explained only approximately 10% of the variance in the oceanic $SS - S$ observations, indicating that the primary source of the observed $SS - S$ signal must be found deeper in

the mantle. Expansion of the smoothed $SS - S$ residuals in terms of spherical harmonics shows that roughly 90% of the variance in the smoothed $SS - S$ residuals can be explained by spherical harmonics of degrees 0 to 8 (corresponding to wavelengths of roughly 40000 km to 5000 km). These findings motivated the use of low degree spherical harmonics as the basis functions in inversions using $SS - S$ and similar data [Woodward and Masters, 1992; see also section 4 of this paper].

Heterogeneity in the lower mantle can be constrained by the observation of $ScS - S$ differential travel times. Again, inspection of Figure 1 reveals that the S and ScS rays have very similar paths through the upper mantle. Thus, the $ScS - S$ differential time is nearly insensitive to upper mantle structure. Figure 3 shows smoothed $ScS - S$ residuals. The residuals are plotted at the surface projection of the ScS core reflection point. These residuals clearly reveal the pattern of large-scale heterogeneity in

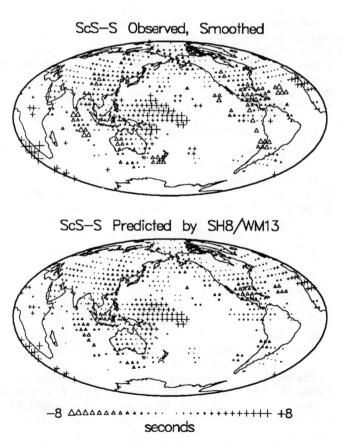

Fig. 3.   (Top) Observed $ScS - S$ differential travel times, from the study of Woodward and Masters [1991b], averaged in 5° spherical caps. (Bottom) Averaged $ScS - S$ predictions of the model $SH8/WM13$ derived in this study. The averaged model predictions account for roughly 80% of the variance in the averaged observations. Notice that a very long-wavelength pattern predominates both observations and model predictions.

the lower mantle. Woodward and Masters [1991b] show that, in a global sense, the $ScS - S$ residuals correlate to ScS residuals, indicating that the dominant signal in the $ScS - S$ observations is due to heterogeneity in the lowermost mantle which is sampled by the ScS phase.

Ratios of compressional and shear wave travel time residuals can be used, through Fermat's principle, to estimate logarithmic velocity derivative ratios. Examination of the correlation between the co-located cap-averaged values of $PP - P$ and $SS - S$ residuals yields a $\delta t_{SS-S}/\delta t_{PP-P}$ ratio of roughly 3 (although it is consistent with a range of 2 to 4) [Woodward and Masters, 1991a], where δt_{SS-S} and δt_{PP-P} are the travel time residuals. This results in a logarithmic derivative $dln(v_s)/dln(v_p)$ in the range of roughly 1.2 to 2.3 for the upper mantle. However, if the $PP - P$ and $SS - S$ residuals are averaged within tectonic regions and these regional averages are compared, the ratio $\delta t_{SS-S}/\delta t_{PP-P}$ is roughly 2.2 ($dln(v_s)/dln(v_p) \sim 1.4$), which is perhaps not inconsistent with the values obtained by comparing cap averages. Similarly, Pulver and Masters [1990] have compared co-located cap-averaged values of $ScS - S$ and $PcP - P$ residuals and found $dln(v_s)/dln(v_p)$ to be roughly 1.7 for the lower mantle. More recently, comparisons of long-period absolute P and S travel time residuals have suggested that the logarithmic derivative $dln(v_s)/dln(v_p)$ may be increasing with depth in the mantle [Bolton and Masters, 1991].

3. THE SPECTRUM OF HETEROGENEITY

Su and Dziewonski [1991] have investigated the spectrum of lateral heterogeneity in the mantle using globally distributed measurements of travel time residuals of the SS phase. These observations are quite similar in nature to the $SS - S$ observations described above and, in fact, are processed in essentially the same way. It has been shown [Kuo et al., 1987; Woodward and Masters, 1991a] that SS residuals are well correlated to $SS - S$ residuals, indicating that the SS residuals are dominantly sensitive to structure in the vicinity of the surface bounce point of SS. SS observations can be obtained for a broader range of epicentral distances than $SS - S$, thus improving global coverage. Interpretation of the SS residuals in terms of near-bounce-point structure then allows constraints to be placed on the character of heterogeneity in the mantle, particularly in the upper mantle.

The smoothed observations of Su and Dziewonski are shown in Figure 4a. The cap averaging procedure (described above) is used to perform the smoothing. Individual SS observations which share a common bounce point location sample roughly the same upper mantle structure beneath the cap location but may have different azimuths and different epicentral distances, thus sampling different source and receiver regions and different turning point locations in the lower mantle. By averaging all the observations which fall within a given cap the signal from the common upper mantle structure beneath the cap should

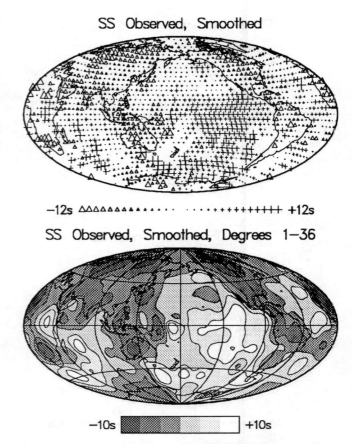

Fig. 4. (Top) SS residuals, from the study of Su and Dziewonski [1991]. The individual residuals are first mapped to the SS surface bounce point, then averaged in 5° spherical caps. Sign conventions are the same as Figure 2. The epicentral distance range for SS observations is 50° to 150°, which allows excellent global coverage. (Bottom) The spherical harmonic expansion of the residuals shown in the top panel.

be reinforced and the incoherent signal from the different source/receiver regions and turning point locations should be minimized.

Figure 4b shows the spherical harmonic expansion of the residuals shown in Figure 4a and yields the spectrum shown in Figure 5. Experiments performed by tracing SS phases through Earth models with random structures have shown that the coverage of SS phases is adequate to properly recover the spectrum of heterogeneity to beyond degree 20. Su and Dziewonski have concluded that the spectrum is 'red', with the energy in each of the harmonic degrees from 1 through 6 being significantly greater than that for degrees greater than 6. Supporting evidence, for both upper and lower mantle, is found in several studies [Inoue et al., 1990; Woodward and Masters, 1991a,b; Zhang and Tanimoto, 1991].

These results should not, however, be interpreted to mean that small-scale structures do not exist or are

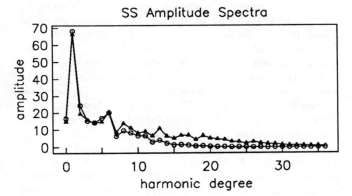

SS Amplitude Spectra

Fig. 5. Spectrum of SS residuals, obtained from the spherical harmonic expansion of the SS residuals which is shown in the bottom of Figure 4. Note how the power beyond degree 6 drops off significantly.

not important.   This clearly is not the case.   But the spectrum of velocity anomalies associated with the subducted lithosphere (e.g.   100 km wide) has a shape that is entirely different from that caused by large-scale convection. This spectrum is spread rather uniformly over a range of harmonic degrees that reach as high as 200. The reason why the local amplitude of such a velocity anomaly can be large (several percent) is that all the appropriate coefficients are in phase, even though their individual numerical values are small.

4. MODELING MANTLE HETEROGENEITY

Since the mid-1970's a number of models of the 3-D structure of shear and compressional velocities of the upper and lower mantle have been presented [e.g. Dziewonski et al., 1977; Masters et al., 1982; Dziewonski, 1984; Woodhouse and Dziewonski, 1984; Nataf et al., 1984, 1986; Tanimoto, 1990; Inoue et al., 1990; Romanowicz, 1990; Montagner and Tanimoto, 1991; Morelli and Dziewonski, 1991; Su and Dziewonski, 1991; Li et al., 1991; Woodward and Masters, 1992; Dziewonski and Woodward, 1992]. The data used to derive these models span three orders of magnitude in frequency and wavelength.   Included were short period arrival times extracted from the Bulletins of the International Seismological Centre, measurements of phase and group velocities of mantle waves, shifts of eigenfrequencies of normal modes, the waveforms of long-period body waves and mantle waves, and the fine structure of the split spectra of normal modes.   In most cases, the structure of the upper and lower mantle have been obtained separately, using different model parameterizations and data.   Although some of the early models differed significantly, they have begun to converge more recently in the sense that a number of large-scale structures are now being reliably mapped.

Heterogeneity in the mantle may be due to thermal, compositional, or phase variations of mantle materials. The

thermal effect is generally assumed to predominate, so that images of velocity heterogeneity in the mantle are typically interpreted as "snapshots" of convective flow in the mantle. Of course, this is only one possible interpretation of these images and it is important to remember that seismic tomography maps seismic velocity variations, whatever their cause, and does not indicate if material is in motion or not.

Below we describe the results of a joint inversion of long-period body wave and mantle wave waveforms and $SS - S$ and $ScS - S$ differential travel times.   We pay particular attention to the parameterization of the upper and lower mantle, which allows us to obtain improved constraints on mantle structure, particularly near the 670 km discontinuity.

4.1 Parameterization

We use low-degree spherical harmonics as the basis functions in which we expand the 3-D structure.   In this method, the 3-D variation in velocity, $\delta v(r, \vartheta, \varphi)$, is expressed as:

$$\delta v(r, \vartheta, \varphi)/v = \sum_{k=0}^{K} \sum_{\ell=0}^{L} \sum_{m=0}^{\ell} f_k(r) \, p_\ell^m(\vartheta)$$
$$(_k A_\ell^m \cos m\varphi +_k B_\ell^m \sin m\varphi); \quad (1)$$

where p_ℓ^m is the normalized associated Legendre polynomial:

$$p_\ell^m(\vartheta) = \left[\frac{(2\ell + 1)(2 - \delta_{m,0})(\ell - m)!}{(\ell + m)!} \right]^{1/2} P_{\ell m}(\vartheta). \quad (2)$$

Such a model is described by relatively few parameters. For example, the parameterization of the lower mantle model $L02.56$ [Dziewonski, 1984] ($K = 4$, $L = 6$, for a total of 245 parameters) requires 48 coefficients to describe the P-velocity anomalies at any particular depth. On the other hand, Inoue et al. [1990] used 2,048 cells for each of their 16 layers spanning the mantle.   Yet when this model is expanded in spherical harmonics its spectrum has a significant decrease in amplitude for degrees greater than 6 (see Figure 5 in Su and Dziewonski, 1991).   Thus the principal heterogeneity at any given depth in Inoue et al.'s model could be described by 48 coefficients. If the spectrum of heterogeneity has significant power at high harmonic degrees then using only low-degree spherical harmonics will lead to results biased by the abrupt truncation of the expansion and aliasing of higher-degree structure into the lower harmonic degrees. Direct examination of the data, as discussed in sections 2 and 3, indicates that the signal in the data is well represented by low-degree spherical harmonics. In addition, the results of Su and Dziewonski indicate that the power in the spectrum of heterogeneity does indeed decrease at higher harmonic degrees. From such results, we conclude that using a low-degree expansion approach, for the data utilized in these inversions, is reasonable.

Dziewonski et al. [1977], Dziewonski [1984], and Woodhouse and Dziewonski [1984], as well as a number of later studies that adopted similar parameterization, used Legendre polynomials for the basis functions $f_k(r)$ in eq. (1). The envelope of the Legendre functions increases significantly, particularly for large k, near the ends of the interval over which they are defined (we use the normalized interval [-1,1]). Such behavior may contribute to instabilities in the solution. Chebyshev polynomials, T_k [e.g. Abramowitz and Stegun, 1964], on the other hand, have a constant envelope in the same interval, making them a better choice for our purpose here. Thus, our functions $f_k(r)$ are defined:

$$f_k(r) = \left[\frac{(4k^2 - 1)}{(4k^2 - 2)}\right]^{1/2} T_k(x) \qquad (3)$$

where r is the radius and x is the normalized radius such that $-1.0 \leq x \leq 1.0$.

In studies treating the entire mantle, the existence of the 670 km discontinuity was explicitly stated and separate sets of coefficients were derived for each region [Woodhouse and Dziewonski, 1986, 1989; Su and Dziewonski, 1991; Dziewonski and Woodward, 1992]. This assumption makes investigation of the change in the pattern of lateral heterogeneity across the discontinuity difficult: it is impossible to tell whether the differences on both sides of the discontinuity are real or an artifact of the adopted parameterization.

We circumvent this limitation by using basis functions that are continuous throughout the mantle. In this case, the radius r is mapped into x:

$$x = (2r - r_{moho} - r_{CMB})/(r_{moho} - r_{CMB}); \qquad (4)$$

where r_{moho} is the radius of the Mohorovičić discontinuity and r_{CMB} the radius of the core-mantle boundary. In view of the findings of Su and Dziewonski [1991], an increase in the order of the spherical harmonic expansion is not essential and we retain $L = 8$, the same as in several earlier studies. But we increase significantly the order of the radial expansion to $K = 13$. This corresponds, roughly, to a resolving length of 220 km throughout the mantle. It is nearly identical to the earlier expansions of the upper mantle, but is significantly better than the 550 km used for the lower mantle in earlier studies [Dziewonski, 1984; Woodhouse and Dziewonski, 1986, 1989]. What is most important, however, is that we can obtain an objective view of the change in the pattern of lateral heterogeneity across the 670 km discontinuity.

4.2 Data

All of the data considered in the modeling described here are taken from the long-period recordings of the Global Digital Seismograph Network [Peterson et al., 1976], which consists of several subnetworks, and the International Deployment of Accelerometers [Agnew et al.,

1976]. The waveform data set used here was collected some time ago and its preliminary interpretation can be found in Woodhouse and Dziewonski [1986, 1989] and Dziewonski and Woodhouse [1987]. Model SH425.2 [Su and Dziewonski, 1991] was also obtained from these data. This data set consists of two fundamental parts. First, the mantle wave records, which are low-passed seismograms (corner frequency of 1/135 Hz) of up to 4.5 hours duration. These recordings are dominated by the fundamental mode Rayleigh and Love surface waves which make one or more orbits around the globe. Surface waves enable a good mapping of the upper mantle, but do not have sufficient sampling of the lower mantle to resolve features in this region. There are approximately 6,000 mantle wave recordings in this part of the data set.

The second portion of the waveform database consists of roughly 9,000 recordings of body waves. These are low passed (corner frequency of 1/45 Hz) recordings of the seismic phases which travel through the interior of the Earth. Aliasing of signal from 'small-scale' structure into the large-scale structure recovered in the global inversion is minimized by using waves with wavelengths of the order of several hundred kilometers. This effectively low-pass filters the short-wavelength structure. This is also why one high quality long-period datum may be 'worth' many more short-period P-wave arrival time readings. In addition, the approximations used to interpret the body-wave waveform data are most appropriate at long periods.

The second major class of data we have used are data sets of $SS - S$ and $ScS - S$ differential travel times observed by Woodward and Masters [1991a,b]. These differential times are robust in the sense that they are insensitive to the source time and have reduced sensitivity to source mislocation and small-scale structure in the source and receiver regions (see section 2). It has been shown that the $SS - S$ times are particularly sensitive to upper mantle structure [Kuo et al., 1987; Woodward and Masters, 1991a], while the $ScS - S$ times are dominantly sensitive to structure in the lowermost mantle [Woodward and Masters, 1991b].

An important aspect of the joint inversions we report here is that the waveform and differential travel times are treated with completely different theories. The theory for interpretation of waveform data has been developed by Woodhouse and Dziewonski [1984]. This theory allows for a recovery of structure characterized by spherical harmonics of both even- and odd-degree. It invokes the 'path average approximation'; a statement that the observed waveforms can be explained by perturbing the average structure along the minor arc path and, for mantle waves, also along the complete great circle. In this approximation the three-dimensional structure is collapsed into one dimension: average radial structure along the great circle path. A much different theory is used for the travel times. The rays are traced through a spherically symmetric reference Earth but the travel time anomaly due to aspherical velocity

Table 1. Coefficients of the three-dimensional model, SH8/WM13, of shear velocities in the mantle defined in eqs. (1) to (4). Units are $10^3 \times \delta v/v$.

ℓ	m	k=0 A	k=0 B	k=1 A	k=1 B	k=2 A	k=2 B	k=3 A	k=3 B	k=4 A	k=4 B	k=5 A	k=5 B	k=6 A	k=6 B	k=7 A	k=7 B	k=8 A	k=8 B	k=9 A	k=9 B	k=10 A	k=10 B	k=11 A	k=11 B	k=12 A	k=12 B	k=13 A	k=13 B
0	0	-0.66		0.62		0.80		0.41		0.22		0.27		0.13		0.01		0.05		0.03		0.11		0.05		-0.01		-0.03	
1	0	1.91		1.27		1.57		1.04		0.72		0.12		-0.18		-0.12		-0.04		-0.02		-0.01		-0.07		-0.09		-0.03	
1	1	0.90	1.04	2.07	0.81	1.00	0.98	0.13	0.18	-0.14	0.18	-0.78	0.04	-0.53	0.08	-0.07	0.06	-0.01	-0.06	0.02	-0.07	0.00	-0.13	0.05	-0.08	0.00	-0.03	-0.03	0.03
2	0	0.97		0.28		1.40		0.82		0.24		-0.17		-0.31		-0.31		-0.23		-0.03		0.13		0.11		0.04		0.00	
2	1	-0.09	-0.56	-0.13	-0.65	0.17	-1.13	0.04	-1.11	0.26	-1.10	0.16	-1.09	-0.02	-0.83	-0.15	-0.46	-0.21	-0.06	-0.04	0.19	0.05	0.23	0.09	0.18	0.03	0.13	-0.03	0.10
2	2	-1.23	-2.62	-2.87	-0.31	0.43	-0.56	0.69	0.04	0.73	0.45	0.64	0.50	0.45	0.19	0.29	-0.05	0.20	-0.09	-0.02	-0.04	-0.18	0.06	-0.13	-0.02	0.02	-0.02	0.01	-0.02
3	0	-0.35		0.79		0.22		0.08		0.14		0.07		0.07		0.12		0.03		0.01		-0.05		-0.03		0.03		-0.02	
3	1	-0.27	-0.15	1.22	0.23	0.25	0.14	0.30	-0.10	0.12	-0.27	0.21	-0.36	0.16	-0.15	0.10	-0.11	0.02	0.12	-0.12	0.04	-0.14	-0.02	-0.09	0.05	0.02	0.05	0.03	0.03
3	2	-0.02	2.01	-0.45	0.23	-0.09	0.83	-0.09	0.43	-0.25	0.49	0.21	0.23	0.37	0.23	0.27	-0.23	0.08	-0.16	-0.03	0.07	-0.03	-0.06	-0.03	0.00	0.01	0.01	0.03	0.01
3	3	-0.62	0.25	-0.08	0.72	-0.40	-0.03	-0.72	-0.25	-0.58	-0.05	-0.38	-0.17	-0.22	0.08	0.11	0.10	0.18	0.09	0.10	0.06	0.06	-0.06	-0.04	0.02	-0.07	0.00	0.00	-0.04
4	0	0.67		1.22		0.87		0.81		0.67		0.38		0.01		0.00		-0.11		-0.08		-0.05		-0.05		0.00		-0.04	
4	1	-0.01	0.13	-0.05	-0.01	0.19	0.40	0.58	0.50	0.26	0.01	0.31	0.05	0.13	0.14	0.06	0.02	0.04	0.05	-0.07	0.00	-0.09	0.04	-0.09	0.00	-0.01	-0.01	0.02	-0.04
4	2	-0.45	-0.04	-1.35	-0.25	-0.37	0.32	-0.51	0.18	0.05	0.09	0.28	-0.49	0.15	-0.49	0.10	-0.20	0.05	-0.11	0.05	0.00	0.04	0.06	-0.01	0.04	-0.08	0.02	-0.04	0.06
4	3	-0.06	-0.56	-0.59	-0.17	-0.11	0.26	-0.82	-0.08	-0.57	0.15	-0.18	0.14	-0.10	0.13	0.10	0.12	0.14	-0.02	0.04	-0.04	-0.02	-0.07	0.00	0.00	0.01	-0.01	0.01	-0.03
4	4	0.05	1.92	0.36	1.54	-0.29	1.62	0.44	0.74	0.31	0.45	0.12	0.52	-0.01	-0.04	-0.25	-0.27	-0.03	-0.29	0.04	-0.13	0.02	0.02	0.00	-0.02	-0.03	-0.04	-0.01	0.00
5	0	-0.84		-0.97		-0.73		-0.46		-0.06		0.12		0.22		0.27		-0.02		-0.06		-0.01		-0.05		0.00		-0.01	
5	1	1.37	-0.02	0.34	0.17	1.11	0.18	0.41	0.10	-0.14	-0.09	0.32	0.12	0.05	0.21	-0.14	0.03	-0.04	-0.05	0.00	-0.07	0.00	-0.06	0.02	-0.02	0.05	0.00	0.02	0.04
5	2	-0.89	-0.22	-0.94	-0.11	-0.84	-0.01	-0.92	-0.33	-0.30	-0.02	-0.15	-0.09	0.12	-0.21	0.13	-0.08	0.01	0.04	-0.03	0.13	0.00	0.04	-0.06	-0.06	0.00	-0.04	0.02	0.03
5	3	-0.75	1.39	-0.14	-0.64	-0.36	0.92	0.03	0.45	0.07	0.43	0.10	0.39	0.03	0.04	0.04	-0.06	0.10	-0.18	0.05	0.02	-0.03	0.06	0.01	0.02	-0.01	0.00	-0.05	-0.04
5	4	1.91	-0.15	1.47	-1.67	1.20	-1.34	0.97	-0.62	0.35	0.18	0.22	0.02	0.06	0.07	-0.31	0.05	-0.08	0.15	0.00	0.02	-0.03	0.01	0.01	0.01	0.02	0.01	-0.01	-0.02
5	5	0.45	0.45	0.75	0.59	0.42	0.04	0.35	0.44	0.24	0.04	0.02	-0.01	-0.05	0.06	0.01	-0.01	0.03	0.07	0.01	0.03	-0.04	-0.12	-0.02	-0.07	0.00	0.05	-0.05	0.04
6	0	0.37		0.41		0.46		0.47		0.29		0.14		0.03		0.00		-0.02		0.02		-0.06		-0.07		0.02		0.01	
6	1	-0.34	-0.67	-0.31	-0.13	-0.13	-0.43	-0.19	-0.26	-0.19	0.40	0.07	0.31	0.17	0.09	-0.07	0.10	-0.07	-0.09	0.05	0.04	-0.10	-0.03	0.01	-0.04	0.01	-0.01	0.03	-0.01
6	2	-0.67	0.64	-0.85	0.25	-0.34	0.53	-0.69	0.60	-0.44	0.02	-0.31	-0.28	-0.01	-0.29	0.02	-0.07	0.04	-0.09	0.08	-0.06	-0.02	-0.04	0.06	0.04	0.00	0.02	-0.01	0.00
6	3	0.21	0.67	-0.08	1.07	0.04	0.36	0.01	0.37	0.03	0.02	0.09	0.03	0.07	-0.11	0.07	-0.11	-0.10	0.09	0.04	0.00	-0.07	-0.02	0.00	-0.05	0.04	0.00	0.00	0.00
6	4	0.46	0.28	0.22	0.27	0.28	0.64	0.45	0.33	0.40	-0.17	0.32	0.24	0.14	-0.16	-0.04	-0.16	-0.11	0.07	-0.09	0.03	-0.07	0.03	-0.04	0.01	0.00	0.00	0.00	-0.02
6	5	0.41	0.18	1.04	0.14	0.67	0.46	0.46	0.46	0.39	0.23	0.24	0.03	0.06	-0.07	-0.07	-0.16	-0.02	-0.02	-0.10	0.12	0.06	0.03	0.01	0.03	-0.03	0.00	0.05	0.01
6	6	0.11	0.08	-0.15	0.58	-0.28	0.21	-0.28	0.08	-0.34	-0.01	-0.23	0.05	-0.07	-0.07	-0.06	-0.07	-0.03	0.00	0.04	-0.01	0.06	0.07	0.03	0.01	-0.03	0.05	0.05	-0.01
7	0	-0.60		-0.44		-0.35		-0.23		-0.16		-0.05		0.13		0.03		-0.01		0.01		-0.01		0.00		0.03		0.02	
7	1	-0.51	0.72	0.03	0.58	-0.15	0.26	-0.14	0.30	-0.09	-0.07	-0.13	0.07	0.08	0.05	-0.05	-0.05	-0.05	0.01	0.03	-0.03	0.01	-0.02	-0.02	-0.02	0.01	0.00	0.05	0.02
7	2	-0.27	0.58	0.22	0.72	0.22	0.46	-0.02	0.50	-0.14	0.04	-0.12	-0.16	0.05	0.01	-0.03	0.01	-0.07	0.06	-0.06	-0.03	0.04	0.00	0.05	0.02	0.02	0.03	-0.02	-0.03
7	3	-0.36	0.24	-0.26	0.53	-0.24	0.13	-0.58	0.32	-0.28	0.00	0.09	0.29	0.10	0.04	0.08	0.08	0.06	-0.06	0.04	0.02	-0.01	0.02	-0.05	0.01	-0.02	-0.03	0.01	-0.04
7	4	0.13	0.58	-0.08	-0.28	-0.10	-0.35	-0.13	0.33	-0.29	-0.01	-0.15	0.04	0.07	-0.02	0.19	-0.02	0.12	0.09	-0.03	0.06	-0.06	0.06	0.02	0.01	0.07	-0.04	0.07	-0.06
7	5	-0.21	-0.98	0.56	-0.40	0.20	-0.38	0.26	-0.10	0.23	-0.53	-0.14	-0.06	0.04	0.00	0.01	-0.01	-0.01	-0.01	-0.02	0.12	-0.04	0.07	-0.01	-0.05	0.05	0.00	-0.02	-0.02
7	6	-0.17	-0.17	-0.05	-0.95	-0.06	-0.08	0.05	-0.09	0.05	-0.01	-0.10	0.13	-0.05	0.00	-0.01	0.00	0.00	-0.10	0.03	0.05	0.04	0.06	0.03	-0.02	-0.02	-0.06	-0.02	-0.02
7	7	-1.10	-0.09	0.11	-0.01	-0.34	0.07	-0.02	-0.02	-0.06	-0.01	-0.16	-0.09	0.09	0.07	0.10	-0.08	0.05	0.04	-0.03	-0.01	0.00	-0.02	-0.01	0.00	0.01	0.01	0.01	0.01
8	0	-0.55		-0.47		-0.70		-0.45		0.01		-0.08		0.00		0.01		0.04		0.06		0.01		-0.05		-0.01		0.03	
8	1	-0.21	0.14	-0.26	-0.35	-0.63	-0.29	-0.18	-0.17	-0.04	-0.16	-0.17	0.04	0.09	0.03	0.01	-0.05	-0.10	0.01	-0.02	0.01	0.08	-0.01	0.05	0.02	0.03	-0.01	0.02	-0.01
8	2	-0.30	0.27	-0.60	-0.02	-0.36	-0.08	-0.37	0.17	-0.21	-0.07	-0.14	-0.09	-0.15	0.03	0.09	0.01	-0.08	0.01	-0.06	-0.02	0.04	0.02	0.00	0.01	0.03	0.00	0.01	-0.01
8	3	0.42	0.52	0.21	0.23	0.31	0.06	0.48	-0.02	0.25	0.04	0.18	-0.05	0.16	0.10	0.12	0.08	-0.04	-0.05	-0.07	0.02	-0.07	-0.02	-0.06	-0.02	0.04	0.02	-0.02	-0.02
8	4	0.19	0.29	-0.12	0.60	0.11	0.21	0.55	-0.05	-0.12	0.02	0.11	-0.01	0.12	-0.01	0.04	-0.01	-0.01	0.00	0.02	-0.01	-0.12	-0.02	0.00	-0.01	-0.12	-0.03	0.06	0.03
8	5	-0.38	0.13	-0.41	-0.18	-0.06	0.14	-0.15	0.02	-0.50	-0.01	0.05	-0.06	0.08	0.02	-0.02	-0.01	-0.03	-0.03	-0.10	-0.08	-0.03	-0.03	-0.03	0.03	0.01	0.07	-0.01	0.06
8	6	-0.34	0.26	0.09	-0.06	-0.42	0.17	0.18	-0.28	-0.01	-0.01	-0.25	-0.03	0.04	-0.09	-0.05	0.00	0.08	0.08	0.10	0.03	-0.09	0.03	0.02	0.02	0.07	0.01	0.00	-0.01
8	7	0.31	0.13	0.05	-0.33	0.18	0.15	-0.22	0.19	-0.01	-0.01	-0.03	-0.05	-0.18	-0.08	-0.01	-0.10	0.08	0.04	0.05	0.02	0.00	0.02	-0.02	0.07	0.00	-0.01	0.01	0.00
8	8	0.25	0.29	-0.17	-0.36	-0.03	-0.09	0.01	-0.16	-0.01	-0.01	-0.26	-0.05	-0.11	-0.08	0.16	0.04	0.04	0.04	0.05	0.00	-0.05	0.02	-0.02	0.08	0.00	-0.01	0.01	-0.04

perturbations is integrated in two dimensions along this path [Dziewonski, 1984].

4.3   Inversion and Data Fit

The model is obtained by solving a non-linear weighted least squares problem stabilized by a penalty function which attempts to minimize the squared gradient of relative velocity perturbations integrated over the mantle volume. The weights of different subsets of data and of the penalty function were chosen empirically; some of the experiments involved a systematic parameter space search. Because the problem is nonlinear, the solution was obtained in three iterations. Corrections for the thickness of the crust have been considered in the same way as in Woodhouse and Dziewonski [1984]. With the parameterization described in the previous section there are 1134 coefficients and these are listed in Table 1 for the model which we call $SH8/WM13$.

The model provides an excellent fit to both the waveform data and the differential travel times. Figure 3a shows a smoothed version of the 2605 $ScS - S$ observations which we used in the inversion. Figure 3b shows the $ScS - S$ predictions of the model. The variance reduction for the individual data is 53% ($\chi^2 = 1694$ for 2605 observations). Comparison of the smoothed predictions to the smoothed observations yields a variance reduction of 80%. The variance reduction for the individual $SS - S$ data is 55% ($\chi^2 = 8877$ for 5388 observations) and for the smoothed data is 78%. The χ^2 values indicate that the $ScS - S$ data are being fit to better than 1 standard deviation, while the $SS - S$ data are being fit to within roughly 1.3 standard deviations.

Variance reductions for the waveform data, relative to PREM [Dziewonski and Anderson, 1981], are roughly 40% for the mantle wave data. The top of Figure 6a shows the comparison of an observed mantle wave record to the synthetic predicted by the spherically symmetric model PREM. The bottom of Figure 6a shows the same observed seismogram compared to the synthetic predicted by our model (the PREM synthetics explain roughly 25% of the total variance in the observed seismograms, the 3-D synthetics explain an additional 40% of the total data variance). Figure 6b shows a similar comparison for a body wave record. For the body wave waveforms the variance reduction, relative to PREM, averages roughly 20% (the PREM synthetics account for roughly 20% of the total data variance, the 3-D synthetics explain an additional 20% of the total variance).

Experiments with different combinations of subsets of data and, in particular, investigation of how much the final model changes if either the waveform data or differential travel times are removed, lead us to the conclusion that these two subsets of data are compatible, despite the different theories employed (see above). Thus, there is empirical evidence that the path average approximation does not lead to systematic errors in the interpretation of

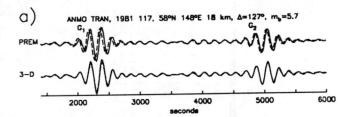

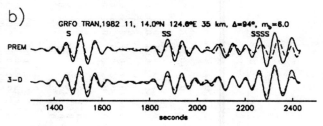

Fig. 6. Comparison of observed (solid line) and synthetic (dashed line) seismograms. (a) Comparison of observed mantle waves (Love waves) which have periods in excess of 135 s with synthetic seismograms. In the top pair of traces the observed seismogram is compared to the synthetic seismogram produced by the spherically symmetric model PREM. In the bottom pair of traces the data is compared to the synthetic seismogram produced by the 3-D model $SH8/WM13$. (b) Comparison of body waves, identified by the appropriate codes, which have periods greater than 45 s with synthetic seismograms. Same details as in (a).

the structure from a large set of overlapping paths. This, of course, does not mean that individual waveforms would not be better matched if a more advanced theory [e.g. Li and Tanimoto, 1992] were used.

4.4   Discussion of Modeling Results

Figure 7a shows the radial variation of the cross-correlation between models $SH8/WM13$ and $SH8/U4L8$. Model $SH8/U4L8$ [Dziewonski and Woodward, 1992] was obtained using the same data set as $SH8/WM13$ and satisfies the data equally well. However, $SH8/U4L8$ has an explicit discontinuity between the upper and lower mantle. The maximum harmonic degree is 8 for both the upper and lower mantle but these regions are parameterized with separate sets of Chebyshev polynomials, with $K = 4$ in the upper mantle and $K = 8$ in the lower mantle.

Figure 7b shows the radial variation of the root-mean-square (rms) amplitude of heterogeneity for the models $SH8/WM13$ and $SH8/U4L8$. For comparison, the rms curve for the model $MDLSH$ of Tanimoto [1990] is also shown. The total rms variation is defined as:

$$G(r) = \left[\sum_\ell g_\ell^2(r) \right]^{1/2} \tag{5}$$

where

$$g_\ell^2(r) = \sum_m \left[\left(\sum_k f_k(r)_k A_\ell^m \right)^2 + \left(\sum_k f_k(r)_k B_\ell^m \right)^2 \right] \tag{6}$$

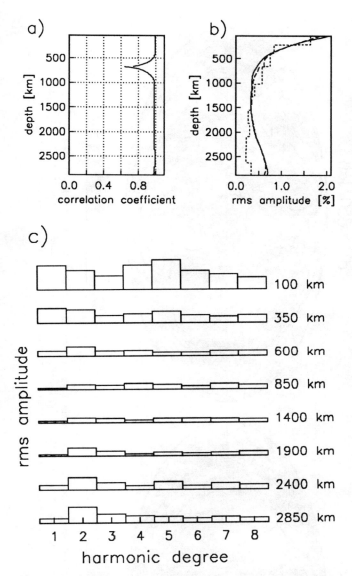

Fig. 7. (a) The cross-correlation between the model $SH8/WM13$, which uses a continuous parameterization of the mantle (from Moho to CMB), and the model $SH8/U4L8$, which uses separate parameterizations for the upper and lower mantle. The correlation is nearly 1.0 throughout the mantle and only drops to roughly 0.6 in the immediate vicinity of the 670 km

$g_\ell(r)$ is the rms amplitude, as a function of depth, for the harmonic degree ℓ. Figure 7c shows the amplitude spectra, $g_\ell(r)$, at eight discrete depths in the mantle (note also that these eight depths are the same as the maps in Figure 8).

Examination of Figures 7b and 7c shows that the heterogeneity is largest near the surface, with a value of 2.0% at the Moho. The amplitude of heterogeneity decreases with depth throughout the upper mantle. In addition, the spectrum is changing, so that in the transition zone the spectrum is dominated by degrees 1 and 2 (Figure

7c). The latter is consistent with the observation of Masters et al. [1982], though their data could not constrain the odd-ℓ part of the structure. By 900 km depth the heterogeneity is reduced to a rather consistent level of 0.35%. In this portion of the mid-mantle, from roughly 900 km depth to 1900 km, the spectrum is relatively uniformly distributed among harmonic degrees from 1 to 8, as Figure 7c demonstrates. At roughly 1900 km depth the amplitude begins rapidly increasing, mostly at harmonic degree 2, but to a lesser extent also 3 and 4, and reaches a value of roughly 0.7% at the CMB.

The general character of the whole mantle rms curve—largest in the upper mantle, smallest in the mid-mantle, increasing in the lowermost mantle—has been known for some time [e.g. Woodhouse and Dziewonski, 1984; Dziewonski, 1984; Tanimoto, 1990, 1991a]. Figure 7b shows that there is a general agreement on the character of the total rms for the models $SH8/WM13$ and $SH8/U4L8$ and for Tanimoto's model $MDLSH$. The agreement is quite good for the upper mantle and mid-mantle (down to roughly 1500 km). However, while $MDLSH$ does show a slight increase in the amplitude of heterogeneity at the base of the mantle, this increase is much smaller than what we observe. Overall, the data used in deriving MDLSH were probably not as sensitive to the amplitude of heterogeneity in the lowermost mantle as the data used here (particularly the $ScS - S$ data) and this may explain some of the discrepancy in Figure 7b.

Figures 8a–h are maps of shear velocity anomalies obtained by synthesizing, according to eqs. 1–4, the A and B coefficients of Table 1. The maps are 'split' along 60°E, because few significant 3-D features span this longitude. The heterogeneity near the surface (Figure 8a) is related to tectonics. The dominant fast features are the continental shields and the dominant slow features are the mid-ocean spreading centers; back arc basins also show as slow regions. The old (and fast) oceanic lithosphere in the west Pacific shows up as relatively fast feature.

At a depth of 350 km in the upper mantle (Figure 8b) the signal is dominated by the difference between continents and oceans. At this depth there is no indication that the relatively slow velocities under the oceans vary with the age of the oceanic lithosphere. At the same time, the mantle under the continents is substantially faster; this lends support to Jordan's hypothesis of the continental tectosphere [Jordan, 1975, 1978].

A gradual change takes place in the transition zone. At 600 km depth (Figure 8c) most of Asia is slow, but there appears a strong velocity high centered on the Philippine plate and extending meridionally far north and south. At this depth there are still high velocities associated with North and South America and Africa while most of the Pacific remains slow. There is a shift in the spectrum towards lower degrees, particularly degree 2 (Figure 7c).

A dramatic change takes place between 600 and 850 km depth (Figure 8d) as the 670 km discontinuity is crossed. There is a significant shift of power towards the higher

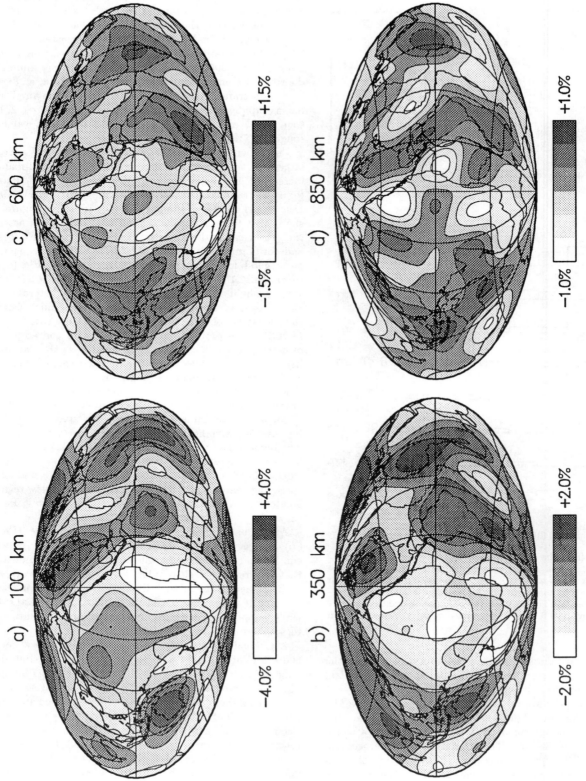

Fig. 8. Maps of relative deviations from the average shear wave velocity at eight depths in the mantle, obtained by the synthesis of the coefficients of model $SH8/WM13$ listed in Table 1. The depth and scale bar are shown with each map. The slowest and fastest values of each map are represented by the white and darkest gray shades, respectively.

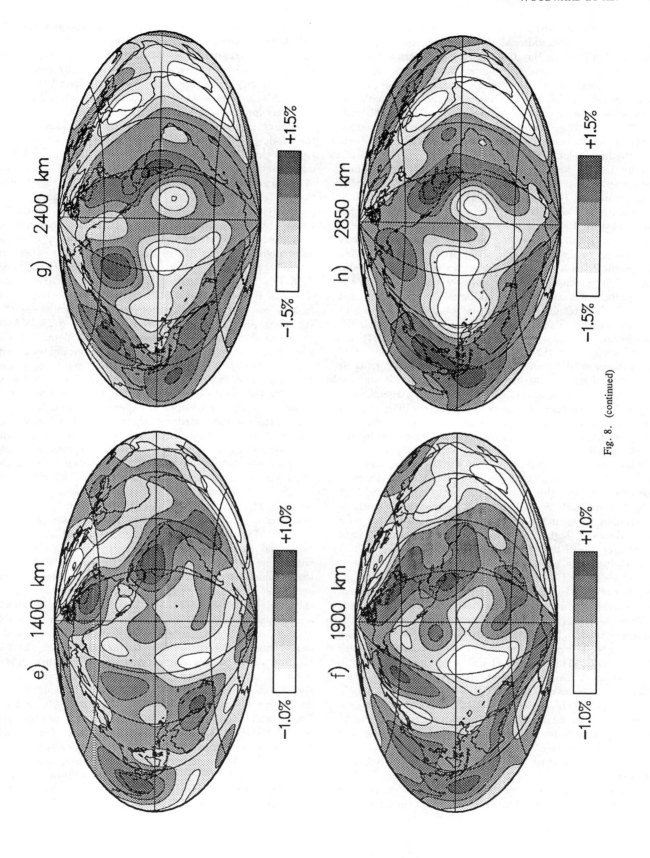

Fig. 8. (continued)

harmonic degrees, although the total power is decreasing. A strong, nearly meridional linear feature develops from the north of Hudson Bay to the south of Tierra del Fuego. Another linear anomaly develops between the Arabian Peninsula and Macquarie triple junction. The pattern of anomalies in the Pacific changes significantly with the development of a velocity high in its north-central part. At the same time, there are certain features that remain continuous: for example the low velocities under Siberia and the high velocities under Africa.

Our results are consistent with the recent regional tomographic inversions of P-wave travel time anomalies [e.g. van der Hilst et al. 1991; Fukao et al., 1992]. The width and amplitude of the velocity highs under the north-western Pacific and easternmost Asia are too large to be the result of low-pass filtering a narrow velocity increase associated with subducting lithosphere (assuming this would have a thickness of roughly 100 km). However, these high velocity regions are entirely consistent with the interpretation that the flow becomes horizontal in that region. Our results are also consistent with the result of Fukao et al. [1992] that the flow associated with the Indonesian subduction zone extends through the 670 km discontinuity. All this is consistent with the observation of Ekström et al. [1990] of deep earthquakes occurring at distances of up to 200 km away from subduction zones. One such earthquake occurred in 1989 in Argentina, a region for which there are no regional tomographic studies, but for which our global inversion predicts a significant horizontal width of a positive velocity anomaly in the transition zone.

There is a significant degree of similarity between the map at 850 km and at 1400 km (Figure 8e), although some features disappear (low velocities beneath Siberia and high velocities beneath Africa) and new ones develop such as the Indian high, although this feature is continuous with the Indonesian high in Figure 8d. There is also a slight rearrangement of the pattern in the central Pacific.

A shift in the spectrum towards longer wavelengths is clearly visible at 1900 km (Figures 8f and 7c). At this depth the ring of high velocities around the Pacific is already well pronounced. This trend is reinforced at 2400 km depth (Figure 8g), where the amplitude of the anomalies has increased by roughly 50%. The amplitude of heterogeneity increases still further towards the core-mantle boundary (Figure 8h). This overall character of heterogeneity in the lowermost mantle has been observed in compressional velocity [Dziewonski, 1984; Inoue et al., 1990] as well as in other studies of shear velocity [Woodhouse and Dziewonski, 1989; Tanimoto, 1990]. The slow features under the Pacific are easily traced up to the spreading center in the east Pacific and similarly, the slow region under Africa may be traced to spreading centers in the Atlantic and Indian Oceans.

It is interesting to compare Figures 8f-h with those of both observed and predicted $ScS - S$ travel time residuals (Figure 3a and b). These $ScS - S$ observations directly reflect the lowermost mantle structure because the ScS phase has a long path through the lowermost 1000 km of the mantle, which is characterized by spatially consistent, strong velocity anomalies, whereas the S phase bottoms in the middle mantle where the amplitude of heterogeneity is relatively low and the spectrum is nearly flat.

Figures 9 a–c show great-circle cross-sections through the model $SH8/WM13$. Figures 9 d and e show cross-sections through the model $SH8/U4L8$, which has separate upper and lower mantle parameterizations. Comparison of Figures 9a and 9b to Figures 9d and 9e shows that the models we obtain are essentially independent of the parameterization used. This is also clear from Figure 7a, which shows the cross-correlation between these models as a function of depth. With the exception of some differences right at 670 km depth, these two models are virtually the same. Both the models SH8/U4L8 and $SH8/WM13$ were obtained subject to smoothness constraints. However, model SH8/U4L8 placed no constraint on how continuous/discontinuous structure across the 670 km discontinuity could be, whereas model $SH8/WM13$ penalized sharp changes in structure at all depths in the mantle, including 670 km. Clearly, the parameterization of model $SH8/U4L8$ contains significant extra degrees of freedom. The fact that the structure in both models is virtually the same demonstrates that the structure which we observe at 670 km depth is required by the data.

Perhaps the most striking feature seen in the images of Figure 9 is the change in structure near 670 km. One can easily discern the location of the 670 km discontinuity in the cross-sections of Figure 9 by the changes in the character of heterogeneity at this depth. Clearly there are features which extend vertically through the 670 km discontinuity, such as the slow features beneath the east Pacific, Indian Ocean, and mid-Atlantic spreading centers. The slow features in the lowermost mantle can be traced to these spreading centers. This is consistent with a scenario in which the major slow features in the lower mantle represent upwellings which are feeding the spreading centers. There are also fast features which extend through the 670 discontinuity, such as beneath Australia and Central America (Figure 9b) or Indonesia (Figure 9c).

It is important to note that not all structures in the cross-sections of Figure 9 are continuous through the 670 km discontinuity. There are a number of structures which seem to be blocked or deflected at this depth: under the North Pacific, Europe and North Africa in Figure 9a, Indian Ocean and Tasman Sea in Figure 9b, and the Pacific and Atlantic oceans in Figure 9c. These blocked and/or deflected features are very important. The models were obtained subject to lateral and radial smoothness constraints. These constraints penalize sharp changes in radial structure. Thus, the modeling procedure will favor smearing structure radially. The fact that we observe structures which are blocked and/or deflected at 670 is

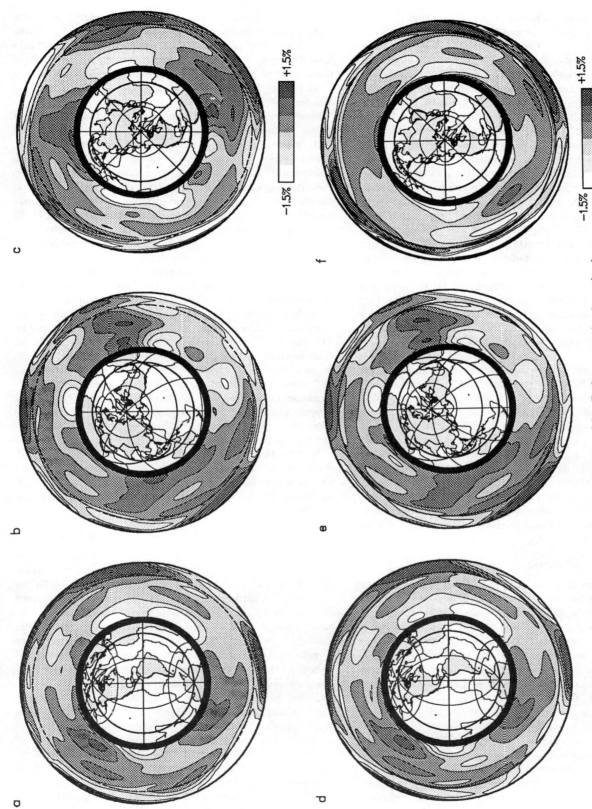

Fig. 9.   Great-circle cross-sections through various 3-D velocity models.   Each cross-section is made along a particular great circle (the heavy line in the inset map) and passes through the center of the Earth.   The outermost ring is closest to the Earth's surface, the innermost corresponds to the CMB.   In the continuous model $SH8/WM13$ (panels a–c) the depth of the 670 km discontinuity is indicated by a dashed line.   Panels d–e show the discontinuous model $SH8/U4L8$, with the 670 km discontinuity shown by a solid black line.   Panel f is a composite of the upper mantle model $M84C$ [Woodhouse and Dziewonski, 1984] and the lower mantle model $L02.56$ (scaled by a factor of two) [Dziewonski, 1984].   The scale represented by the shading ranges from -1.5% (white) to $+1.5\%$ (darkest gray).   Significant saturation of the scale is possible in the upper mantle.

thus very significant. This does not mean that features which are continuous across the 670 km discontinuity should be discounted. These features make sense when interpreted as signatures of mantle flow, with continuity being observed between mid-ocean spreading centers and low velocity upwellings in the lower mantle, and continuity between subduction zones and high velocity features in the lowermost mantle. These points are addressed further in the subsequent section, where we observe that the geodynamic calculations yield the best predictions when a whole mantle style of flow is assumed.

We also note that the pattern of convection indicated by our model may have important bearing on large-scale patterns of isotopic signatures in oceanic basalts. The similarity in the pattern of the Dupal anomaly [Dupre and Allegre, 1983; Hart, 1984] and the pattern of low velocities in the lower mantle has been noted by Castillo [1988]. More recent results indicate that it may be difficult to establish correlations between the isotopic and velocity anomaly patterns [Ray and Anderson, 1991]. However, the cross-sections of Figures 9a and 9c show that the upwelling beneath Africa is apparently connected to the spreading centers in both the South Atlantic and Indian Oceans, implying that the Dupal anomaly is consistent with a lower mantle source.

Figure 9f is a composite cross-section through the upper mantle of model M84C [Woodhouse and Dziewonski, 1984] and the lower mantle of model L02.56 [Dziewonski, 1984]; in the latter, the P-velocity anomalies are scaled by a factor of two [Giardini et al., 1987]. The details of L02.56 near the 670 km discontinuity should be ignored, since some of the upper mantle heterogeneity might have been mapped into this part of the model. The upper mantle images in Figures 9c and f are difficult to tell apart. Parameterization of model L02.56 is much coarser than that of $SH8/WM13$, hence Figure 9c contains much more detail in the lower mantle, but the similarity of large-scale features is unmistakable.

5. GEODYNAMIC IMPLICATIONS

The cross-sections of the model shown in Figures 9 bear a strong resemblance to features which characterize thermal convective flow. On a very large scale, one sees relatively narrow upwellings in the lower mantle which mushroom out into much larger heads in the upper mantle. Return flow is perhaps less distinctive, and primarily consists of large features which descend straight down (e.g. under Australia, Central America, and the Canadian shield) or with some lateral deflection.

By scaling the velocity perturbations of the seismic models to density perturbations, it is possible to calculate the density-driven viscous flow in the Earth's mantle. Extensive discussions of buoyancy-induced flow in spherical shells, in which it is assumed that the fluid is incompressible, may be found in Ricard et al. [1984], Richards and Hager [1984], and Forte and Peltier [1987]. A complete formalism

which also takes into account the effects arising from the finite compressibility of the mantle may be found in Forte and Peltier [1991a]. From such calculations of the mantle flow driven by seismically inferred density contrasts it is possible to make predictions of the main convection-related observables: the nonhydrostatic geoid, CMB topography, and tectonic plate motions.

However, such calculations as described above require several assumptions. First, we find that the geoid and plate motion predictions only provide good fits if a whole-mantle style of convection is assumed. If the 670 km discontinuity is assumed to be a barrier to radial flow, the fits degrade significantly. Further support for this assumption comes from the observation that the long-wavelength nonhydrostatic geoid (especially at degree 2) is strongly correlated to shear velocity heterogeneity near 670 km depth. If the 670 km seismic discontinuity were a barrier to flow, the density contrasts near this depth would tend to be compensated by the deflection of this boundary and thus would produce little geoid signal (density perturbations right at 670 km would be entirely compensated and produce no geoid signal). The strong correlation between geoid and $\delta v_s/v_s$ which we observe at 670 km depth suggests that the density perturbations at this depth do have significant impact on the geoid, which, in turn, implies that mantle flow is whole-mantle in style.

A second important assumption required by the geo-dynamic calculations is a knowledge of the $\delta ln\rho/\delta lnv_s$ proportionality factors throughout the mantle. By fitting geoid data we obtain new inferences of depth-dependent $\delta ln\rho/\delta lnv_s$ proportionality factors.

Finally, the calculation of plate motions and the nonhy-drostatic geoid requires a knowledge of both the relative and absolute mantle viscosities and thus, based on our modeling, we infer a new radial viscosity profile for the mantle.

5.1 Nonhydrostatic Geoid

Three-dimensional density perturbations give rise to buoyancy forces which drive 3-D flow in the mantle. From the theoretical description of this flow it is possible to calculate the kernels which relate the flow-related surface observables to density perturbations in the mantle. The viscous flow models we employ here assume that the mantle viscosity is only a function of radius and consequently these kernels are only sensitive to radial variations in mantle viscosity. Further, the kernels for the nonhydrostatic geoid are only sensitive to the relative radial viscosity profile and are not sensitive to the actual value of viscosity. Constraining the absolute value of mantle viscosity requires other types of data, such as plate motions, which we will discuss in the next section.

We are able to predict well the observed nonhydrostatic geoid by using a mantle viscosity profile which we have obtained by performing forward modeling experiments which take advantage of the known sensitivity of the geoid to the depth dependent variation of viscosity. The viscosity profile we propose is illustrated in Figure 10 and contains two important features: a thin zone of low viscosity at the base of the upper mantle and a zone of higher viscosity in the middle of the lower mantle. The experiments which motivated these inferences are described briefly in Forte et al. [1992] and will be the subject of a future contribution.

At this point it is important to stress the non-uniqueness associated with viscosity profiles inferred from the geoid. It is possible to alter the details of the viscosity profile without noticeably changing the geoid predictions. However, the major features of the viscosity profile (a thin low viscosity zone at the base of the upper mantle and a zone of higher viscosity in the lower mantle) are required in order to obtain adequate fits to the geoid. In Forte et al. [1992] we discuss further refinements of the viscosity profile used here. Such non-uniqueness suggests the need for obtaining a viscosity profile which satisfies the data while also being as simple (e.g. smooth) as possible. Inversions to obtain such a viscosity profile will be the subject of a future contribution.

We must also assume values for the scaling of shear velocity perturbations to density perturbations. We assume that this scaling, $\delta ln\rho/\delta lnv_s$, is only depth dependent and does not vary laterally. If we further assume that this ratio is constant within four depth ranges in the mantle,

0–400 km, 400–670 km, 670–1000 km, 1000–2891 km, we can then solve for the scalings in these depth ranges which give the best fit between the observed nonhydrostatic geoid and that predicted by the 3-D shear velocity model. The values we infer from this procedure are:

$$\delta ln\rho/\delta lnv_s = \begin{cases} -0.02, & 0\text{–}400 \text{ km}, \\ +0.33, & 400\text{–}670 \text{ km}, \\ +0.20, & 670\text{–}1000 \text{ km}, \\ +0.10, & 1000\text{–}2891 \text{ km} \end{cases} \quad (7)$$

and are illustrated in Figure 10.

The $\delta ln\rho/\delta lnv_s$ value in the transition zone (400-670 km depth) is only slightly smaller than the value $\delta ln\rho/\delta lnv_s \approx 0.4$ expected on the basis of laboratory measurements [e.g. Anderson et al., 1968; Isaak et al., 1989] and this reduction, if significant, may be due to the effect of the high ambient pressure at these depths (the laboratory studies referred to here are conducted at 1 atm pressure). The effect of the high-pressure environment of the deep mantle on the temperature derivatives of the seismic velocities has been discussed in detail by Anderson [1987] who concludes, among other things, that the in situ derivatives will be significantly smaller in magnitude than those obtained in standard-pressure laboratory measurements.

An important feature of the inferences in (7) is the marked decrease in the value of $\delta ln\rho/\delta lnv_s$ from the upper mantle to the bottom of the lower mantle. This decrease may again reflect the effect of increasing pressure with depth and is consistent with the recent high-pressure data of Chopelas [1988,1990] which has been used by Yuen et al. [1991] to show a marked decrease, with increasing pressure, of the $\delta ln\rho/\delta lnv_s$ values for MgO. Since the temperature derivatives of the seismic wave speeds are expected to decrease with increasing depth, it is clear from (7) that the coefficient of thermal expansion must decrease even more rapidly. The recent measurements by Chopelas and Boehler [1989] show that the coefficient of thermal expansion will decrease significantly (perhaps by as much as an order of magnitude) across the depth-range of the mantle.

The $\delta ln\rho/\delta lnv_s$ value we infer for the top 400 km of the mantle is anomalous, both in sign and amplitude, with respect to that predicted by thermal variations alone. One possible explanation for this anomalous value is that we are inferring an effective horizontal average of a laterally-varying $\delta ln\rho/\delta lnv_s$ which may reflect the effects of partial-melting below mid-ocean ridges and/or the effects of the continent-ocean differences envisaged by Jordan [1975, 1978]. Such an explanation is consistent with the results of Tanimoto [1991b], who found that at shallow depths (top 200 km of the mantle) higher than average shear velocities beneath continental shields are associated with lower than average densities, thus implying a negative value of $\delta ln\rho/\delta lnv_s$ in these regions. It is important to remember, however, that the $\delta ln\rho/\delta lnv_s$ value we infer for the top 400 km of the mantle is sensitive to the details of

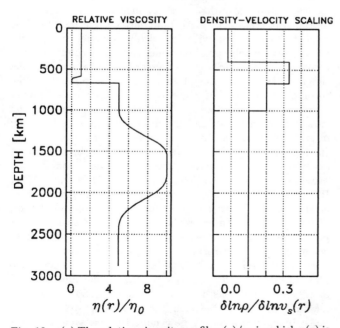

Fig. 10.   (a) The relative viscosity profile $\eta(r)/\eta_0$ in which $\eta(r)$ is the actual mantle viscosity and η_0 is a reference mantle viscosity. This viscosity profile is inferred on the basis of the nonhydrostatic geoid data (see text). (b) The scaling used to convert shear velocity perturbations to density perturbations. This profile is obtained by inverting for the scaling which provides the optimal fit between the observed geoid and that predicted from the model $SH8/U4L8$.

a) GEM–T2 NONHYDROSTATIC GEOID (L=2–8)

b) PREDICTED GEOID (L=2–8)

c) ADJUSTED PREDICTED GEOID (L=2–8)

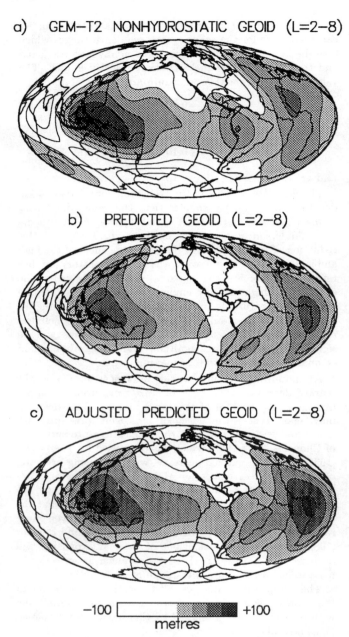

−100 [ ] +100
metres

Fig. 11. (a) The GEM-T2 geoid [Marsh et al., 1990], filtered by removal of the hydrostatic flattening [e.g. Jeffreys, 1963; Nakiboglu, 1982], in the degree range $\ell = 2-8$. (b) The nonhydrostatic geoid, in the degree range $\ell = 2-8$, predicted with model $SH8/U4L8$ using the relative viscosity and $\delta ln \rho / \delta ln v_s$ profiles of Figure 10. (c) The nonhydrostatic geoid of (b) which has been adjusted by setting its Y_2^0 coefficient equal to the value of the observed Y_2^0 coefficient in (a). The contour interval in all cases is 25 m.

the assumed radial viscosity profile and is therefore rather uncertain [Forte et al., 1992].

Figure 11 shows the observed nonhydrostatic geoid and that predicted from model $SH8/U4L8$ by using

the viscosity and velocity–density scaling profiles shown in Figure 10. The predicted nonhydrostatic geoid in Figure 11b explains 65% of the variance in the observed nonhydrostatic geoid shown in Figure 11a. Structure in the vicinity of the 670 km discontinuity is extremely important. We find that density variations in the depth range of 400 to 1000 km account for 65% of the total variance reduction obtained with the geoid prediction of Figure 11b.

The greatest source of misfit between the observed and predicted geoids is the Y_2^0 geoid coefficient. To illustrate the importance of this misfit, Figure 11c shows the predicted geoid of Figure 11b which has now been adjusted by setting its Y_2^0 coefficient to be exactly equal to that in the observed nonhydrostatic geoid. It is evident that this adjustment has considerably improved the the fit to the observed geoid in Figure 11a (variance reduction now equals 82%). This Y_2^0 misfit is of significant geodynamic interest. A brief discussion of the importance and consequences of the Y_2^0 misfit is provided in Forte et al. [1992] and a more complete discussion will be deferred to a later contribution.

5.2   Tectonic Plate Motions

The motion of the tectonic plates is a clear expression of convective flow in the mantle. The plate velocity field may be completely described in terms of its horizontal divergence and radial vorticity. The horizontal divergence field describes the rate of divergence of plates at ridges and trenches, and the radial vorticity field describes the plate motions at transform plate boundaries. The mantle flow predicted by models with spherically symmetric viscosity profiles is entirely poloidal, and cannot account for the toroidal flows which are necessary to characterize the radial vorticity component of actual plate motions. The toroidal component of plate velocities may be explained in terms of a mantle and lithosphere possessing lateral variations of rheology and the most extreme manifestation of such variations is likely provided by the tectonic plates themselves. More than one approach has been used to calculate convective mantle flow which accounts for the presence of rigid plates [e.g. Hager and O'Connell, 1981; Ricard and Vigny, 1989; Forte and Peltier; 1991a,b; Gable et al., 1991]. The approach we use here [Forte and Peltier, 1991a,b] is based on an explicit treatment of surface flows permitted by the requirement that plate motions may only occur by rigid body rotations around Euler poles.

Using the scaling relationships of (7) we calculate the plate-like surface flow driven by the seismically inferred density perturbations derived from model $SH8/U4L8$. By maximizing the fit between the predicted and observed plate-divergence fields we determine the reference viscosity η_0 which normalizes the relative viscosity profile in Figure 10. The value $\eta_0 = 1.2 \times 10^{21}$ Pa s thus obtained from the plate velocities agrees closely with the upper-mantle viscosity $\eta_0 = 1 \times 10^{21}$ Pa s which is usually inferred from glacial isostatic adjustment data [e.g. Cathles, 1975; Peltier and Andrews, 1976]. The predicted plate motions are shown in Figure 12 in which we observe that the

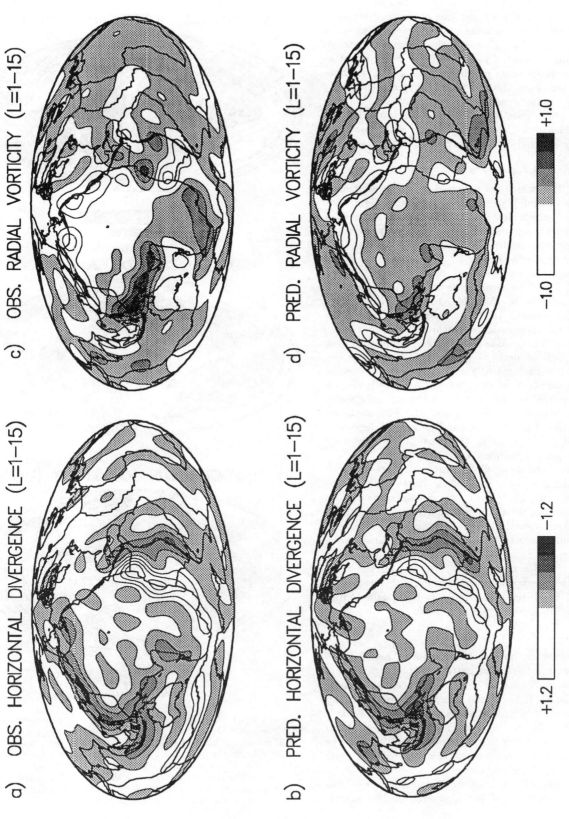

a) OBS. HORIZONTAL DIVERGENCE (L=1–15)

b) PRED. HORIZONTAL DIVERGENCE (L=1–15)

c) OBS. RADIAL VORTICITY (L=1–15)

d) PRED. RADIAL VORTICITY (L=1–15)

Fig. 12. (a) The horizontal divergence of the Minster and Jordan [1978] tectonic plate velocities [Forte and Peltier, 1987] in the degree range $\ell = 1 - 15$. (b) The horizontal divergence, in the degree range $\ell = 1 - 15$, predicted with model $SH8/U4L8$. The contour interval in (a) and (b) is 0.3×10^{-7} rad/yr. (c) The radial agreement between the predicted and observed divergence fields is excellent. The predicted plate divergence provides a 66% variance reduction. However, the agreement between the observed and predicted radial vorticity fields is not as good, with the predictions providing only a 20% variance reduction. We have found that the radial vorticity predictions are quite sensitive to any mismatch between the observed plate geometry (which is an input for our plate-like surface flow calculations) and the geometry of the seismically inferred density contrasts in the mantle.

5.3 Dynamic CMB Topography

The mantle flow, which is calculated in the above discussion of the geoid and plate motions, produces significant normal stresses which act on the core-mantle boundary (CMB). These stresses deflect the CMB away from its hydrostatic reference position. Using the $\delta ln\rho/\delta lnv_s$ and radial viscosity profiles of Figure 10, we calculate the CMB topography expected on the basis of model $SH8/U4L8$. The predicted CMB topography is shown in Figure 13, where it is compared to the CMB topography model of Morelli and Dziewonski [1987]. There is a reasonable agreement of the largest-scale features, with both the observations and predictions showing the prominent ring of depressed CMB around the circum-Pacific and the elevated CMB below the central Pacific and Atlantic oceans. However, the amplitude of the observed and predicted topography are somewhat different (note the different scales for Figure 13a versus Figures 13b and c). The predicted CMB topography has an rms amplitude (for degrees 1 through 8) of 1.0 km while the rms of the observed topography (degrees 1 through 4) is 2.5 km.

Additional constraints on CMB topography are provided by studies of the forced nutation of the Earth. Such observations [Gwinn et al., 1986] place very strong constraints on the Y_2^0 component of the topography. Given the uncertainties in the seismic models, and the assumptions involved, the seismic and geodetic inferences of CMB topography can be easily made to agree. A more detailed discussion of the dynamic CMB topography predicted on the basis of the seismic tomographic models of mantle heterogeneity may be found in Hager et al. [1985], Forte and Peltier [1989,1991c], Forte et al. [1992].

6. CONCLUSIONS

We have reviewed the use of differential travel times for placing constraints on the pattern of heterogeneity in the upper and lower mantle. We have seen that these observations clearly indicate the presence of heterogeneity in the upper mantle which is related to tectonic province and oceanic age. In the lower mantle, these observations indicate that there is significant heterogeneity near the CMB, and that the variations in this heterogeneity are dominantly of long wavelength.

From the study of SS absolute travel times we have seen how the spectrum of heterogeneity in the mantle

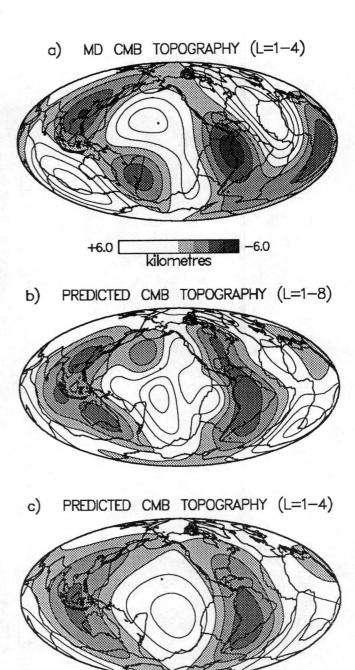

Fig. 13. (a) The seismically-inferred CMB topography model of Morelli and Dziewonski [1987] in the degree range $\ell = 1 - 4$. The contour interval is 1.5 km. (b) The flow-induced CMB topography, in the degree range $\ell = 1 - 8$, predicted with model $SH8/U4L8$ using the relative viscosity and $\delta ln\rho/\delta lnv_s$ profiles of Figure 10. The contour interval is 0.75 km. (c) The flow-induced CMB topography in (b) shown for the truncated degree range $\ell = 1 - 4$. The contour interval is 0.75 km.

may be constrained. These studies show that the overall spectrum of heterogeneity in the mantle is dominanted by the longest wavelengths (i.e. the longest wavelengths of heterogeneity have the largest amplitudes). These studies should not, however, be interpreted to mean that smaller-scale structure does not exist or is not important. For example, shorter wavelength structure is clearly needed to produce the sharp lateral variations in structure which we observe in the upper mantle. In addition, the 3-D models we observe have a fairly 'white' low-amplitude spectrum in the mid-mantle, which is a reasonable expectation for a region of upwelling and downwelling material [Jarvis and Peltier, 1986].

We have obtained improved resolution of structure at all depths in the mantle by combining two major classes of seismic data in a joint inversion. We observe a number of what are, by now, relatively familiar features in the lower-most mantle. What is striking is that these features can be traced through the lower mantle and many of them continue on through the upper mantle as well. The continuity of such features through the 670 km discontinuity implies that convective transport of material between the upper and lower mantle is occurring. However, there are other regions where material is either blocked or deflected at depths of 600 to 800 km or more. In general, there are significant changes in the character of heterogeneity which occur at depths of 600 to 800 km. The amplitude of heterogeneity begins increasing above this depth, and both the spectrum and pattern of heterogeneity change.

Finally, we have seen how this new generation of models may be used to predict completely independent observations. The calculation of geoid anomalies, plate velocities, and CMB topography all represent important tests of any new model. Of course, these calculations require assumptions about mantle viscosities and velocity-to-density scalings. Much useful information can therefore be gained by exploring the range of viscosity models and velocity-density scalings which improve the fit to the convection-related observables. This information can be obtained by additional forward modeling experiments, such as those described here, as well as by formal inversions for viscosity and velocity-to-density scaling profiles. Another logical step along these lines is to combine the inversion of the seismic data and, for example, the geoid. This will allow a more precise quantification of the level of agreement between such different types of data, as well as improving the knowledge of variations in mantle structure. Such experiments will be the subject of a future contribution.

Acknowledgments. The waveform portion of the data set and several programs used in this study were developed in collaboration with John Woodhouse, now at Oxford University. We thank T. Tanimoto and an anonymous referee for constructive reviews. We thank the staffs of the GDSN and IDA seismographic networks for their efforts in collecting the data used in the studies described here. This work has been supported by a grant EAR90-05013 from the National Science Foundation.

REFERENCES

Abramowitz, M., and I. A. Stegun, *Handbook of Mathematical Functions with Formulas, Graphs and Mathematical Tables*, Nat. Bur. Stand., Appl. Math. Ser. *55*, 1964.

Agnew, D. C., J. Berger, R. Buland, W. Farrell, and F. Gilbert, International deployment of accelerometers: A network for very long period seismology, *EOS Trans. AGU*, *57*, 180–188, 1976.

Anderson, D. L., A seismic equation of state II. Shear properties and thermodynamics of the lower mantle, *Phys. Earth Planet. Inter.*, *45*, 307–323, 1987.

Anderson, O. L., E. Schreiber, R. C. Lieberman, and N. Soga, Some elastic constant data on minerals relevant to geophysics, *Rev. Geophys. Space Phys.*, *6*, 491–524, 1968.

Bolton, H., and G. Masters, Long period absolute P times and lower mantle structure, *EOS Trans. AGU*, *72*, 339, 1991.

Butler, R., Shear-wave travel times from SS, *Bull. Seismol. Soc. Am.*, *69*, 1715–1732, 1979.

Castillo, P. R., The Dupal anomaly–low velocity regions of the lower mantle correlation: implications for mantle convection, *EOS Trans. AGU*, *69*, 490–491, 1988.

Cathles, L. M., *The Viscosity of the Earth's Mantle*, Princeton University Press, Princeton, N.J., 1975.

Chopelas, A., New accurate sound velocity measurements of lower mantle materials at very high pressures, *EOS Trans. AGU*, *69*, 1460, 1988.

Chopelas, A., Thermal expansion, heat capacity, and entropy of MgO at mantle pressures, *Phys. Chem. Minerals*, *17*, 249–257, 1990.

Chopelas, A., and R. Boehler, Thermal expansion measurements at very high pressure, systematics, and a case for a chemically homogeneous mantle, *Geophys. Res. Lett.*, *16*, 1347–1350, 1989.

Dupre, B. and C. J. Allegre, Pb-Sr isotope variation in Indian Ocean basalts and mixing phenomena, *Nature*, *303*, 142–146, 1983.

Dziewonski, A. M., Mapping the lower mantle: determination of lateral heterogeneity in P velocity up to degree and order 6, *J. Geophys. Res.*, *89*, 5929–5952, 1984.

Dziewonski, A. M., and D. L. Anderson, Preliminary reference Earth model, *Phys. Earth Planet. Inter.*, *25*, 297–356, 1981.

Dziewonski, A. M., B. H. Hager, and R. J. O'Connell, Large-scale heterogeneities in the lower mantle, *J. Geophys. Res.*, *82*, 239–255, 1977.

Dziewonski, A. M., and J. H. Woodhouse, Global images of the earth's interior, *Science*, *236*, 37–48, 1987.

Dziewonski, A. M. and R. L. Woodward, Acoustic imaging at the planetary scale, *Acoustical Imaging*, *19*, 785–797, 1992.

Ekström, G., A. M. Dziewonski, and J. Ibañez, Deep earthquakes outside slabs, *EOS Trans. AGU*, *71*, 1462, 1990.

Forte, A. M., and W. R. Peltier, Plate tectonics and aspherical Earth structure: The importance of poloidal-toroidal coupling, *J. Geophys. Res.*, *92*, 3645–3679, 1987.

Forte, A. M., and W. R. Peltier, Core-mantle boundary topography and whole-mantle convection, *Geophys. Res. Lett.*, *16*, 621–624, 1989.

Forte, A. M., and W. R. Peltier, Viscous flow models of global geophysical observables. I. Forward problems, *J. Geophys. Res.*, *96*, 20,131–20,159, 1991a.

Forte, A. M., and W. R. Peltier, Gross Earth data and mantle convection: New inferences of mantle viscosity, in *Glacial Isostasy, Sea-Level and Mantle Rheology*, edited by R. Sabadini, K. Lambeck, and E. Boschi, pp. 425–444, NATO ASI Series, Kluwer, Boston, 1991b.

Forte, A. M., and W. R. Peltier, Mantle convection and core-mantle boundary topography: Explanations and implications, *Tectonophysics*, *187*, 91–116, 1991c.

Forte, A. M., A. M. Dziewonski, and R. L. Woodward, Aspherical Structure of the Mantle, Tectonic Plate Motions, Nonhydrostatic Geoid, and Topography of the Core-Mantle Boundary, in *Dynamics of the Earth's Deep Interior and Earth Rotation*, edited by J.-L. Le Mouël, pp. , AGU Geodyn. Ser., in press, 1992.

Fukao, Y., M. Obayashi, H. Inoue, and M. Nenbai, Subducting slabs stagnant in the mantle transition zone, *J. Geophys. Res.*, **97**, 4809–4822, 1992.

Gable, C. W., R. J. O'Connell, and B. J. Travis, Convection in three dimensions with surface plates, *J. Geophys. Res.*, **96**, 8391–8405, 1991.

Giardini, D., X.-D. Li, and J. H. Woodhouse, Three dimensional structure of the Earth from splitting in free oscillation spectra, *Nature*, **325**, 405–411, 1987.

Girardin, N., Travel time residuals of *PP* waves reflected under oceanic and continent platform regions, *Phys. Earth Planet. Inter.*, **23**, 199–206, 1980.

Grand, S. P., and D. V. Helmberger, Upper-mantle shear structure beneath Asia from multi-bounce *S*-waves, *Phys. Earth Planet. Inter.*, **41**, 154–169, 1985.

Gwinn, C. R., T. A. Herring, and I. I. Shapiro, Geodesy by radio interferometry: Studies of the forced nutations of the Earth, 2, Interpretation, *J. Geophys. Res.*, **91**, 4755–4765, 1986.

Hager, B. H., and R. J. O'Connell, A simple global model of plate dynamics and mantle convection, *J. Geophys. Res.*, **86**, 4843–4867, 1981.

Hager, B. H., R. W. Clayton, M. A. Richards, R. P. Comer, and A. M. Dziewonski, Lower mantle heterogeneity, dynamic topography and the geoid, *Nature*, **313**, 541–545, 1985.

Hart, S. R., A large-scale isotope anomaly in the Southern Hemisphere mantle, *Nature*, **309**, 753–757, 1984.

Isaak, D. G., O. L. Anderson, T. Goto, and I. Suzuki, Elasticity of single-crystal forsterite measured up to 1700 K, *J. Geophys. Res.*, **94**, 10,637–10,646, 1989.

Inoue, H., Y. Fukao, K. Tanabe, and Y. Ogata, Whole mantle *P*-wave travel time tomography, *Phys. Earth Planet. Inter.*, **59**, 294–328, 1990.

Jarvis, G. T., and W. R. Peltier, Lateral heterogeneity in the convecting mantle, *J. Geophys. Res.*, **91**, 435–451, 1986.

Jeffreys, H., On the hydrostatic theory of the figure of the Earth, *Geophys. J. R. Astr. Soc.*, **8**, 196–202, 1963.

Jordan, T. H., The continental tectosphere, *Rev. Geophys. Space Phys.*, **13**, 1–12, 1975.

Jordan, T. H., Composition and development of the continental tectosphere, *Nature*, **274**, 544–548, 1978.

Kuo, B.-Y., D. W. Forsyth, and M. Wysession, Lateral heterogeneity and azimuthal anisotropy in the north Atlantic determined from *SS − S* differential travel times, *J. Geophys. Res.*, **92**, 6421–6436, 1987.

Li, X.-D., D. Giardini, and J. H. Woodhouse, Large-scale three-dimensional even-degree structure of the Earth from splitting of long-period normal modes, *J. Geophys. Res.*, **96**, 551–577, 1991.

Li, X.-D., and T. Tanimoto, Waveforms of long-period body waves in a slightly aspherical Earth model, *Geophys. J. Int.*, submitted, 1992.

Marsh, J. G., F. J. Lerch, B. H. Putney, T. L. Felsentreger, B. V. Sanchez, S. M. Klosko, G. B. Patel, J. W. Robbins, R. G. Williamson, T. L. Engelis, W. F. Eddy, N. L. Chandler, D. S. Chinn, S. Kapoor, K. E. Rachlin, L. E. Braatz, and E. C. Pavlis, The GEM-T2 gravitational model, *J. Geophys. Res.*, **95**, 22,043–22,071, 1990.

Masters, G., T. H. Jordan, P. G. Silver, and F. Gilbert, Aspherical earth structure from fundamental spheroidal-mode data, *Nature*, **298**, 609–613, 1982.

Minster, J. B., and T. H. Jordan, Present-day plate motions, *J. Geophys. Res.*, **83**, 5331–5354, 1978.

Montagner, J.-P., and T. Tanimoto, Global upper mantle tomography of seismic velocities and anisotropies, *J. Geophys. Res.*, **96**, 20,337–20,351, 1991.

Morelli, A., and A. M. Dziewonski, Topography of the core-mantle boundary and lateral homogeneity of the liquid core, *Nature*, **325**, 678–683, 1987.

Morelli, A., and A. M. Dziewonski, Joint determination of lateral heterogeneity and earthquake location, in *Glacial Isostasy, Sea-Level and Mantle Rheology*, edited by R. Sabadini, K. Lambeck, and E. Boschi, pp. 515–534, NATO ASI Series, Kluwer, Boston, 1991.

Nakiboglu, S. M., Hydrostatic theory of the Earth and its mechanical implications, *Phys. Earth Planet. Int.*, **28**, 302–311, 1982.

Nataf, H.-C., I. Nakanishi, and D. L. Anderson, Anisotropy and shear-velocity heterogeneities in the upper mantle, *Geophys. Res. Lett.*, **11**, 109–112, 1984.

Nataf, H.-C., I. Nakanishi, and D. L. Anderson, Measurements of mantle wave velocities and inversion for lateral heterogeneities and anisotropy, 3. Inversion, *J. Geophys. Res.*, **91**, 7261–7307, 1986.

Olson, P., P. G. Silver, and R. W. Carlson, The large-scale structure of convection in the Earth's mantle, *Nature*, **344**, 209–215, 1990.

Peltier, W. R., and J. T. Andrews, Glacial-isostatic adjustment-I. The forward problem, *Geophys. J. R. Astr. Soc.*, **46**, 605–646, 1976.

Peterson, J., H. M. Butler, L. G. Holcomb, and C. R. Hutt, The seismic research observatory, *Bull. Seismol. Soc. Am.*, **66**, 2049–2068, 1976.

Pulver, S., and G. Masters, *PcP − P* travel times and the ratio of *P* to *S* velocity variations in the lower mantle, *EOS Trans. AGU*, **71**, 1465, 1990.

Ray, T. W., and D. L. Anderson, Correlations between isotopic data and mantle tomography, *EOS Trans. AGU*, **72**, 526, 1991.

Ricard, Y., L. Fleitout, and C. Froidevaux, Geoid heights and lithospheric stresses for a dynamic Earth, *Ann. Geophys.*, **2**, 267–286, 1984.

Ricard, Y., and C. Vigny, Mantle dynamics with induced plate tectonics, *J. Geophys. Res.*, **94**, 17,543–17,559, 1989.

Richards, M. A., and B. H. Hager, Geoid anomalies in a dynamic Earth, *J. Geophys. Res.*, **89**, 5987–6002, 1984.

Romanowicz, B. A., The upper mantle degree 2: constraints and inferences from global mantle wave attenuation measurements, *J. Geophys. Res.*, **95**, 11051–11071, 1990.

Romanowicz, B., Seismic Tomography of the Earth's Mantle, *Annu. Rev. Earth Planet. Sci.*, **19**, 77–99, 1991.

Sheehan, A. F., and S. C. Solomon, Joint inversion of shear wave travel time residuals and geoid and depth anomalies for long-wavelength variations in upper mantle temperature and composition along the Mid-Atlantic Ridge, *J. Geophys. Res.*, **96**, 19981–20009, 1991.

Stark, M., and D. W. Forsyth, The geoid, small-scale convection, and differential travel time anomalies of shear waves in the central Indian Ocean, *J. Geophys. Res.*, **88**, 2273–2288, 1983.

Su, W.-J., and A. M. Dziewonski, Predominance of long-wavelength heterogeneity in the mantle, *Nature*, **352**, 121–126, 1991.

Tanimoto, T., Long-wavelength *S*-wave velocity structure throughout the mantle, *Geophys. J. Int.*, **100**, 327–336, 1990.

Tanimoto, T., Predominance of large-scale heterogeneity and the shift of velocity anomalies between the upper and lower mantle, *J. Phys. Earth*, **38**, 493–509, 1991a.

Tanimoto, T., Waveform inversion for three-dimensional density and *S* wave structure, *J. Geophys. Res.*, **96**, 8167–8189, 1991b.

van der Hilst, R., R. Engdahl, W. Spakman, and G. Nolet, Tomographic imaging of subducted lithosphere below northwest Pacific island arcs, *Nature*, **353**, 37–43, 1991.

Woodhouse, J. H., and A. M. Dziewonski, Mapping the upper mantle: three-dimensional modeling of earth structure by inversion of seismic waveforms, *J. Geophys. Res.*, **89**, 5953–5986, 1984.

Woodhouse, J. H., and A. M. Dziewonski, Three dimensional mantle models based on mantle wave and long period body wave data, *EOS Trans. AGU*, **67**, 307, 1986.

Woodhouse, J. H., and A. M. Dziewonski, Seismic modeling of the Earth's large-scale three-dimensional structure, *Phil. Trans. R. Soc. Lond.*, **328**, 291–308, 1989.

Woodward, R. L., and G. Masters, Global upper mantle structure from long-period differential travel times, *J. Geophys. Res.*, **96**, 6351–6377, 1991a.

Woodward, R. L., and G. Masters, Lower mantle structure from $ScS - S$ differential travel times, *Nature*, **352**, 231–233, 1991b.

Woodward, R. L., and G. Masters, Upper mantle structure from long-period differential traveltimes and free oscillation data, *Geophys. J. Int.*, **109**, 275–293, 1992.

Yuen, D. A., A. M. Leitch, and U. Hansen, Dynamical influences of pressure-dependent thermal expansiity on mantle convection, in *Glacial Isostasy, Sea-Level, and Mantle Rheology*, edited by R. Sabadini, K. Lambeck, and E. Boschi, pp. 663–701, NATO ASI Series, Kluwer, Boston, 1991.

Zhang, Y.-S., and T. Tanimoto, Global Love wave phase velocity variation and its significance to plate tectonics, *Phys. Earth Planet. Int.*, **66**, 160–202, 1991.

A. M. Dziewonski, A. M. Forte, W.-J. Su, and R. L. Woodward, Department of Earth and Planetary Sciences, Harvard University, 20 Oxford St., Cambridge, MA, 02138, USA.

Density and Elasticity of Model Upper Mantle Compositions and Their Implications for Whole Mantle Structure

JOEL ITA

Department of Geology and Geophysics, University of California, Berkeley

LARS STIXRUDE[1]

Geophysical Laboratory and Center for High Pressure Research, Carnegie Institution of Washington, Washington, DC

We compare the predictions of compositional models of the mantle to properties inferred from seismological data by constructing phase diagrams in the $MgO - FeO - CaO - Al_2O_3 - SiO_2$ system and estimating the elasticity of the relevant minerals. Mie-Grüneisen theory is combined with the Birch-Murnaghan or the Universal equation of state to extrapolate experimental measurements of thermal and elastic properties to high pressures and temperatures. The resulting semi-empirical thermodynamic potentials combined with the estimated phase diagrams predict self-consistently the density, seismic parameter and mantle adiabats for a given compositional model. The transition zone and upper mantle likely have similar compositons. We find that both olivine rich (pyrolite-like) and olivine poor (piclogite-like) compositions agree well with the observed properties of the upper mantle and transition zone. Compositions low in olivine substantially underestimate the magnitude of the 400 km discontinuity unless they have enough Al to cause the pyroxene-garnet transition to coincide approximately with this boundary. The temperature jump associated with a compositional change at 400 km leads to poor agreement with seismic observations for all compositions. The composition of the lower mantle most likely differs from that of the upper mantle. The candidate upper mantle compositions considered here provide acceptable fits to lower mantle observations, but only along adiabats which are significantly cooler than expected. A thermal boundary layer near 670 km depth is highly unlikely in an isochemical mantle.

INTRODUCTION

Seismological observations reveal that the Earth's mantle consists of several layers. The thermal and chemical evolution of the Earth, the dynamics of its interior and thus the forces which drive plate tectonics depend critically on the nature of this layering. Some have argued that these layers are separated by phase changes alone while others have argued that they differ in composition as well [see *Silver et al.*, 1988]. While the former is consistent with whole mantle convection, the latter requires convection in multiple layers, which greatly reduces the efficiency of chemical mixing and heat transport. Multi-layer convection implies a slowly evolving Earth, one which readily preserves large scale heterogeneities and retains most of its primordial heat.

Many types of geophysical observations bear on the nature of mantle layering, but none have unambiguously resolved whether the Earth's mantle is isochemical or compositionally layered. Deep seismicity patterns, used to infer the presence of subducting slabs, show that many slabs penetrate into and in some cases, through the transition zone [*Isacks and Molnar*, 1971; *Jarrard*, 1986]. Travel time and wave-form studies indicate that slabs in the Kurile, Mariana and Tonga trenches penetrate well into the lower mantle [*Creager and Jordan*, 1984, 1986; *Fischer et. al.*, 1991; *Silver and Chan*, 1986], although these results are controversial [*Zhou and Clayton*, 1990; *Gaherty et al.*, 1991; *Schwartz et al.*, 1991]. Isochemical mantle models easily accomodate these observations, but slab penetration is also consistent with a compositionally layered mantle in which the layers are separated by permeable (leaky) boundary layers, or boundaries that are only intermittently breached [*Silver et al.*, 1988; *Machetel and Weber*, 1991]. Travel time observations [*Zhou and Clayton*, 1990] and analysis of seismicity patterns [*Giardini and Woodhouse*, 1984, 1986] also indicate that at least some subducting slabs are substantially deformed by the boundary between the upper and lower mantle, suggesting a

[1]Now at School of Earth and Atmospheric Science, Georgia Institute of Technology, Atlanta, Georgia.

Evolution of the Earth and Planets
Geophysical Monograph 74, IUGG Volume 14

change in material properties at 670 km depth. Slab deformation is easily explained by a change in composition between upper and lower mantles, but is also consistent with an isochemical mantle which exhibits a 10- to 30-fold increase in viscosity at 670 km depth [Vassiliou et al., 1984; Gurnis and Hager, 1988], a viscosity structure which has been proposed on the basis of geoid observations [Hager, 1984; Hager and Richards, 1989]. Many geochemical observations are most easily explained by the survival of two or more physically distinct reservoirs, such as multiple convective layers [Depaolo, 1983; Allegre and Turcotte, 1985; Zindler and Hart, 1986; Silver et al., 1988]. However, inefficient mixing may allow the long-term survival of chemical heterogeneity on many length scales even in single-layer convection [Gurnis and Davies, 1986].

Comparison of seismological observations of the Earth's interior with known material properties is the most direct way of determining the composition of the mantle. The petrologic complexity of the transition zone and the extreme pressures and temperatures of the lower mantle, however, have severely hindered past efforts. Recent advances in high pressure experimental petrology have disentangled the stability relations among the more than 25 phases which may exist in the transition zone, while Brillouin spectroscopy has determined the elastic constants of many of these phases. In situ measurements at combined high pressures and temperatures have placed the first experimental constraints on the thermal expansivity of likely lower mantle phases at high pressure. Previous studies, forced to rely on estimates of these crucial quantities, have come to very different conclusions regarding the composition of the transition zone [Weidner, 1985; Irifune, 1987; Duffy and Anderson, 1989] the lower mantle [Jeanloz and Knittle, 1989; Bina and Silver, 1990; Bukowinski and Wolf, 1990], and the existence of chemical stratification in the Earth's mantle.

We analyze three models of mantle composition by combining a self-consistent thermodynamic formulation with recent measurements of mineral elasticity, phase equilibria and high pressure thermal expansivity. The models include the pyrolite composition, and two olivine poor compositions similar to piclogite. Pyrolite and piclogite were originally proposed in the context of one- and three-layer models of mantle convection which have very different consequences for the state of the Earth's interior. The development of these two models is discussed by Ringwood [1975], Anderson and Bass [1984], Irifune and Ringwood [1987a] and Duffy and Anderson [1989].

We define the compositional models below and synthesize experimental petrologic data by constructing summary phase diagrams for each composition. We then describe the semiempirical thermodynamic potential formalism which allows self-consistent determination of density, bulk modulus, and adiabats throughout the pressure and temperature regime of the Earth's mantle. We focus on comparisons with the seismologically determined density and bulk sound velocity profiles, since these properties are independent of the shear modulus, the pressure dependence of which is unknown for nearly all mantle phases. Careful attention is given to uncertainties in the seismic profiles and in the extrapolation of material properties to mantle pressure

temperature conditions. We estimate the range of compositions permitted by seismic observations of the upper mantle, transition zone and lower mantle and discuss the geophysical implications.

COMPOSITION AND PHASE EQUILIBRIA

The major oxide contents of the three upper mantle compositions considered are shown in Table 1. Composition A is that of pyrolite as defined by Irifune [1987], B is the model deduced from the preferred phase assemblage of Duffy and Anderson [1989] and C is an aluminum enriched version of B and is similar to piclogite compositions defined by Bass and Anderson [1984] and Anderson and Bass [1984]. To fully exploit phase equilibrium data on psuedo-binary systems, we assume that these bulk compositions can be divided into two subsystems, olivine and the remainder (residuum) which do not affect each other's phase equilibria [Jeanloz and Thompson, 1983; Irifune, 1987]. This is consistent with the results of Akaogi and Akimoto [1979] which show no change in the relative proportions of these two subsystems to at least 20 GPa.

Summary phase diagrams for the two subsystems are shown in Figures 1 and 2. For the olivine subsystem, the diagram is based on the results of Katsura and Ito [1989] and Ito and Takahashi [1989]. The phase diagram for the residuum system is based on experimental data along the Mg, Fe, and (Mg,Ca) pyroxene-garnet joins [Akaogi et al., 1987; Kanzaki, 1987; Irifune and Ringwood, 1987a] and results on pyrolite residuum [Irifune and Ringwood, 1987b] [see Ita and Stixrude, 1992 for a complete discussion]. The summary diagrams are in excellent agreement with thermodynamic calculations [Akaogi et al., 1989; Wood, 1990; Fei et al., 1991; Stixrude and Bukowinski, 1992] and with in situ observations

TABLE 1. Oxide Composition

	Composition		
	A	B	C
		mole %	
SiO_2	38.59	41.94	39.82
MgO	49.09	42.32	42.77
FeO	6.24	5.30	5.35
CaO	3.25	8.67	8.76
Al_2O_3	2.20	1.78	3.3
Na_2O	0.33	-	-
Cr_2O_3	0.15	-	-
Ti_2O	0.14	-	-
$\dfrac{100 \times MgO}{MgO + FeO}$	88.72	88.88	88.88
$(Mg,Fe)_2SiO_4$ wt%	62	40	40
Si#	3.23	3.46	3.26

Si# is number of Si atoms per every 12 O atoms

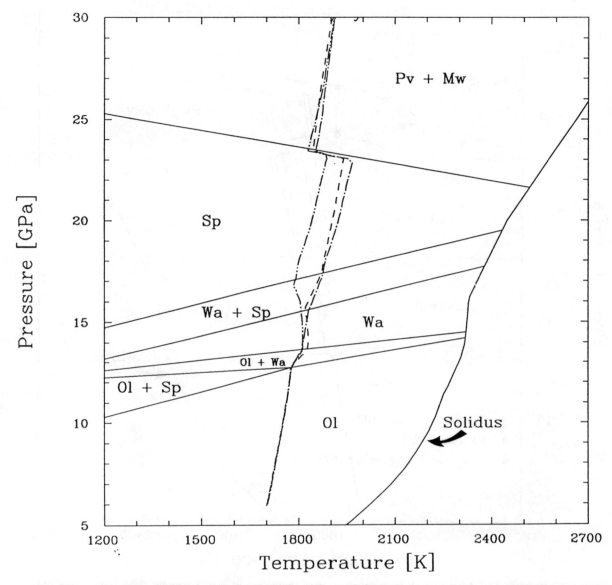

Fig. 1. $(Mg,Fe)_2SiO_4$ phase diagram. The mantle solidus is from *Ito and Takahashi* [1987] and *Gasparik* [1990]. The 1700 K adiabats of compositions A (dashed line), B (dash - double dotted line), and C (dash - dotted line) are superimposed.

vations of peridotite phase equilibria at lower mantle pressures [*O'Neill and Jeanloz*, 1990]. In the upper mantle and transition zone, Fe and Al partition coefficients between coexisting phases are taken from the experimental results quoted above. In the lower mantle, the partitioning of Fe between perovskite and magnesiowüstite is taken from the results of *Stixrude and Bukowinski* [1992] while that between garnet and perovskite is assumed to be unity.

THERMODYNAMICS AND ELASTICITY

Calculations of mineral elasticity and mantle adiabats are self-consistently based on semi-empirical thermodynamic poten-

tials. We express the Gibbs free energy per formula unit, G, of a phase composed of a solid solution of N species as a function of pressure and temperature (P and T) as:

$$G_i(P,T) = \sum_{i=1}^{N} x_{ij}[g_{ij}(P,T) + s_i RT \ln a_{ij}] \tag{1}$$

where g_{ij}, x_{ij}, and a_{ij} are the Gibbs free energy, mole fraction, and activity of species j in phase i, s_i is the stoichiometric coefficient of the mixing site and R is the gas constant. We assume symmetric solid solution behavior [e.g. *Guggenheim*, 1952]

$$RT \ln a_{ij} = RT \ln x_{ij} + W_i(1-x_{ij})^2 \tag{2}$$

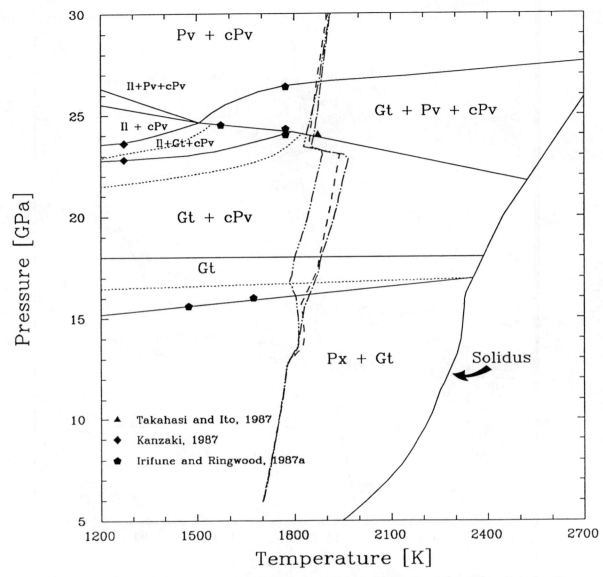

Fig. 2. Residuum phase diagram. Symbols present experimental observations of phase boundaries. Solid lines represent phase boundaries in compositions A and C. Where composition B differs from the other two, its phase boundaries are represented by dotted lines. Mantle solidus and the superimposed adiabats are the same as in Figure 1. Px = Opx + Cpx; Gt = Py + Gr + Maj + Ca-Maj.

where W_i is the interaction parameter assumed to be independent of pressure and temperature. We write the Gibbs free energy as $g_{ij}(P,T) = F_{ij}(V_{ij},T)+PV_{ij}$ where V_{ij} is the molar volume. We assume that the Helmholtz free energy, F_{ij}, can be divided into a reference term, a volume dependent part given by Birch-Murnaghan or Universal equations of state, and a thermal part given by Debye theory (subscripts ij understood) [*Stixrude and Bukowinski*, 1990]

$$F(V,T) = F_o + F_C(V,T_o) + [F_{TH}(V,T) - F_{TH}(V,T_0)] \quad (3)$$

where the subscript o refers to zero pressure and temperature

T_o. The cold part is given by

$$F_C(V) = 9K_oV_o(f^2/2 + a_1f^3/3) \quad (4a)$$

where

$$f = (1/2) [(V_o/V)^{2/3} - 1] \quad (4b)$$

$$a_1 = (3/2)(K_o' - 4), \quad (4c)$$

for the Birch Murnaghan equation of state and by integration of the pressure volume relation proposed by *Vinet et al.* [1987] for the universal equation of state:

$$F_C(V) = \frac{9K_oV_o}{\eta^2}[1 + e^{\eta(1-X)}(\eta - \eta X - 1)] \qquad (4d)$$

where

$$X = (V/V_o)^{1/3} \qquad (4e)$$

$$\eta = 3(K_o'-1)/2 \qquad (4f)$$

Here, K_o and K_o' are the isothermal bulk modulus and its first pressure derivative. The thermal part is given by:

$$F_{TH} = 9nRT(T/\theta)^3 \int_0^{\theta/T} ln(1 - e^t)t^2 \, dt \qquad (5a)$$

where n is the number of atoms per formula unit and θ is the Debye temperature, whose negative logarithmic volume derivative is the Grüneisen parameter, γ, which is assumed to have the form:

$$\gamma = \gamma_o(V/V_o)^q \qquad (5b)$$

where we assume q=1 ± 2 throughout, except in the case of (Mg,Fe)SiO$_3$ perovskite where this quantity has been measured [Mao et al., 1991; Stixrude et al., 1992].

The molar volume (V), isothermal bulk modulus (K), thermal expansivity (α) , heat capacity (C_V), entropy (S), and adiabatic bulk modulus (K_S), of a mineral solid solution at a given pressure and temperature are given directly by the Gibbs free energy (G) and the Helmholtz free energy (F):

$$V = (\partial G/\partial P)_T = \sum_{i=1}^N x_iV_i \qquad (6)$$

$$K = -V/[(\partial^2 G/\partial P^2)_T] = V/(\sum_{i=1}^N x_iV_i/K_i) \qquad (7)$$

$$\alpha = (\partial^2 G/\partial P \partial T)/V = (\sum_{i=1}^N x_iV_i\alpha_i)/V \qquad (8)$$

$$C_V = -T(\partial^2 F/\partial T^2)_V = \sum_{i=1}^N x_iC_{Vi} \qquad (9)$$

$$S = -(\partial G/\partial T)_P = \sum_{i=1}^N x_iS_i - sRx_i lnx_i \qquad (10)$$

$$K_s = K(1 + KTV\alpha^2/C_V) \qquad (11)$$

Mineral properties are combined to calculate ρ, entropy and bulk sound velocity, $V_\phi = \sqrt{K_s/\rho}$, of mantle compositions. We use the Voight-Reuss-Hill method [Watt et al., 1976] to determine the bulk modulus of the poly-phase aggregate, resulting in tight bounds on V_ϕ (± 0.35% on average). The assumption that aggregates are isotropic and homogeneous, inherent in this, as well as other, more restrictive bounding schemes [e.g. Hashin and Shtrikman, 1963; see Watt et al., 1976 and Salerno and Watt, 1986] is unlikely to significantly bias our comparisons with seismic data. Tomographic studies of the transition zone and lower mantle show that anisotropy is small or unresolvable in these regions [Montagner and Tanimoto, 1991] and studies of scattering attenuation show that inhomogeneities on the scale

length of long period seismic waves (10-100 km) are small below 200 km [Korn, 1988].

The model parameters for mantle species are given in Table 2. Reference free energies, F_o, and interaction parameters, W_{ij}, of lower mantle minerals are determined from phase equilibria data and are given in Table 3 of Stixrude and Bukowinski [1992]. The molar volumes of most minerals are directly measured by x-ray diffraction, while those of fictive Fe end-members (Fe-wadsleyite, Fe-perovskite) are determined by measurements along the Mg-Fe join. Using the known volumes of pyrope, almandine and grossular, we simultaneously inverted diffraction data on a suite of garnets for the molar volumes of magnesium, iron, and calcium majorites. The 30 garnet solid solutions span the range of compositions considered here [Akaogi and Akimoto, 1977; Jeanloz, 1981; Irifune et al., 1986; Irifune, 1987; Irifune and Ringwood, 1987a]. Derived end-member majorite values are constrained to within 0.1% (68% confidence level) and agree well with previous determinations of Mg and Fe majorite volumes [Jeanloz, 1981]. Garnet volumes are explained to within 0.2% using the ideal solution model.

Where available, K_o values are taken from ultrasonic or Brillouin scattering measurements. When experimental constraints were unavailable, we estimated the bulk and shear moduli by:

$$M_oV_o = M_{oR}V_{oR} \qquad (12)$$

where M_o is the unknown modulus of a mineral, V_o is its molar volume and M_{oR} and V_{oR} are the corresponding known properties of a reference mineral [see also Duffy and Anderson, 1989]. Assumed values of K_o' were chosen to provide the best fit to compression data for the values of K_o used here. We prefer this approach, also advocated by Bass et al. [1981], to determinations of K_o' based on relatively low pressure ultrasonic measurements of K [e.g. Gwanmesia et al., 1990], since it is based on direct measurement of an observable mantle property (ρ) at mantle pressure conditions.

For most minerals, values of θ_o and γ_o are constrained by heat capacity and thermal expansivity data as described by Stixrude and Bukowinski [1990]. When heat capacity data are not available, we estimated θ_o by:

$$\theta_o/\theta_o^e = \theta_{oR}/\theta_{oR}^e \qquad (13)$$

where θ_o is the unknown thermal Debye temperature of a mineral, θ_o^e is its known elastic Debye temperature and θ_{oR} and θ_{oR}^e are the corresponding known properties of a reference mineral [see also Watanabe, 1982].

Due to the lack of a proper analog, the K_o and K_o' of majorites are treated as variable within the range of current estimates and adjusted to provide the best fit with seismic properties. Thus they also represent the greatest source of uncertainty in the upper mantle. The effect of this uncertainty in modeling mantle properties is discussed in detail below.

The effects of the uncertainties in the lower mantle parameters on calculated properties tend to be larger, but are well characterized. The recent abundance of data relevant to the desired thermodynamic properties allow a quantitative error analysis. Experimental measurements of thermal expansion and

TABLE 2. Parameters of the Thermodynamic Potential

Name	Formula	Abbrv	V_0[a] (cc/mol)	K_0 (GPa)	K_0'	θ_0 (°K)	γ_0
Forsterite	Mg_2SiO_4	Ol	43.76	$128^{1,2}$	$5.0^{1,3}$	$924^{4,5}$	$1.14^{4,5}$
Fayalite	Fe_2SiO_4		46.27	127^6	$5.2^{6,7}$	$688^{8,9}$	$1.08^{8,9}$
Wadsleyite	Mg_2SiO_4	Wa	40.52	174^{10}	$4.0^{11}(4.8)^{12}$	$974^{5,13}$	$1.32^{5,13}$
	Fe_2SiO_4		43.22	$174^{14,b}$	$4.0^{14,b}(4.8)^b$	771^c	1.32^b
Spinel	Mg_2SiO_4	Sp	39.65	183^{15}	$4.1^{16}(5.0)^{17}$	$1017^{5,18}$	$1.21^{5,18}$
	Fe_2SiO_4		42.02	192^{19}	$4.1^{20}(5.0)^b$	$805^{9,18}$	$1.52^{9,18}$
Enstatite	$MgSiO_3$	Opx	31.33	106^{21}	$5.0^{22,23}$	$935^{24,25}$	$0.97^{24,25}$
Ferrosilite	$FeSiO_3$		32.96	101^{26}	5.0^b	$676^{9,27}$	$0.98^{9,27}$
Diopside	$CaMgSi_2O_6$	Ca-cpx	66.11^{28}	114^{29}	4.5^{30}	$941^{25,28}$	$1.06^{25,28}$
Hedenbergite	$CaFeSi_2O_6$		67.84^{31}	120^{32}	4.5^b	$845^{d,21,32}$	0.95^{31}
Pyrope	$Mg_3Al_2Si_3O_{12}$	Py	113.19	173^{33}	3.8^{34}	$981^{35,36}$	$1.24^{35,36}$
Almandine	$Fe_3Al_2Si_3O_{12}$		115.23	177^{37}	4.0^e	$909^{f,37}$	1.06^{38}
Grossular	$Ca_3Al_2Si_3O_{12}$	Gr	125.30^{38}	168^{37}	4.5^{39}	904^{40}	1.05^{38}
Majorite	$Mg_4Si_4O_{12}$	Maj	114.15^g	$151^{41}(173)^{42}$	$4.0^e(4.9)^{22}$	$949^{f,37,41}$	1.24^h
	$Fe_4Si_4O_{12}$		117.70^g	$146^i(168)^i$	$4.0^e(4.9)^{22}$	$820^{f,i,37}$	1.24^h
Ca-Majorite	$Ca_4Si_4O_{12}$	Ca-maj	127.57^g	$135^i(155)^i$	$4.0^e(4.9)^{22}$	$850^{f,i,37}$	1.24^h
Perovskite	$MgSiO_3$	Pv	24.46	$263^{43}\pm4$	$3.9^{43}\pm0.3$	$1017^{44,45}\pm4$	$1.96^{44,45}\pm0.05$
	$FeSiO_3$		25.49	$263^{b,46}\pm6$	$3.9^{b,46}\pm0.4$	$749^{j,k,47}\pm50$	$1.96^b\pm.25$
Al-perovskite	$Mg_{3/4}Al_{1/2}Si_{3/4}O_3$	Al-Pv	24.84^{48}	$259^l\pm10$	$3.9^b\pm0.6$	$1010^{k,m,47}\pm50$	$1.96^b\pm0.25$
	$Fe_{3/4}Al_{1/2}Si_{3/4}O_3$		25.87^{48}	$248^l\pm10$	$3.9^b\pm0.6$	$833^{k,m,47}\pm50$	$1.96^b\pm0.25$
Ca-perovskite	$CaSiO_3$	Ca-Pv	27.27^{49}	$301^{49}\pm12$	$3.8^{49}\pm0.4$	$917^{k,m,47}\pm50$	$1.96^b\pm0.25$
Periclase	MgO	Mw	11.25	$160.4^{50}\pm0.2$	$4.1^{50}\pm0.1$	$777^{4,40}\pm5$	$1.47^{4,40}\pm0.01$
Wustite	FeO		12.25	$152.3^{51}\pm0.2$	$4.9^{51}\pm0.1$	$434^{n,50,51}\pm50$	$1.48^{52}\pm0.25$

1 Graham and Barsch, 1969; 2 Issak et al., 1989; 3 Olinger, 1977; 4 Suzuki, 1975a; 5 Ashida et al., 1987; 6 Graham et al., 1988; 7 Williams et al., 1990; 8 Suzuki et al., 1981; 9 Watanabe, 1982; 10 Sawamoto et al., 1984; 11 Fei et al., 1992; 12 Gwanmesia et al., 1990; 13 Suzuki et al., 1980; 14 Hazen et al., 1990; 15 Weidner et al., 1984; 16 Sawamoto et al., 1986; 17 Rigden et al., 1991; 18 Suzuki, 1979; 19 Liebermann, 1975; 20 Bass et al., 1981; 21 Weidner et al., 1978; 22 Duffy and Anderson, 1989; 23 Watt and Ahrens, 1986; 24 Suzuki, 1975b; 25 Krupka et al., 1985; 26 Bass and Weidner, 1984; 27 Sueno et al., 1976; 28 Finger and Ohashi, 1976; 29 Levien et al., 1979; 30 Levien and Prewitt, 1981; 31 Cameron et al., 1973; 32 Kandelin and Weidner, 1988; 33 O'Neill et al., 1989; 34 Leger et al., 1990; 35 Suzuki and Anderson, 1983; 36 Robie et al., 1976; 37 Bass, 1989; 38 Skinner, 1956; 39 Weaver et al., 1976; 40 Krupka et al., 1979; 41 Bass and Kanzaki, 1990; 42 Yeganeh-Haeri et al., 1990; 43 Knittle and Jeanloz, 1987; Mao et al., 1991; see Hemley et al., 1992 44 Knittle et al., 1986; Mao et al., 1991; see Stixrude et al., 1992 45 Ito and Takahashi, 1989; 46 Mao et al., 1991; 47 Yeganeh-Haeri et al., 1989; 48 Weng et al., 1982; 49 Mao et al., 1989; 50 Jackson, 1982; 51 Jackson, 1990; 52 Touloukian et al., 1977.

a. Unless otherwise noted, volumes are from Jeanloz and Thompson, 1983; b. Assumed to be the same as the Mg end-member; c. Dependence on Mg/Fe ratio assumed the same as for spinel; d. Calculated from Eq. 13 and the properties of diopside; e. Assumed similar to pyrope; f. Calculated from Eq. 13 and the properties of pyrope; g. This work, see text; h. Assumed to be the same as pyrope; i. Bulk and shear moduli calculated from Eq. 12 and the properties of Mg-Majorite; j. Effect of Fe on shear modulus assumed the same as for enstatite; k. Calculated from Eq. 13 and the properties of Mg-perovskite; l. Calculated from Eq. 12 and the properties of Mg-perovskite; m. Shear modulus calculated from Eq. 12 and the properties of Mg-perovskite; n. Elastic value.

heat capacity constrain θ and γ, compression data determine the errors associated with the values of K and K', and phase equilibria measurements provide constraints on the interaction parameters and the reference free energies. The best fitting values and their associated errors are found by minimizing the value of χ^2 between the experimental and calculated values [see *Stixrude and Bukowinski*, 1992 and *Stixrude and Bukowinski*, 1990 for details and results]. We use the variance and covariance of the thermodynamic parameters determined in this way as generating functions in a Monte Carlo algorithm to simulate the probability distribution of the density and bulk modulus of mineral assemblages at elevated pressures and temperatures. The simulation

results show that the distributions of ρ and K are, to a good approximation gaussian. Thus the non-linearities inherent in the thermodynamic and mechanical mixing of end-member species do not cause the errors in the predicted lower mantle ρ and V_Φ estimates to significantly deviate from a normal distribution.

SEISMOLOGICAL PROFILES

We estimate the expected density profile of the mantle and its uncertainty with the mean and standard deviation of 14 published models [*Derr*, 1969; *Haddon and Bullen*, 1969; *Mizutani and Abe*, 1971; *Wang*, 1972; *Jordan and Anderson*, 1974; *Gil-*

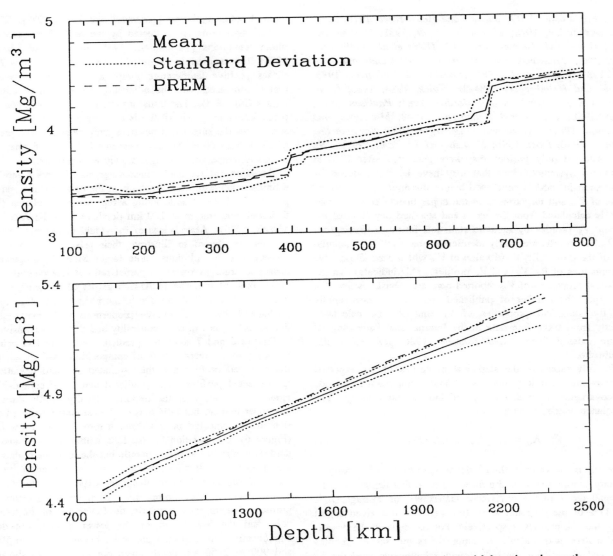

Fig. 3. Comparison of the expected mean and standard deviation of density with the PREM density model throughout the mantle.

bert and Dziewonski, 1975; Kind and Muller, 1975; Hart et al., 1976; Nakada and Hashizume, 1983; Lerner-Lam and Jordan, 1987; Montagner and Anderson, 1989; Dost, 1990]. Some of the older models may contain systematic errors due to the effects of attenuation [Hart et al., 1976], but this error does not appear to significantly bias the mean in the upper mantle. A comparison is made with the Earth model PREM [Dzeiwonski and Anderson, 1981] which was purposely left out of the average to provide an independent test of the mean. The mean and PREM are in excellent agreement throughout the upper mantle except just above the 400 and 670 km discontinuities (see Figure 3). Even there, PREM falls within the uncertainty of the mean. It appears that any error introduced by the older models has only increased the standard deviation in this region, making it a conservative estimate. A significant bias does appear between PREM and our

mean curve in the lower mantle. The standard deviation we estimate in this region should overestimate the actual uncertainty given the increased disparity between profiles with and without this bias. In order to account for this bias, we assume that PREM represents the true mean in the lower mantle and has the same standard deviation as the biased curve. The average standard deviation is 0.6% which is in good agreement with other estimates of the error [Silver et al., 1988].

We base our estimate of the expected bulk sound velocity profile of the upper mantle and its uncertainty on 31 P-wave velocity (V_P) profiles and 30 S-wave velocity (V_S) profiles generated from high resolution travel-time studies or waveform modeling studies of body waves or surface waves [Gilbert and Dziewonski, 1975; Hart, 1975; Kind and Muller, 1975; Dey-Sarkar and Wiggins, 1976; Hart et al., 1976; King and Calcagnile, 1976; *Fukao*, 1977; *Burdick and Helmberger*, 1978; *Sengupta and Julian*, 1978; *McMechan*, 1979; 1981; *Uhrhammer*, 1979; *Given and Helmberger*, 1980; *Hales et al.*, 1980; *Burdick*, 1981; *Dziewonski and Anderson*, 1981; *Vinnik and Ryaboy*, 1981; *Fukao et al.*, 1982; *Nakada and Hashizume*, 1983; *Grand and Helmberger*, 1984a,b; *Walck*, 1984, 1985; *Lyon-Caen*, 1986; *Lerner-Lam and Jordan*, 1987; *Paulssen*, 1987; *Grad*, 1988; *Graves and Helmberger*, 1988; *Montagner and Anderson*, 1989; *Lefevre and Helmberger*, 1989; *Bowman and Kennett*, 1990; *Dost*, 1990; *Kennett and Engdahl*, 1991]. We have selected only profiles that were generated after 1974 to avoid any systematic bias that may have been introduced by attenuation in models developed before this time. The expected value of V_ϕ and its variance in the upper mantle and transition zone is calculated from the mean and standard deviation of the V_P and V_S profiles by standard statistical formulae [*Bevington*, 1969]. The results are nearly identical to the results of a simulation of the probability distribution of V_ϕ which used all possible combinations of the V_P and V_S profiles. This indicates that the statistical properties of V_ϕ reported here are robust. Below 790 km depth, the number of published profiles decreases rapidly and the statistical properties of V_P and V_S are calculated directly from traveltime data in the International Seismological Centre catalog [*Tralli and Johnson*, 1986; *Ita and Tralli*, unpublished data].

To fully represent the statistical properties of the expected lower mantle seismic profiles, we found it important to include the covariance for both density and bulk sound velocity. The covariance matrix, Λ, is given by

$$\Lambda_{ij} = \frac{1}{N} \sum_{k=1}^{N} (x_{ik} - \bar{x}_i)(x_{jk} - \bar{x}_j), \tag{14}$$

where x_{ik} is the value at the ith depth point of the kth density or velocity profile and $\bar{x}_i$ is the mean value at that depth. The off-diagonal terms in the covariance matrix are, on average, 78% and 11% of the diagonal terms (the variance at a given depth) for ρ and V_ϕ profiles, respectively. For the density, the covariance matrix was calculated using the same set of profiles described above. For the bulk sound velocity, we used the 11 V_P and V_S profiles available in the lower mantle. We have also calculated the full covariance matrices for the predicted profiles and taken full account of covariance in our comparisons. Comparisons between predicted and expected lower mantle profiles exclude the upper 120 km of the lower mantle because of possible non-adiabatic temperature gradients, which we do not attempt to explain here, and the lower 560 km because of contamination of the S wave traveltime curve by SKS arrivals.

RESULTS

Isentropes of the three compositions are determined from the same fundamental thermodynamic relations used in the mineral elasticity calculations. This is accomplished by finding contours of constant entropy in pressure - temperature space. Isentropes should closely approximate geotherms since, for mantle conditions, the effects of viscous dissipation are only significant in

thermal boundary layers [*Machetel and Yuen*, 1989]. The calculated geotherms are deflected by the heat of reaction during phase transitions [*Verhoogen*, 1965; *Jeanloz and Thompson*, 1983]. The exothermic nature of phase transitions below 21 GPa causes positive temperature jumps and high gradients in this region compared with the adiabatic compression of individual phases (Figure 4). The transformation of garnet and spinel into perovskite + magnesiowüstite is a strongly endothermic reaction and causes the substantial negative jump seen between pressures of 21 and 25 GPa. Higher pressures induce no further transitions and temperature changes slowly with pressure.

Based on our summary phase diagrams (Figures 1 and 2), we estimate relative mineral fractions in the upper mantle and upper part of the lower mantle along a representative adiabat (Figure 5, initial temperature at 180 km depth of 1700 K; *Jeanloz and Morris*, 1986). Abundances of individual garnet and perovskite species are shown to illustrate their relative effects on their respective solid solutions. The larger Si # in composition B causes a larger pyroxene to garnet ratio at low pressures and a larger ratio of majorite to Al-rich garnet components at higher pressures. Below 900 km depth, the phase assemblage remains unchanged, but their relative proportions will change slightly due to differences in compressibility and thermal expansivity.

Figures 6 and 7 show the predicted bulk sound velocity and density profiles, respectively, of compositions A-C together with the expected profiles and the estimated standard deviations in the expected profiles. The predicted density of composition A agrees very well with the expected density profile throughout the upper mantle, but falls outside the estimated standard deviation in the expected profile through most of the lower mantle (Figure 6). Composition C also falls within the estimated standard deviation in the upper mantle but the agreement above 400 km is marginal. It also predicts densities that differ by more than one standard deviation from the expected profile for most of the lower mantle. Composition B matches the estimated ρ profile of the uppermost mantle, the lower part of the transition zone and the lower part of the lower mantle, but departs significantly from expected properties between 400 and 500 km and 900 to 2150 km depth given our estimates of the uncertainty.

In the uppermost mantle, the predicted bulk sound velocity (Figure 7) of all assemblages falls within one standard deviation of the expected velocity profile. Near the 400 km discontinuity, the character of B differs significantly from the other two. Composition B predicts a jump that is only two-thirds of the expected value. The velocity remains consistently low for the next 50 km and then rises to the expected value within a 30 km depth range. A rise in velocity occurs near the proposed 520 km seismic discontinuity [*Shearer*, 1990] in all assemblages due to the dissolution of Ca-perovskite (Figure 5). It is more pronounced in B and C due to their higher calcium contents. The rise is then followed by predicted values that agree with the mean velocity in the rest of the transition zone. In the lower mantle, the velocities predicted by A fall within one standard deviation of the expected profile except in the lowermost regions while B and C fall systematically above the uncertainty.

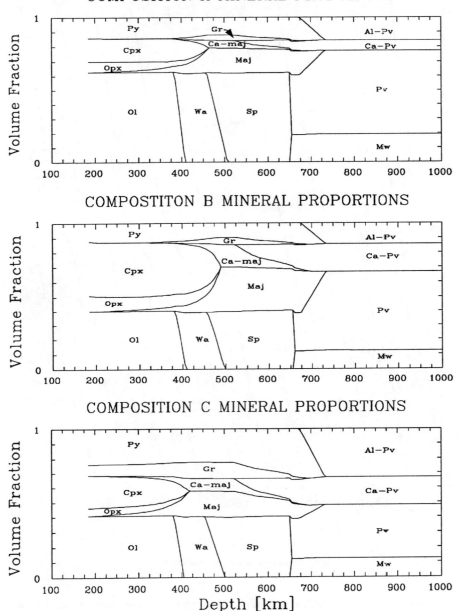

Fig. 4.   Volume percentage of the minerals present in compositions A, B, and C as a function of depth.

DISCUSSION

Our results show that it is possible to account for the anomalously high velocity gradients in the transition zone with a uniform composition. In excellent agreement with *Birch's* [1952] explanation for the seismic properties of this region, we find that a series of phase transitions, including pyroxene to garnet, wadsleyite to spinel and the exsolution of Ca-perovskite in

compositions A and C, readily account for the expected density and velocity in this region. A compositional gradient through the transition zone is not required in order to predict the expected properties in this region.

We find that pyrolite (composition A) provides an excellent account of the expected properties in the upper mantle and the transition zone. This is in agreement with the results of *Weidner* [1985] and *Weidner and Ito* [1987] who also con-

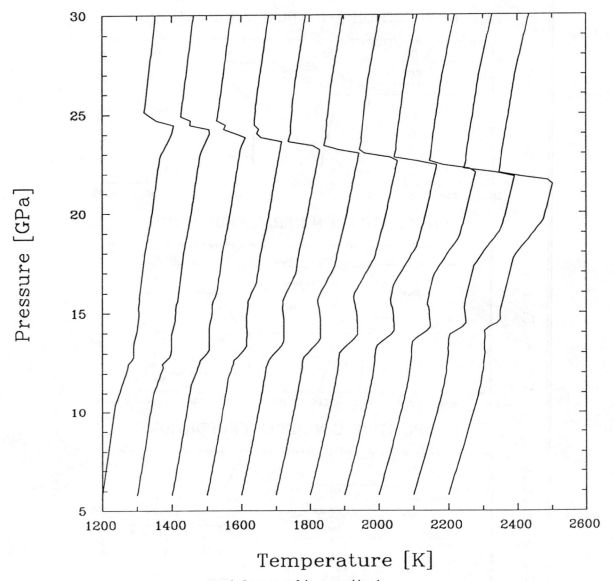

Fig. 5. Isentropes of the composition A.

cluded that seismic observations are consistent with a pyrolite composition to at least 670 km depth.

Olivine-poor compositions, however, are also able to account for the expected profiles. In composition A, the 400 km discontinuity is caused by the transition of olivine to wadsleyite. In composition C, the effect of this transition is muted because of its lesser olivine content. However, its greater Al content, relative to compositions A and B, lead to a shallower pyroxene to garnet transition. The combined effect of the olivine to wadsleyite and pyroxene to garnet transitions in composition C leads to good agreement with the expected magnitude of the discontinuity. In composition B, the magnitude of the discontinuity is too small due to the delay of the pyroxene to garnet

transition to approximately 450 km because of its relatively low Al content coupled with it's low olivine fraction.

To account for the uncertainties in the thermodynamic parameters in the upper mantle and transition zone, we consider alternative estimates of K_o for the majorite species and K_o' for the majorites, wadsleyite, and spinel. Examination of the variation of other parameters within their uncertainties demonstrates that they have a comparatively small effect on derived mantle velocities and densities as does the form of the equation of state (Figure 8). As one can see from Figure 9, average values of K_o and K_o' (Table 2) yield the best fit to the expected profiles for composition B. Lower and higher majorite bulk moduli lower and raise transition zone velocities, respectively, and magnify

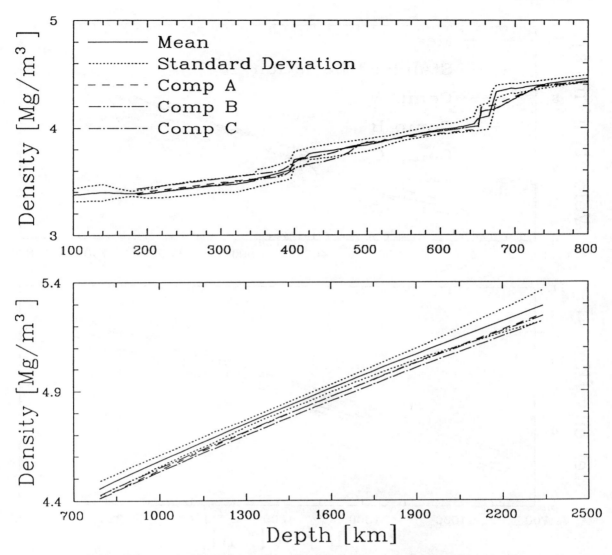

Fig. 6. Comparison of the expected density profile and its standard deviation with the predicted densities of compostitions A, B, and C.

the disparity between predicted and expected bulk sound velocity profiles. Raising or lowering mantle temperatures can produce acceptable calculated velocities but leads to disagreement with expected densities. Regardless of variations in the parameters or temperature, piclogite does not provide an acceptable fit to both bulk sound velocity and density profiles if one accepts our estimates of the uncertainty.

For composition A, values of K_o and $K_o{}'$ in the lower half of the range considered here lead to the best agreement between calculated and expected properties. Higher majorite bulk moduli lead to higher bulk sound velocities in the transition zone, causing the predictions to deviate significantly from the expected profile. Again, agreement with velocities can be improved by raising or lowering assumed mantle temperatures

but only at the expense of the ρ comparison. Similar behavior is seen in the calculations for composition C.

For the range of compositions considered here, a chemically layered upper mantle is unlikely. Adiabats initiating at 1700-1800 K at 185 km lead to the best agreement with expected transition zone properties regardless of composition. These temperatures agree very well with independent determinations from geothermometry [see *Jeanloz and Knittle*, 1989]. A chemical boundary at 400 km, however, requires layered convection and a thermal boundary at this depth of 500-1000 K [*Jeanloz and Richter*, 1979]. Retaining a 1700 K adiabat in the uppermost mantle, we find that transition zone properties along a 2200 K adiabat are in severe disagreement with expected densities (Figure 10). Furthermore, these high temperatures may

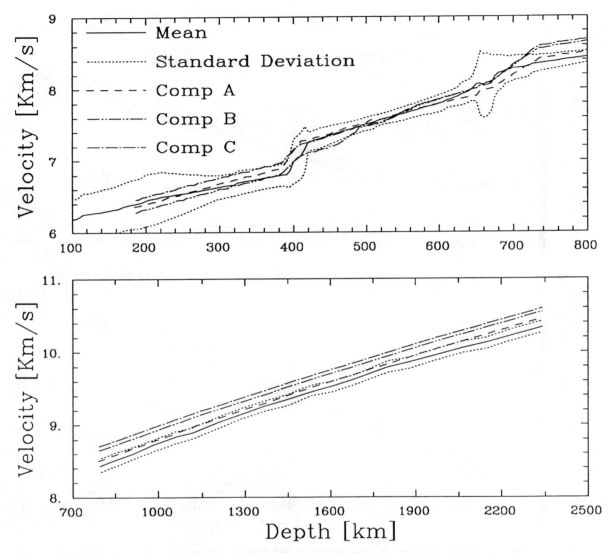

Fig. 7. Comparison of the expected V_Φ profile and its standard deviation with the predicted profiles for compositions A, B, and C.

cause wide spread partial melting of the transition zone (Figures 1 and 2) leading to shear velocities much lower than observed. Alternatively, if we assume a 1700 K adiabat in the transition zone, the uppermost mantle follows a 1200 K adiabat which is marginally consistent with the expected properties for some compositions but inconsistent with geothermometry.

In the lower mantle, the bounds on the predicted and expected profiles of both ρ and V_Φ overlap. Over the depth range of the lower mantle, the average relative error of the expected ρ and V_Φ is approximately 0.7% and 0.8%, respectively. The predicted ρ and V_Φ have an average relative error of 0.6% and 1.8%, respectively. The errors are clearly of the same order and both should be taken into account when making these comparisons.

A formal measure of the significance of the difference between the predicted and expected profiles is given by the χ^2 statistic:

$$\chi^2 = (\vec{p} - \vec{e}) \cdot (\Lambda_p + \Lambda_e)^{-1} \cdot (\vec{p} - \vec{e}) \qquad (15)$$

where $\vec{p}$ and $\vec{e}$ are vectors of length N whose elements are the predicted and expected values, respectively, of ρ or V_Φ at N depth points and Λ_p and Λ_e are their associated covariance matrices. We compare predicted and expected profiles at sixteen depth points located 100 km apart for V_Φ and five depth points located 300 km apart for ρ. The spacing was chosen based on the expected resolution of ρ and V_Φ in the lower mantle appropriate for our estimated uncertainties in the expected profiles [Tralli and Johnson, 1986; Gilbert et al., 1973].

For the purpose of illustration, we note that χ^2 is less than the number of degrees of freedom (N) for all compositions, as

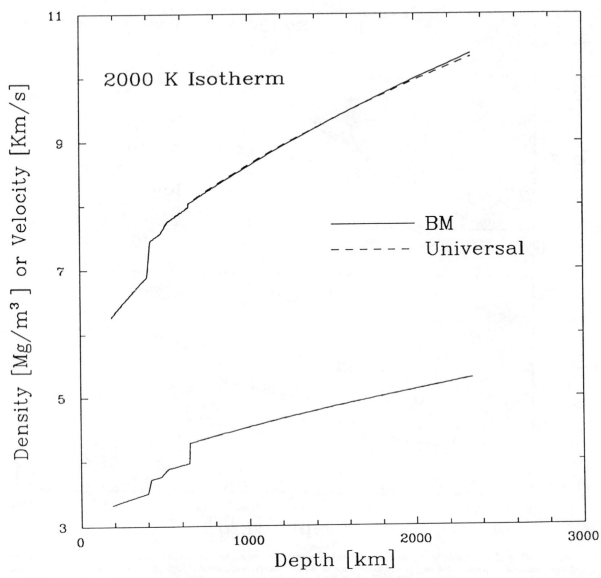

Fig. 8. Composite value of the density (ρ) and acoustic velocity (V_ϕ) along the 2000 K adiabat for a pure olivine composition using the Birch-Murnaghan finite strain equation of state (solid line) and the Universal equation of state (dashed line).

suggested by the overlap of error bounds shown in Figure 11. This means that with at least 68% confidence the following hypothesis (H1) can be rejected: the predicted and expected profiles are drawn as random samples from different parent probability distributions. This test would be relevant, for example, in the case of comparing a single experimental measurement of density at a given pressure and temperature with a seismological determination of density at the corresponding depth. However, our expected and predicted profiles are based on many measurements so that we have some confidence in their mean values and the shapes of their probability distributions. In our case, it is appropriate to test the following hypothesis (H2): the means

of the predicted and expected probability distributions are determined by repeated sampling of different parent probability distributions. Clearly this is a more restrictive test since the range of values within which the mean of a finite sample of the parent distribution is expected to fall for a given level of confidence is smaller than the expected range for any single event in that sample at the same confidence level.

We have adopted the modelling procedure of Press et al., [1986, pp. 534-5] to determine the temperature at the foot of the lower mantle adiabat (670 km depth) which minimizes χ^2 for each composition and the uncertainty in this best-fitting temperature. This is equivalent to determining the range of adiabats

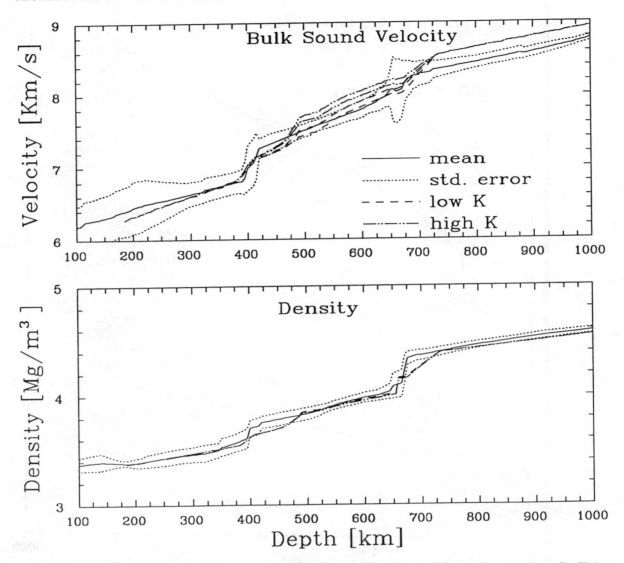

Fig. 9. Effect of uncertainty in the elastic parameters of wadsleyite, spinel, and the majorites on the composite profile. High and low values of K produce the higher and lower pairs of dashed curves, respectively. High and low values of K' produce the higher and lower curves for each pair. The mean and standard deviation of expected velocities are represented by the solid and dotted lines, respectively.

which do not satisfy H2 in the case that the best fitting temperature yields a perfect fit between predicted and expected profiles (minimum value of $\chi^2=0$). The results of this procedure are shown in Table 3.

The adiabat which produced the best fits to expected properties of the upper mantle and transition zone (1700 K at 185 km depth, e.g. Figure 1) implies a temperature of 1822 to 1841 K at the top of the lower mantle. Table 3 shows that temperatures which produce the best fits to lower mantle observations are 250 to 260 K colder. This difference is not significant at the 95% confidence level, but is significant at greater than the 68% confidence level in the case of compositions A and C, the two

which produced acceptable fits to the upper mantle and transition zone.

None of the compositional models considered here match expected profiles over the entire depth range of the mantle at the 68% confidence level. The extent to which this result is affected by possible, as yet undiscovered, phase transitions in the perovskite species and magnesiowüstite or the possible significance of higher order thermodynamic parameters (e.g. volume dependence of q) at very high pressures and temperatures will require further experimental and theoretical investigation. Seismological observations, in particular the expected density profile, are also not free of systematic uncertainty. If the

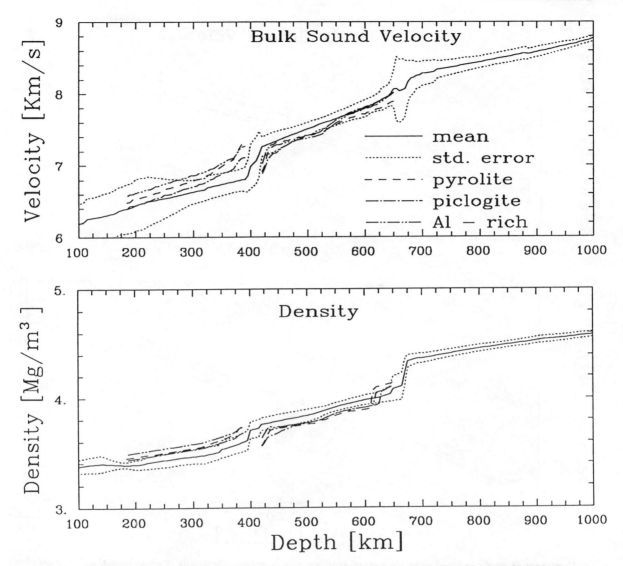

Fig. 10. Effect of a thermal boundary layer on calculated mantle properties. For the three compositions considered here, upper mantle properties are shown along the 1200 K adiabat. Transition zone properties are shown along the 2200 K adiabat. The mean and standard deviation of acoustic velocity and density are represented by the solid and dotted lines (see text for discussion).

14 density profiles we have considered here are not significantly biased by attenuation and we assume that their mean best represents the expected density profile, rather than PREM, best fitting temperatures for compositions A-C are raised by approximately 100 K, making them consistent with our best fitting upper mantle adiabat.

The effect of neglecting covariance and uncertainties in the predicted profiles is also shown in Table 3. If the off-diagonal terms in Λ are ignored, H2 and the less restrictive hypothesis, H1, are satisfied at the 95% confidence level for both compositions B and C regardless of temperature. The best fitting temperature for composition A is higher if covariance is ignored, but its upper bound is still well below what one would expect

from the best fitting upper mantle adiabat. Ignoring uncertainties in the predicted profiles also causes B and C to satisfy H2 and H1 at the 95% confidence level. The best fitting temperature for A is higher in this case because higher temperatures lead to better agreement with the expected V_Φ profile at the expense of the ρ profile. The V_Φ profile is weighted more strongly if predicted errors are ignored since its error is reduced by a greater percentage than that of density.

Our best estimates of predicted and expected profiles and their uncertainties indicate that the composition of the lower mantle differs from that of the upper mantle. This conclusion is considerably strengthened if a thermal boundary layer exists at 670 km. Recent studies of mantle convection indicate that the

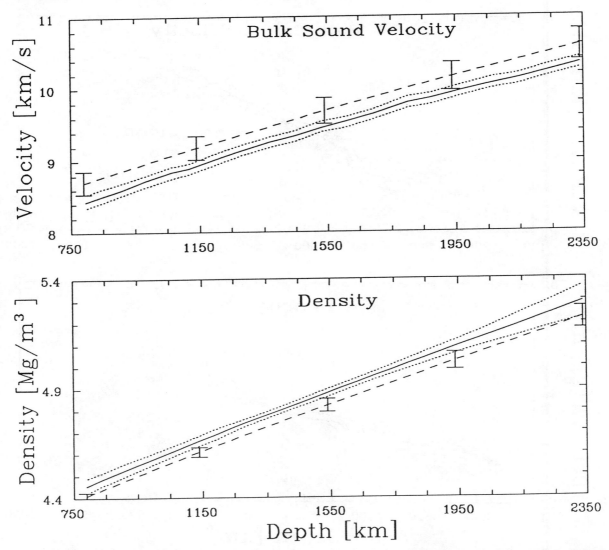

Fig. 11. Comparison of the the expected mean and standard deviation with the predicted properties of composition C and their associated uncertainties.

thermal energy of perovskite forming reactions may be sufficient, even in an isochemical mantle, to induce two-layer convection [*Machetel and Weber*, 1991]. This would raise lower mantle temperatures by 500-1000 K, [*Jeanloz and Richter*, 1979], increasing the discrepancy between compositions A-C and seismic observations. The lower mantle is most likely enriched in SiO_2, and possibly FeO components, which would lead to inherently denser and faster, perovskite-dominated assemblages which would match lower mantle observations along the relatively hot mantle adiabats required by a thermal boundary layer [*Jeanloz and Knittle*, 1989; *Bina and Silver*, 1990; *Stixrude et al.*, 1992].

These results are in excellent agreement with those of *Stixrude et al.* [1992] who found that upper mantle compositions matched seismic observations of the lower mantle only

along adiabats which were unreasonably cold. The best fitting temperatures in that study were slightly less than those reported here primarily because the effects of Ca-perovskite, the densest and fastest (highest V_Φ) mineral in our predicted lower mantle assemblage, was not included. The presence of Ca- and Al-bearing perovskites also account primarily for the larger uncertainties in this study since the thermal properties of these minerals have been estimated, and thus assigned generous uncertainties (Table 2).

CONCLUSIONS

Self-consistent temperature, ρ and V_Φ profiles show that pyrolite-like compositions (A) agree very well with seismically observed properties throughout the upper mantle and transition

TABLE 3. Error Analysis: Temperatures at the Foot of the Lower Mantle Adiabat (K)

Composition	Best Fit	68% Confidence		95% Confidence	
		Low	High	Low	High
A	1575	1410	1750	1260	1880
no covariance	1675	1570	1780	1465	1885
no predict. errors	1780	1690	1875	1580	1975
B	1690	1530	1870	1375	2050
C	1570	1385	1765	1215	1970

zone. A piclogite-like compositions (C) also matches the seismic data primarily because its high Al content causes the pyroxene to garnet transition to occur near the 400 km discontinuity. Olivine-poor, pyroxene-rich compositions (B) significantly underestimate velocities and densities between 400 and 500 km. We have shown that a thermal boundary layer at 400 km is highly unlikely, so that the bulk composition of the upper mantle and transition zone are probably similar. Adiabats which yield the best agreement above 670 km lead to disagreement with expected profiles in the lower mantle. This discrepancy is significant at the 68% confidence level and indicates that the composition of the lower mantle differs from that above 670 km. A thermal boundary layer at 670 km considerably strengthens this conclusion, so that a dynamically stratified isochemical mantle is highly unlikely. The mantle is most likely dynamically and compositionally stratified, with a lower mantle enriched in SiO_2 and possibly FeO components.

Acknowledgements. We thank D. L. Anderson and two anonymous referees for their constructive reviews and C. Lithgow-Bertelloni, and L. R. Johnson for comments which improved the manuscript. We are also grateful to P. J. Stark for many enlightening discussions. This research was supported by Grant EAR-9105515 of the National Science Foundation and by the Director, Office of Energy Research, Division of Basic Energy Sciences, Engineering, and Geosciences, of the U.S. Department of Energy under contract DE-ACO3-76SF0098. All computations were carried out at the Center for Computational Seismology of the Lawrence Berkeley Laboratory.

REFERENCES

Akaogi, M., and S. Akimoto, Pyroxene-garnet solid-solution equilibria in the systems $Mg_4Si_4O_{12} - Mg_3Al_2Si_3O_{12}$ and $Fe_4Si_4O_{12} - Fe_3Al_2Si_3O_{12}$ at high pressures and temperatures, *Phys. Earth Planet. Inter., 15*, 90-106, 1977.

Akaogi, M., and S. Akimoto, High-Pressure phase equilibria in a garnet lherzolite, with special reference to $Mg^{2+} - Fe^{2+}$ partitioning among constituent minerals, *Phys. Earth Planet. Inter., 19*, 31-51, 1979.

Akaogi, M., A. Navrotsky, T. Yagi, and S. Akimoto, Pyroxene-garnet transition: Thermochemistry and elasticity of garnet solid solutions, an application to a pyrolite mantle, in *High Pressure Research in Mineral Physics*, edited by M. H. Manghnani and Y. Syono, pp. 427-438, TERRAPUB/AGU, Tokoyo/Washington, 1987.

Akaogi, M., E. Ito, and A. Navrotsky, Olivine - modified spinel - spinel transitions in the system $Mg_2SiO_4 - Fe_2SiO_4$: Calorimetric measurements, Thermochemical calculation, and geophysical application, *J. Geophys. Res., 94*, 15671-15686, 1989.

Allègre, C. J., and D. L. Turcotte, Geodynamic mixing in the mesosphere boundary layer and the origin of oceanic islands, *Geophys. Res. Lett., 12*, 207-210, 1985.

Anderson, D. L., and J. D. Bass, Mineralogy and composition of the upper mantle *Geophysical Research Letters, 11*, 637-640, 1984.

Anderson, D. L., and J. D. Bass, Transition region of the Earth's upper mantle, *Nature, 320*, 321-328, 1986.

Ashida, T., S. Kume, and E. Ito, Thermodynamic aspects of phase boundary among α, β, and γ Mg_2SiO_4 , in *High Pressure Research in Mineral Physics*, edited by M. H. Manghnani and Y. Syono, pp. 427-438, TERRAPUB/AGU, Tokoyo/Washington, 1987.

Bass, J. D., Elasticity of grossular and spessartite garnets by Brillouin spectroscopy, *J. Geophys. Res., 94*, 7621-7628, 1989.

Bass, J. D., and D. L. Anderson, Composition of the upper mantle: Geophysical tests of two petrological models, *Geophysical Research Letters, 11*, 229-232, 1984.

Bass, J. D., R. C. Liebermann, D. J. Weidner, and S. J. Finch, Elastic Properties from acoustic and volume compression experiments, *Phys. Earth Planet. Int., 25*, 140-158, 1981.

Bass, J. D., and M. Kanzaki, Elasticity of a majorite-pyrope solid solution, *Geophys. Res. Lett., 17*, 1989-1992, 1990.

Bass, J. D. and D. J. Weidner, Elasticity of single-crystal orthoferrosilite, *J. Geophys. Res., 89*, 4359-4371, 1984.

Bevington, P. R., *Data Reduction and Error Analysis for the Physical Sciences*, 336 pp., McGraw-Hill, New York, 1969.

Bina, C. R. and P. G. Silver, Constraints on lower mantle composition and temperature from density and bulk sound velocity profiles, *Geophys. Res. Lett., 17*, 1153-1156, 1990.

Birch, F., Elasticity and constitution of the Earth's interior, *J. Geophys. Res., 57*, 227-286, 1952.

Bowman, J. R., and B. L. N. Kennett, An investigation of the upper mantle beneath NW Australia using a hybrid seismograph array, *Geophys. J. Int., 101*, 411-424, 1990.

Bukowinski, M. S. T., and G. H. Wolf, Thermodynamically consistent decompression: Implications for lower mantle composition, *J. Geophys. Res., 95*, 12583-12593, 1990.

Burdick, L. J., A comparison of the upper mantle structure beneath north America and Europe, *J. Geophys. Res., 86*, 5926-5936, 1981.

Burdick, L. J., and D. V. Helmberger, The upper mantle P velocity structure of the western United States, *J. Geophys. Res., 83*, 1699-1712, 1978.

Cameron, M., S. Sueno, C. T. Prewitt and J. J. Papike, High-temperature crystal chemistry of acmite, diopside, hedenbergite, jadeite, spodumene, and ureyite, *Am. Mineral., 58*, 594-618, 1973.

Creager, K. C., and T. H. Jordan, Slab penetration into the lower mantle, *J. Geophys. Res., 89*, 3031-3049, 1984.

Creager, K. C., and T. H. Jordan, Slab penetration into the lower mantle beneath the Mariana and other island arcs of the northwest pacific, *J. Geophys. Res., 91*, 3573-3589, 1986.

DePaolo, D. J., Geochemical evolution of the crust and mantle, *Revs. Geophys. Space Phys., 21*, 1347-1358, 1983.

Derr, J. S., Internal structure of the Earth inferred from free oscillations, *J. Geophys. Res., 74*, 5202-5220, 1969.

Dey-Sarkar, S. K., and R. A. Wiggins, Upper mantle structure in western Canada, *J. Geophys. Res., 81*, 3619-3632, 1976.

Dost, B., Upper mantle structure under western Europe from fundamental and higher mode surface waves using the NARS array, *Geophys. J. Int., 100*, 131-151, 1990.

Duffy, T. S., and D. L. Anderson, Seismic velocities in mantle minerals and the mineralogy of the upper mantle, *J. Geophys. Res., 94*, 1895-1912, 1989.

Dziewonski, A. M., and D. L. Anderson, Preliminary reference Earth model, *Phys. Earth Planet. Inter., 25*, 297-356, 1981.

Fei, Y., H. K. Mao, B. O. Mysen, Experimental determination of element partitioning and calculation of phase relations in the

MgO–FeO–SiO$_2$ system at high pressure and high temperature, *J. Geophys. Res., 96*, 2157-2170, 1991.

Fei, Y., H. K. Mao, J. Shu, G. Parthasarathy, W. Bassett, and J. Ko, Simultaneous high-P, high-T X-ray diffraction study of β-(Mg,Fe)$_2$SiO$_4$ to 26 GPa and 900 K, *J. Geophys. Res., 97*, 4489-4496, 1992.

Finger, L. W., and Y. Ohashi, The thermal expansion of diopside to 800 C and a refinement of the crystal structure at 700 C. *Am. Mineral., 61*, 303-310, 1976.

Fischer, K. M., K. C. Creager, and T. H. Jordan, Mapping the Tonga slab, *J. Geophys. Res., 96*, 14403-14427, 1991.

Fukao, Y., Upper mantle P structure on the ocean side of the Japan-Kurile arc, *Geophys. J. R. astr. Soc., 50*, 621-642, 1977.

Fukao, Y., T. Nagahashi, and S. Mori, Shear velocity in the mantle transition zone, in *High Pressure Research in Geophysics*, by S. Akimoto and M. H. Manghnani, Center for Academic Publications, Tokyo, Japan, 1982.

Gaherty, J. B., T. Lay, and J. E. Vidale, Investigation of deep slab structure using long-period S waves, *J. Geophys. Res., 96*, 16349-16367, 1991.

Gasparik, T., Phase relations in the transition zone, *J. Geophys. Res., 95*, 15751-15769, 1990.

Giardini, D., and J. H. Woodhouse, Deep seismicity and modes of deformation in Tonga subduction zone, *Nature, 307*, 505-509, 1984.

Giardini, D., and J. H. Woodhouse, Horizontal shear flow in the mantle beneath the Tonga arc, *Nature, 319*, 551-555, 1986.

Gilbert, F., and A. M. Dziewonski, An application of normal mode theory to the retrieval of structural parameters and source mechanisms from seismic spectra, *Phil. trans. Royal Soc., London A, 278*, 187-269, 1975.

Gilbert, F., A. M. Dziewonski, and J. Brune, An information solution to a seismological inverse problem, *Proc. Natl. Acad. Sci. USA, 70*, 1410-1413, 1973.

Given, J. W., and D. V. Helmberger, Upper mantle structure of northwestern Eurasia, *J. Geophys. Res., 85*, 7183-7194, 1980.

Grad, M., Seismic model of the Earth's crust and upper mantle for the east European platform, *Phys. Earth Planet. Inter., 51*, 182-184, 1988.

Graham, E., and G. Barsch, Elastic constants of single-crystal forsterite as a function of temperature and pressure, *J. Geophys. Res., 74*, 5949-5960, 1969.

Graham, E., J. Schwab, S. Sopkin, and H. Takei, The pressure and temperature dependence of the elastic properties of single-crystal fayalite Fe$_2$SiO$_4$, *Phys. Chem. Miner., 16*, 186-198, 1988.

Grand, S. P., and D. V. Helmberger, Upper mantle shear structure of North America, *Geophys. J. R. astr. Soc., 76*, 399-438, 1984a.

Grand, S. T., and D. V. Helmberger, Upper mantle shear structure beneath the northwest Atlantic ocean, *J. Geophys. Res., 89*, 11465-11475, 1984b.

Graves, R. W., and D. V. Helmberger, Upper mantle cross section from Tonga to Newfoundland, *J. Geophys. Res., 93*, 4701-4711, 1988.

Guggenheim, E. A., *Mixtures*, 270 pp., Clarendon, Oxford, 1952.

Gurnis, M., and G. F. Davies, Mixing in numerical models of mantle convection incorporating plate kinematics, *J. Geophys. Res., 91*, 6375-6395, 1986.

Gurnis, M., and B. H. Hager, Controls of the structure of subducted slabs, *Nature, 335*, 317-321, 1988.

Gwanmesia, G., S. Rigden, I. Jackson, and R. Liebermann, Pressure dependence of elastic wave velocity for β–Mg$_2$SiO$_4$ and the composition of the Earth's mantle. *Science, 250*, 794-797, 1990.

Haddon, R. A. W., and K. E. Bullen, An Earth model incorporating free oscillation data, *Phys. Earth Planet. Inter., 2*, 35-49, 1969.

Hager, B. H., Subducted slabs and the geoid: Constraints on mantle rheology and flow, *J. Geophys. Res., 89*, 6003-6015, 1984.

Hager, B. H., and M. A. Richards, Long-wavelength variations in Earth's geoid: Physical models and dynamical implications, *Philos. Trans. R. Soc. London, Ser. A, 328*, 309-327, 1989.

Hales, A. L., K. J. Muirhead, and J. M. W. Rynn, A compressional velocity distribution for the upper mantle, *Tectonophysics, 63*, 309-348, 1980.

Hart, R. S., Shear velocity in the lower mantle from explosion data, *J. Geophys Res., 80*, 4889-4894, 1975.

Hart, R. S., D. L. Anderson, and H. Kanamori, Shear velocity and density of an attenuating Earth, *Earth Planet. Sci. Let., 32*, 25-34, 1976.

Hashin, Z., and S. Shtrikman, A variational approach to the elastic behavior of multiphase materials, *J. Mech. Phys. Solids, 11*, 127-140, 1963.

Hazen, R. M., J. Zhang, and J. Ko, Effects of Fe/Mg on the compressibility of synthetic wadsleyite: β − (Mg$_{1-x}$Fe$_x$)SiO$_4$ (x < 0.25), *Phys. Chem. Min., 17*, 416-419, 1990.

Hemley, R. J., L. Stixrude, Y. Fei, and H. K. Mao, Constraints on lower mantle composition from P-V-T measurements of (Fe,Mg)SiO$_3$ perovskite and (Fe,Mg)O magnesiowüstite, in *High Pressure Research: Application to Earth and Planetary Sciences*, edited by Y. Syono and M. H. Manghnani (in press), 1992.

Irifune, T., An experimental investigation of the pyroxene - garnet transformation in a pyrolite composition and its bearing on the constitution of the mantle, *Phys. Earth Planet. Inter., 45*, 324-336, 1987.

Irifune, T., T. Sekine, A. E. Ringwood, and W.O. Hibberson, The eclogite-garnetite transformation at high pressure and some geophysical implications, *Earth Planet. Sci. Lett., 77*, 245-256, 1986.

Irifune, T., and A. E. Ringwood, Phase transformations in a harzburgite composition to 26 GPa: implications for dynamical behaviour of the subducting slab, *Earth Planet. Sci. Lett., 86*, 365-376, 1987a.

Irifune, T., and A. E. Ringwood, Phase transformations in primitive MORB and pyrolite compositions to 25 GPa and some geophysical implications, in *High Pressure Research in Mineral Physics*, edited by M. H. Manghnani and Y. Syono, pp. 427-438, TERRAPUB/AGU, Tokoyo/Washington, 1987b.

Isaak, D. G., O. Anderson, and T. Goto, Elasticity of single-crystal forsterite measured to 1700 K, *J. Geophys. Res., 94*, 5895-5906, 1989.

Isacks, B. L., and P. Molnar, Distribution of stresses in the descending lithosphere from a global survey of focal mechanism solutions of mantle earthquakes, *Rev. Geophys., 9*, 103-174, 1971.

Ita, J. J., and L. Stixrude, Petrology, Elasticity, and Composition of the Mantle Transition Zone, *J. Geophys. Res., 97*, 6849-6866, 1992.

Ito, E., and E. Takahashi, Melting of peridotite at uppermost lower-mantle conditions, *Nature, 328*, 514-517, 1987.

Ito, E., and E. Takahashi, Postspinel Transformations in the system Mg$_2$SiO$_4$ - Fe$_2$SiO$_4$ and some geophysical implications, *J. Geophys. Res., 94*, 10637-10646, 1989.

Jackson, I., and H. Niesler, The elasticity of periclase to 3 GPa and some geophysical implications, in *High Pressure Research in Geophysics*, edited by S. Akimoto and M. H. Manghnani, pp. 93-113, Center for Academic Publications, Tokoyo, 1982.

Jackson, I., Elasticity and polymorphism of wüstite Fe$_{1-x}$O, *J. Geophys. Res., 95*, 21671-21685, 1990.

Jarrard, R. D., Relations among subduction parameters, *Rev. Geophys., 24*, 217-284, 1986.

Jeanloz, R., Majorite: Vibrational and compressional properties of a high-pressure phase, *J. Geophys. Res., 86*, 6171-6179, 1981.

Jeanloz, R., Effects of phase transitions and possible compositional changes on the seismological structure near 650 km depth, *Geophys. Res. Lett., 18*, 1743-1746, 1991.

Jeanloz, R., and E. Knittle, Density and composition of the lower mantle, *Phil. Trans. R. Soc. Lond. A, 328*, 377-389, 1989.

Jeanloz, R., and S. Morris, Temperature distribution in the crust and mantle, *Ann. Rev. Earth Planet. Sci., 14*, 377-415, 1986.

Jeanloz, R., and F. M. Richter, Convection, composition, and the thermal state of the lower mantle, *J. Geophys. Res., 84*, 5497-5504, 1979.

Jeanloz, R., and A. B. Thompson, Phase transitions and mantle discontinuities, *Rev. Geophys. Space. Phys., 21*, 51-74, 1983.

Jordan, T. H., and D. L. Anderson, Earth structure from free oscillations and travel times, *Geophys. J. R. astr. Soc., 36*, 411-459, 1974.

Kandelin, J., and D. J. Weidner, Elastic properties of hedenbergite, *J. Geophys. Res., 94*, 1063-1072, 1988.

Kanzaki, M., Ultrahigh - pressure phase relations in the system $Mg_4Si_4O_{12}$ - $Mg_3Al_2Si_3O_{12}$, *Phys. Earth Planet. Inter., 49*, 168-175, 1987.

Katsura, T., and E. Ito, The system Mg_2SiO_4–Fe_2SiO_4 at high pressures and temperatures: Precise determination of stabilities of olivine, modified spinel, and spinel, *J. Geophys. Res., 94*, 15663-15670, 1989.

Kennett, B. L. N., and E. R. Engdahl, Traveltimes for global earthquake location and phase identification, *Geophys. J. Int., 105*, 429-465, 1991.

Kind, R., and G. Muller, Computations of SV waves in realistic Earth models, *J. Geophys., 4*, 149-172, 1975.

King, D. W., and G. Calcagnile, P-wave velocities in the upper mantle beneath Fennoscandia and western Russia, *Geophys. J. R. astr. Soc., 46*, 407-432, 1976.

Knittle, E., R. Jeanloz, and G. L. Smith, Thermal expansion of silicate perovskite and stratification of the Earth's mantle, *Nature, 319*, 214-216, 1986.

Knittle, E., and R. Jeanloz, Synthesis and equation of state of $(Mg,Fe)SiO_3$ perovskite to over 100 gigapascals, *Science, 235*, 666-670, 1987.

Korn, M., P-wave coda analysis of short-period array data and the scattering and absorptive properties of the lithosphere, *Geophys. J. R. astr. Soc., 93*, 437-449, 1988.

Krupka, K. M., B. S. Hemingway, R. A. Robie, and D. M. Kerrick, High temperature heat capacities and derived thermodynamic properties of anthophyllite, diopside, dolomite, enstatite, bronzite, talc, tremolite, and wollastonite, *Am. Mineral., 70*, 261- 271, 1985.

Krupka, K. M., R. A. Robie, and B. S. Hemingway, High temperature heat capacities of corundum, periclase, anorthite, $CaAl_2Si_2O_8$ glass, muscovite, pyrophyillite, $KAlSi_3O_8$ glass, grossular, and $NaAlSi_3O_8$ glass, *Am. Mineral., 64*, 86-101, 1979.

Lay, T., and D. V. Helmberger, A lower mantle S-wave triplication and the shear velocity of D'', *Geophys. J. R. astr. Soc., 75*, 799-837, 1983.

Leger, J. M., A. M. Redon, and C. Chateau, Compressions of synthetic pyrope, spessartine and uvarovite garnets up to 25 GPa, *Phys. Chem. Minerals, 17*, 161-167, 1990.

Lefevre, L. V., and D. V. Helmberger, Upper mantle P velocity structure of the Canadian shield, *J. Geophys. Res., 94*, 17749-17765, 1989.

Lerner-Lam, A. L., and T. H. Jordan, How Thick Are the Continents?, *J. Geophys. Res., 92*, 14,007-14026, 1987.

Levien, L. R., and C. T. Prewitt, High pressure structural study of diopside, *Am. Mineral., 66*, 315-323, 1981.

Levien, L., D. J. Weidner, C. T. Prewitt, Elasticity of diopside, *Phys. Chem. Min., 4*, 105-113, 1979.

Liebermann, R. C., Elasticity of olivine (α), beta (β), and spinel (γ) polymorphs of germanates and silicates. *Geophys. J. R. astr. Soc., 42*, 899-929, 1975.

Lyon-Caen, H., Comparison of the upper mantle shear wave velocity structure of the Indian Shield and the Tibetan Plateau and tectonic implications, *Geophys. J. R. astr. Soc., 86*, 727-749, 1986.

McMechan, G. A., An amplitude constrained P-wave velocity profile for the upper mantle beneath the eastern United States, *Bull. Seis. Soc. Am., 69*, 1733-1744, 1979.

McMechan, G. A., Mantle P-wave velocity structure beneath Antarctica, *Bull. Seis. Soc. Am., 71*, 1061-1074, 1981.

Machetel, P., and P. Weber, Intermittent layered convection in a model mantle with an endothermic phase change at 670 km, *Nature, 350*, 55-57, 1991.

Machetel, P., and D. A. Yuen, Penetrative convective flows induced by internal heating and mantle compressibility *J. Geophys. Res., 94*, 10609-10626, 1989.

Mao, H. K., L. C. Chen, R. J. Hemley, A. P. Jephcoat, Y. Wu, and W. A. Basset, Stability and equation of state of $CaSiO_3$ - perovskite to 134 GPa, *J. Geophys. Res., 94*, 17889-17894 , 1989.

Mao, H. K., R. J. Hemley, Y. Fei, J. F. Shu, L. C. Chen, A. P. Jephcoat, Y. Wu and W. A. Bassett, Effect of pressure, temperature and composition on lattice parameters and density of $(Fe,Mg)SiO_3$ -perovskites to 30 GPa, *J. Geophys. Res., 96*, 8069-8080, 1991.

Mizutani, H., and K. Abe, An Earth model consistent with free oscillation and surface wave data, *Phys. Earth Planet. Inter., 5*, 345-356, 1971.

Montagner, J., and D. L. Anderson, Constrained reference mantle model, *Phys. Earth Planet. Inter., 58*, 205-227, 1989.

Montagner, J. P., and T. Tanimoto, Global upper mantle tomography of seismic velocities and anisotropies, *J. Geophys. Res., 96*, 20337-20351, 1991.

Nakada, M., and M. Hashizume, Upper mantle structure beneath the Canadian shield derived from higher modes of surface waves, *J. Phys. Earth, 31*, 387-405, 1983.

Olinger, B., Compression studies of forsterite (Mg_2SiO_4) and enstatite ($MgSiO_3$), in *High-Pressure Research: Applications in Geophysics*, edited by M. H. Manghnani and S. Akimoto, pp. 255-266, Academic, San Diego, Calif. 1977.

O'Neill, B., J. D. Bass, J. R. Smyth, and M. T. Vaughan, Elasticity of grossular-pyrope-almandine garnet, *J. Geophys. Res., 94*, 17819-17824, 1989.

O'Neill, B., and R. Jeanloz, Experimental petrology of the lower mantle: a natural peridotite taken to 54 GPa *Geophys. Res. Lett., 17*, 1477-1480, 1990.

Paulssen, H. Lateral heterogeneity of Europe's upper mantle as inferred from modeling of broad-band body waves, *Geophys. J. R. astr. Soc, 91*, 171-199, 1987.

Press, W. H., B. P. Flannery, S. A. Teukolsky, W. T. Vetterling, *Numerical Recipes*, 818 pp., Cambridge University Press, Cambridge, 1986.

Rigden, S. M., G. D. Gwanmesia, J. D. Fitz Gerald, I. Jackson, and R. C. Liebermann, Spinel elasticity and seismic structure of the transition zone of the mantle, *Nature, 354*, 143-145, 1991.

Ringwood, A. E., *Composition and Petrology of the Earth's Mantle*, 618 pp., McGraw-Hill, New York, 1975.

Robie, R. A., B. S. Hemingway, and J. R. Fisher, Thermodynamic properties of minerals and related substances at 298.15 K and 1 bar (10^5 Pascals) pressure and higher temperatures, *U. S. Geol. Survey, Dept. of the Interior Bull. 1452*, 456 pp, 1976.

Salerno, C. M., and J. P. Watt, Walpole bounds on the effective elastic moduli of isotropic multicomponent composites, *J. Appl. Phys., 60*, 1618-1624, 1986.

Sawamoto, H., D. Weidner, S. Sasaki, and M. Kumazawa, Single-crystal elastic properties of the modified spinel (beta) phase of Mg_2SiO_4 , *Science, 224*, 749-751, 1984

Sawamoto, H., M. Kozaki, A. Jujimura, and T. Akamatsu, Precise measurement of compressibility of $\gamma - Mg_2SiO_4$ using synchrotron radiation, paper presented at 27th High Pressure Conference Sapporo, Japan, 1986; as quoted by *Akaogi et al.* [1989].

Schwartz, S. Y., T. Lay, and S. L. Beck, Shear wave travel time, amplitude and waveform analysis for earthquakes in the Kurile slab: constraints on deep slab structure and mantle heterogeneity, *J. Geophys. Res., 96*, 14445-14460, 1991.

Sengupta, M. K., and B. R. Julian, Radial variation of compressional and shear velocities in the Earth's lower mantle, *Geophys. J. R. astr. Soc., 54*, 185-219, 1978.

Shearer, P. M., Seismic imaging of upper-mantle structure with new evidence for a 520-km discontinuity, *Nature, 344*, 121-126, 1990.

Silver, P. G., and W. W. Chan Observations of body wave multipathing from broadband seismograms: evidence for lower mantle slab penetration beneath the Sea of Okhotsk, *J. Geophys. Res., 91*, 13787-13802, 1986.

Silver, P. G., Carlson, R. W., and P. Olson, Deep slabs, geochemical heterogeneity, and the large-scale structure of mantle convection: Investigation of an enduring paradox, *Ann. Revs Earth Planet. Sci., 16*, 477-541, 1988.

Skinner, B. J., Physical properties of end-members of the garnet group, *Am. Mineral., 41*, 428-436, 1956.

Stixrude L., and M. S. T. Bukowinski, Fundamental thermodynamic relations and silicate melting with implications for the constitution of D", *J. Geophys. Res.* , *95*, 19311-19326, 1990

Stixrude, L., and M. S. T. Bukowinski, Thermodynamic analysis of the system MgO–FeO–SiO$_2$ at high pressure and the structure of the lowermost mantle *this volume*, 1992.

Stixrude, L., R. J. Hemley, Y. Fei, and H. K. Mao, Thermoelasticity of silicate perovskite and magnesiowüstite and stratification of the Earth's lower mantle, *Science*, *257* 1099-1101, 1992.

Sueno, S., M. Cameron, and C. T. Prewitt, Orthoferrosilite: High temperature crystal chemistry, *Am. Mineral.*, *61*, 38-53, 1976.

Suzuki, I., Thermal expansion of olivine and periclase and their anharmonic properties, *J. Phys. Earth* , *25*, 145-159, 1975a.

Suzuki, I., Cell parameters and linear thermal expansion coefficients of orthopyroxenes, *J. Seismol. Soc. Jpn.*, *28*, 1-9, 1975b.

Suzuki, I., Thermal expansion of γ – Mg$_2$SiO$_4$. *J. Phys. Earth*, *27*, 53-61, 1979.

Suzuki, I., and O. L. Anderson, Elasticity and thermal expansion of a natural garnet up to 1000 K, *J. Phys. Earth.*, *31*, 125-138, 1983.

Suzuki, I., E. Ohtani, and M. Kumazawa, Thermal expansion of modified spinel, β – Mg$_2$SiO$_4$, *J. Phys. Earth*, *28*, 273-280, 1980.

Suzuki, I., K. Seya, H. Takei, and Y. Sumino, Thermal expansion of fayalite, *Phys. Chem. Minerals*, *7*, 60-63, 1981.

Tralli, D. M., and L. R. Johnson, Lateral variations in mantle P velocity from tectonically regionalized tau estimates, *Geophys. J. R. astr. Soc.*, *86*, 475-489, 1986.

Uhrhammer, R., Shear-wave velocity structure for a spherically averaged earth, *Geophys. J. R. astr. Soc.*, *58*, 749-767, 1979.

Vassiliou, M. S., B. H. Hager, and A. Raefsky, The distribution of earthquakes with depth and stress in subducting slabs, *J. Geodynam.*, *1*, 11-28, 1984.

Verhoogen, J., Phase changes and convection in the Earth's mantle, *Phil. Trans. R. Soc. London, A*, *258*, 276-283, 1965.

Vinet, P., J. Ferrante, J. H. Rose, and J. R. Smith, Compressibility of solids, *J. Geophys. Res.*, *92*, 9319-9326, 1987.

Vinnik, L. P., and V. Z. Ryaboy, Deep structure of the east European platform according to seismic data, *Phys. Earth Planet. Inter.*, *25*, 27-37, 1981.

Walck, M. C., The P-wave upper mantle structure beneath an active spreading center: The Gulf of California, *Geophys. J. R. astr. Soc.* , *76*, 697-723, 1984.

Walck, M. C., The upper mantle beneath the north-east Pacific rim: a comparison with the Gulf of California, *Gephys. J. R. astr. Soc.*, *81*, 243-276, 1985.

Wang, C., A simple Earth model, *J. Geophys. Res.*, *77*, 4318-4329, 1972.

Watanabe, H., Thermochemical properties of synthetic high-pressure compounds relevant to the Earth's mantle, in *High Pressure Research in Geophysics*, edited by S. Akimoto and M. H. Manghnani, pp. 93-113, Center for Academic Publications, Tokoyo, 1982.

Watt, J. P., and T. J. Ahrens, Shock wave equation of state of enstatite, *J. Geophys. Res*, *91*, 7495-7503, 1986.

Watt, J. P., G. F. Davies, and R. J. O'Connell, The elastic properties of composite materials, *Rev. Geophys. Space. Phys* , *14*, 541-563, 1976.

Weaver, J. S., T. Takahashi, and J. Bass, Isothermal compression of grossular garnets to 250 kbars and the effect of calcium on the bulk modulus, *J. Geophys. Res.*, *81*, 2475-2482, 1976.

Weidner, D. J., A mineral physics test of a pyrolite mantle, *Geophys. Res. Lett.*, *12*, 417-420, 1985.

Weidner, D. J., H. Sawamoto, and S. Sasaki, Single-crystal elastic properties of the spinel phase of Mg$_2$SiO$_4$, *J. Geophys. Res.*, *89*, 7852-7859, 1984

Weidner, D. J., H. Wang, and J. Ito, Elasticity of orthoenstatite, *Phys. Earth Planet. Int.*, *17*, P7-P13, 1978.

Weng, K., H. K. Mao, and P. M. Bell, Lattice parameters of the perovskite phase in the system MgSiO$_3$–CaSiO$_3$–Al$_2$O$_3$, *Carnegie Inst. Wash. Yearbook, 1981*, 273-277, 1982.

Williams, Q., E. Knittle, R. Reichlin, S. Martin, and R. Jeanloz, Structural and electronic properties of Fe$_2$SiO$_4$-fayalite at ultrahigh pressures: amorphization and gap closure, *J. Geophys. Res.*, *95*, 21549-21564, 1990.

Wood, B. J., Postspinel transformations and the width of the 670-km discontinuity: A comment on "Postspinel transformations in the system Mg$_2$SiO$_4$–Fe$_2$SiO$_4$ and some geophysical implications" by E. Ito and E. Takahashi, *J. Geophys. Res.*, *95*, 12681-12685, 1990.

Yeganeh-Haeri, A., D. J. Weidner, and E. Ito, Single crystal elastic moduli of magnesium metasilicate perovskite, in *Perovskite: A Structure of Great Interest to Geophysics and Materials Science*, edited by A. Navrotsky and D. J. Weidner, pp. 13-26, AGU, Washington, 1989.

Yeganeh-Haeri, A., D. J. Weidner, and E. Ito, Elastic properties of the pyrope-majorite solid solution series, *Geophys. Res. Lett.*, *17*, 2453-2456, 1990.

Zhou, H. W., and R. W. Clayton, P and S wave travel-time inversions for subducting slabs under island arcs of the northwest pacific, *J. Geophys. Res.*, *95*, 6829-6851, 1990.

Joel Ita, Department of Geology and Geophysics, University of California, Berkeley, CA 94720.

Lars Stixrude, School of Earth and Atmospheric Science, Georgia Institute of Technology, Atlanta, GA 30332.

Thermodynamic Analysis of the System MgO-FeO-SiO$_2$ at High Pressure and the Structure of the Lowermost Mantle

LARS STIXRUDE[1] AND M.S.T. BUKOWINSKI

Dept. of Geology and Geophysics, University of California at Berkeley

Semi-empirical thermodynamic potentials, based on Debye theory and Birch-Murnaghan finite strain theory are constructed for important mantle phases in the MgO-FeO-SiO$_2$ system. The parameters are constrained by inverting experimental compression, thermal expansion, thermochemical and phase equilibria data on the relevant phases. Calculated pseudo-unary MgSiO$_3$ and Mg$_2$SiO$_4$, and pseudo-binary Mg$_2$SiO$_4$-Fe$_2$SiO$_4$ phase diagrams agree well with available data. Analysis of the complete ternary system shows that (Mg,Fe)SiO$_3$ perovskite is stable throughout the likely pressure-temperature and compositional regime of the Earth's mantle. The breakdown of perovskite to its constituent oxides is unlikely in the Earth, even at the extreme pressure-temperature conditions of the core-mantle boundary. This reaction had been proposed to reconcile proposed geotherms with predicted silicate melting curves and seismic observations of the deep mantle.

INTRODUCTION

The constitution of the D" layer at the base of the mantle is of broad geophysical interest because the layer is expected to strongly modulate and possibly control the geophysical fields which originate in the Earth's core. The surface manifestations of core heat flow and the geodynamo provide essential information for our understanding of the deep interior and how it affects tectonic, igneous and possibly climatic processes. A significant fraction of core heat may be transported to the surface by mantle plumes that likely originate in D" and are responsible for flood basalts, oceanic plateaus, hot spots and their associated climatic effects [*Duncan and Richards*, 1990; *Tarduno et al.*, 1991]. Possible metallic constituents in D" caused by chemical reaction between mantle and core may distort surface observations of the geomagnetic field, with important implications for models of fluid motion in the core and the geodynamo [*Jeanloz*, 1990]. Correlations between flood basalt activity and geomagnetic field reversal frequency [*Courtillot and Besse*, 1987] as well as arguments based on the secular variation of the magnetic field [*Bloxham and Jackson*, 1991] suggest that lateral and temporal variations in temperature in

D", caused by large scale mantle flow, the generation of plumes or the impingement of subducted oceanic crust, may influence the nature of flow in the core and the geodynamo.

The D" layer is a few hundred kilometer thick region of anomalous seismic properties. It is distinguished from the rest of the lower mantle by strong lateral heterogeneity, anomalous, often negative velocity gradients and a velocity discontinuity over much of its upper boundary [*Young and Lay*, 1987; *Weber and Körnig*, 1990]. The mineralogical or compositional changes responsible for these unusual seismic properties remain unknown.

(Mg,Fe)SiO$_3$ perovskite is generally accepted as the most abundant mineral in the lower mantle [*Jeanloz and Knittle*, 1989; *Bukowinski and Wolf*, 1990]. Perovskite dominated assemblages readily account for the seismic properties of the bulk of the lower mantle and phase equilibrium studies over most of the mantle's pressure-temperature regime confirm its stability [*Knittle and Jeanloz*, 1987]. However, the stability of perovskite under D" pressure-temperature conditions has not yet been experimentally addressed.

A comparison of recently proposed geotherms [*Williams and Jeanloz*, 1990; *Knittle and Jeanloz*, 1991b] with predicted silicate melting curves in the deep mantle suggests that perovskite cannot be a constituent of D" [*Stixrude and Bukowinski*, 1990]. Near the core-mantle boundary, its melting point falls more than a thousand degrees below estimated temperatures in the Earth. The proposed core-mantle boundary temperature is primarily determined by the measured melting temperature of iron under shock-loading [*Williams et al.*, 1987] which is found to be significantly higher than model calculations of the iron hugoniot had

[1]Now at School of Earth and Atmospheric Science, Georgia Institute of Technology, Atlanta, Georgia.

Evolution of the Earth and Planets
Geophysical Monograph 74, IUGG Volume 14

predicted [*Brown and McQueen*, 1982]. If the shock wave measurements and the predicted perovskite melting curve are approximately correct, the constitution of D'' must differ substantially from the rest of the mantle, since seismic observations rule out significant amounts of partial melt. If the region's bulk chemistry is dominated, like the rest of the mantle, by Mg, Fe and Si oxide components, perovskite must undergo a transformation to a high pressure phase or assemblage under D'' conditions. A proposed breakdown of perovskite to its constituent oxides permits a solid D'', since oxides are expected to be more refractory than their compounds, and provides a possible explanation of the region's distinctive seismic properties [*Stixrude and Bukowinski*, 1990].

In this report, we present a thermodynamic analysis of the stability of perovskite at high pressures and temperatures and show that the breakdown of perovskite to its constituent oxides is unlikely in the Earth. The reaction does occur at appropriate pressure-temperature conditions but only for iron contents so high that the products disagree severely with the density and bulk modulus of the lowermost mantle. We discuss other possible resolutions of the apparent discrepancy between the perovskite melting curve, the geotherm and seismic observations.

THERMODYNAMIC POTENTIALS

The Gibbs free energy per formula unit, G, of a phase consisting of a solid solution of N species as a function of pressure, P, and temperature, T, is:

$$G_i(P,T) = \sum_{j=1}^{N} x_{ij}[g_{ij}(P,T) + RT \ln a_{ij}] \qquad (1)$$

where x_{ij}, g_{ij} and a_{ij} are the mole fraction, Gibbs free energy and activity of species j in phase i . We assume symmetric regular solution behavior [e.g. *Guggenheim*, 1952]:

$$RT \ln a_{ij} = s_i[RT \ln x_{ij} + W_i(1 - x_{ij})^2] \qquad (2)$$

where W_i is the interaction parameter and s_i is the stoichiometric coefficient of the mixing site. The Gibbs free energy is written as: $g_{ij}(P,T) = F_{ij}(V_{ij},T) + PV_{ij}$, where V is the molar volume. We assume that the Helmholtz free energy, F_{ij}, can be divided into a reference term, a purely volume dependent part given by Birch-Murnaghan finite strain theory, a thermal part, given by Debye theory and an electronic contribution, given by free electron theory (subscripts i,j understood) [*Stixrude and Bukowinski*, 1990]:

$$F(V,T) = F_o + F_C(V,T_o) + [F_{TH}(V,T) - F_{TH}(V,T_o)]$$
$$+ [F_{EL}(V,T) - F_{EL}(V,T_o)] \qquad (3)$$

where the subscript o refers to zero pressure and temperature T_o. The cold part is given by:

$$F_C(V) = 9K_oV_o[\frac{1}{2}f^2 + \frac{1}{3}a_1f^3 + \frac{1}{4}a_2f^4 ...] \qquad (4a)$$

$$f = \frac{1}{2}[(V/V_o)^{-2/3} - 1] \qquad (4b)$$

$$a_1 = \frac{3}{2}(K_o' - 4) \qquad (4c)$$

$$a_2 = \frac{3}{2}[K_oK_o'' + K_o'(K_o' - 7) + 143/9] \qquad (4d)$$

where f is the finite strain and K_o and K_o' are the isothermal bulk modulus and its first pressure derivative. Terms of 4th and higher order are unimportant for the phases and pressure ranges considered here and are neglected. In general, the thermal contribution contains both quasiharmonic terms, for which we assume the Debye model, and anharmonic terms [*Wallace*, 1972]:

$$F_{TH}(V,T) = 9nRT(T/\theta)^3 \int_0^{\theta/T} \ln(1-e^{-t})t^2dt + A_2T^2 \qquad (5)$$

where n is the number of atoms in the formula unit and θ is the Debye temperature, whose negative logarithmic volume derivative is γ, the Grüneisen parameter. Anharmonic terms are expected to be small [*Hardy*, 1980; *Stixrude and Bukowinski*, 1990] at the high pressures of interest here and are neglected ($A_2 = 0$). The electronic term appears only for the high pressure form of wüstite (Wü II), which is known to be metallic:

$$F_{EL} = -\frac{1}{2}\beta_o(V/V_o)^{2/3}T^2 \qquad (6)$$

where $\beta_o = (m^*/m)\beta_{FE}$, m^*/m is the free electron mass ratio, which we assume is the same as for Fe and β_{FE} is the free electron heat capacity coefficient, which depends only on zero pressure conduction electron density [e.g. *Ashcroft and Mermin*, 1976]. The structure of Wü II and its solution properties remain unknown. We assume that Wü II forms a complete solid solution with Pe II and that this phase (Mw II) contains an electronic contribution to its free energy even at very small FeO concentrations. These assumptions lead to a lower bound on the free energy of Mw II and thus a lower bound on the width of the perovskite stability field since the structure of Wü II may differ from that of Pe II, and, if a solid solution exists, it is likely insulating or semi-conducting at small FeO contents.

The parameters are either directly measured or constrained by experimental data or first principles calculations (Table 1). Following the approach of *Bass et al.* [1981], we adopt the directly measured value of K_o, where ultrasonic or brillouin determinations exist, and constrain K_o' by inverting compression data. The reference state Debye temperature, θ_o, and Grüneisen parameter, γ_o, are determined by inverting thermal expansion and calorimetric data as described by *Stixrude and Bukowinski* [1990] (see Figures 1-3) [see also *Ita and Stixrude*, 1992; *Hemley et al.*, 1992]. Reference free energies and interaction parameters are

TABLE 1. Thermodynamic potential parameters.

Phase	Formula	V_o (cm³/mol)	K_o (GPa) Value	K_o (GPa) Ref.	K_o' Value	K_o' Ref	θ_o, γ_o θ_o	θ_o, γ_o γ_o	θ_o, γ_o Ref.
Olivine (Ol)	Mg_2SiO_4	43.76	128(2)	1,2	5.0(5)	17	924(10)	1.14(10)	25,26
	Fe_2SiO_4	46.27	127(2)	3	5.2(5)	18	688(10)	1.08(10)	27,28
Wadsleyite (Wa)	Mg_2SiO_4	40.52	174(2)	4	4.0(5)	19	974(10)	1.32(10)	25,29
	Fe_2SiO_4	43.22	174(2)[a]	5	4.0(5)[a]	5	771(50)[c]	1.32(20)[a]	
Spinel (Sp)	Mg_2SiO_4	39.65	183(3)	6	4.1(5)	20	1017(10)	1.21(10)	25,30
	Fe_2SiO_4	42.02	192(3)	7	4.1(5)	21	781(10)	1.21(10)	27,30
Enstatite (En)	$MgSiO_3$	31.33	106(2)	8	5.0(5)	22,23	935(10)	0.97(10)	31,32
Majorite (Mj)	$Mg_4Si_4O_{12}$	114.15	151(10)	9	4(1)[b]		949(50)[d]	1.24(20)[b]	
Ilmenite(Il)	$MgSiO_3$	26.35	212(4)	10	4.3(5)	22	1026(10)	1.48(10)	33
Perovskite (Pv)	$MgSiO_3$	24.46	263(7)	11,12	3.9(5)	11,12	1017(10)	1.96(10)	34,35
	$FeSiO_3$	25.49	263(7)[a]	12	3.9(5)[a]	12	749(50)[d]	1.96(10)[a]	
Magnesio-wüstite (Mw)	MgO	11.25	160(2)	13	4.1(2)	13	776(10)	1.45(10)	26,36
	FeO	12.25	152(2)	14	4.9(2)	14	434(50)[d]	1.45(10)	37
Magnesio-wüstite II (Mw II)	MgO	10.79	161(10)	15	4.1(1.0)	15	698(50)	1.36(20)	15
	FeO	10.84	205(10)	14	6(1)	14	630(50)	1.45(20)[e]	38
Stishovite (St)	SiO_2	14.01	314(8)	16	8.8(8)	24	1152(10)	1.35(10)	27,39
Liquid	Mg_2SiO_4	50.12	37.7(2.0)	41	6.3(5)	23	595(20)	0.99(20)	40,41
Liquid	$MgSiO_3$	38.73	21.3(2.0)	41	6.4(5)	23	625(20)	0.36(20)	40,41

Estimated standard deviations are shown in parentheses. $q=2.5(1.7)$ for perovskite [Hemley et al., 1992] and is assumed to be 1(2) for all other phases. References: 1, Graham and Barsch [1969]; 2, Isaak et al. [1989]; 3, Graham et al. [1988]; 4, Sawamoto et al. [1984]; 5, Hazen et al. [1990]; 6, Weidner et al. [1984]; 7, Liebermann [1975]; 8, Weidner et al. [1978]; 9, Bass and Kanzaki [1990]; 10, Weidner and Ito [1985]; 11, Knittle and Jeanloz [1987]; 12, Mao et al. [1991]; 13, Jackson and Niesler [1982]; 14, Jackson et al. [1990]; 15, Bukowinski [1985]; 16, Weidner et al. [1982]; 17, Olinger [1977]; 18, Williams et al. [1990]; 19, Fei et al. [1991b]; 20, Sawamoto et al. [1986]; 21, Bass et al. [1981]; 22, Duffy and Anderson [1989]; 23, Stixrude and Bukowinski [1990]; 24, Tsuchida and Yagi [1989]; 25, Ashida et al. [1987]; 26, Suzuki [1975a]; 27, Watanabe [1982]; 28, Suzuki et al. [1981]; 29, Suzuki et al. [1980]; 30, Suzuki [1979]; 31, Krupka et al. [1985]; 32, Suzuki [1975b]; 33, Ashida et al. [1988]; 34, Ito and Takahashi [1989]; 35, Knittle et al. [1986]; 36, Krupka et al. [1979]; 37, Touloukian et al. [1977]; 38, Knittle and Jeanloz [1991]; 39, Ito et al. [1974]; 40, Lange and Carmichael [1987]; 41, Stebbins et al. [1984]. Volumes are from Jeanloz and Thompson [1983] except those of Majorite [Ita and Stixrude, 1992] and the liquid phases [Lange and Carmichael, 1987]. a. Assumed to be the same as the Mg end-member; b. Assumed to be the same as pyrope; c. Dependence on Mg/Fe ratio assumed to be the same as for spinel; d. Ratio of majorite to pyrope thermal debye temperatures assumed to be the same as the ratio of the corresponding elastic debye temperatures.

determined by inverting phase equilibria measurements. We find the set of W_i 's and F_{oij}'s which provide the best simultaneous solution to a set of reactions (Table 2) of the form:

$$\sum_{i=1}^{N} \nu_{ij}\mu_{ij}(P_{exp}, T_{exp}, x_{ij\,exp}) = 0 \qquad (7)$$

where μ_{ij}, the chemical potential of species j in phase i:

$$\mu_{ij} = F_{ij} + PV_{ij} + s_i[RT \ln x_{ij} + W_i(1-x_{ij})^2] \qquad (8)$$

is evaluated at the experimentally observed pressure, temperature and mole fraction with the parameters of Table 1. The coefficients, ν_{ij}, balance the reaction. Unary reactions are represented by single P,T points, while each binary reaction includes several observations of the coexisting phases. Because the set of equations (7) depends linearly on the parameters (F_{oij}, W_i) the best least-squares solution is readily found, e.g. by singular value decomposition (Table 3).

The thermodynamic potential (1) is expressed in terms of its natural variables and thus contains all thermodynamic information about the model system; it is a fundamental thermodynamic relation in the terminology of Callen

[1985]. All thermodynamic properties of the model system are given by pressure and temperature derivatives of (1). The potential formulation has important advantages over alternative schemes which determine G by integrating empirical expressions for the volume and entropy in terms of their conjugate variables (polynomials in pressure and temperature or more complex forms) [see e.g. Wood and Fraser, 1977]. Unless self-consistency is explicitly imposed over the temperature and pressure regime of interest, they lead to path dependent results for G and its derivatives. Thermodynamic identities, such as those between mixed partial derivatives:

$$-\left(\frac{\partial S}{\partial P}\right)_T = \left(\frac{\partial V}{\partial T}\right)_P \qquad (9)$$

are everywhere satisfied by our approach, while they are generally not by empirical integration schemes.

In addition to complete self-consistency, the thermodynamic potentials used here generally require only a small number of parameters since they are physically based (finite strain theory and the Debye model of the vibrational density of states), which allows us to constrain each parameter independently by a large number of experimental

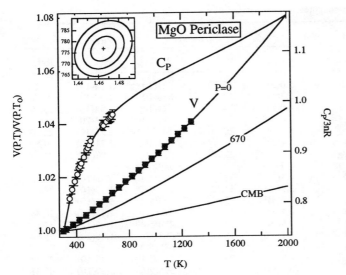

Fig. 1. Inversion of heat capacity and thermal expansion data for the Grüneisen parameter, γ_o, and Debye temperature, θ_o, of Periclase. Lines are calculated using the best fitting parameters, indicated by the cross in the inset (θ_o on the vertical axis and γ_o on the horizontal). The ellipses are 1-, 2- and 3-σ confidence intervals. Thermal expansion at pressures corresponding to the upper mantle-lower mantle boundary (24 GPa), and the core-mantle boundary (136 GPa) is also shown. Sources of the data are given in Table 1.

measurements. The potentials describe a wide variety of thermodynamic properties of mantle minerals at high pressures and temperatures, including their equations of state, phase equilibria and melting behavior [*Stixrude and Bukowinski*, 1990]. In the form of their volume derivative - the Mie-Grüneisen equation of state - they have been extensively tested against shock wave and static compression

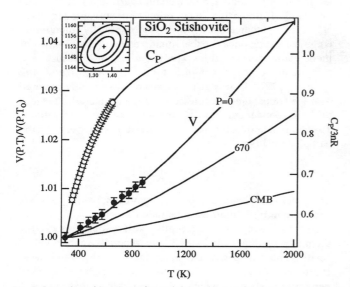

Fig. 2. Inversion of heat capacity and thermal expansion data for γ_o and θ_o of stishovite. The format is the same as in Figure 1.

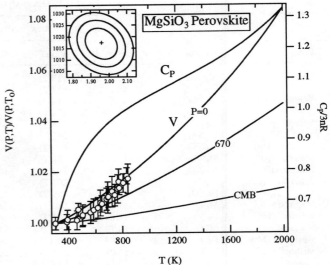

Fig. 3. Inversion of thermal expansion data and experimentally determined dT/dP slopes of the Il=Pv and Sp=Pv+Mw reactions (not shown) for γ_o and θ_o of perovskite. The format is the same as in Figure 1.

data on a wide variety of solids including minerals to lower mantle pressures [*Shapiro and Knopoff*, 1969; *Knopoff and Shapiro*, 1969; *McQueen et al.*, 1970; *Jeanloz*, 1989]. To the extent that anharmonic terms are negligible, the potentials remain sufficiently accurate for our purposes at the extreme

TABLE 2. Reactions used to constrain reference free energies and interaction parameters (Table 2).

Reaction	P (GPa)	T (K)	X_{Fe}	Ref.
Ol=Wa	15.0	1873	0.0	1
Wa=Sp	20.5	1873	0.0	1
Sp=Pv+Mw	24.0	1600	0.0	2
2En=Wa+St	16.1	1573	0.0	3
En=Mj	16.8	1973	0.0	3
2Il=Sp+St	19.5	1273	0.0	3
Il=Pv	22.2	2123	0.0	3
Fo=Liquid	0.0	2163	0.0	4
En=Liquid	0.0	1830	0.0	4
Ol=Sp	6.0	1473	1.0	5
Ol=Sp	12.0	1473	Univ	1
Ol=Sp	13.0	1873	Univ	1
Wa=Sp	12.0	1473	Univ	1
Wa=Sp	13.0	1873	Univ	1
Sp=2Mw+St	22.5-24.5	1373	0.3-0.5	2
Sp=2Mw+St	22-23.1	1873	0.3-0.6	2
Pv=Mw+St	24.5-25.5	1373	0.3-0.4	2
Pv=Mw+St	23-25	1873	0.4-0.5	2

Columns show the pressure and temperature at which the reactions proceed for the indicated bulk iron content (X_{Fe}). For binary reactions, observations over a range of pressure and bulk composition were used. Ol-Wa-Sp binary equilibria were constrained by the observed three phase univariant line. References: 1, *Katsura and Ito* [1989]; 2, *Ito and Takahashi* [1989]; 3, *Gasparik* [1990]; 4, *Stebbins et al.* [1984]; 5, *Akimoto* [1987].

TABLE 3. Thermodynamic potential parameters: reference free energies and interaction parameters. All quantities in kJ/mol.

Species	Global		High T		Low T	
	F_o	W	F_o	W	F_o	W
Forsterite	-127.7	6.1	-133.4	6.3	-124.0	6.5
Fayalite	-93.3	6.1	-96.6	6.3	-89.7	6.5
Mg-Wadsleyite	-99.6	3.9	-105.3	3.9	-95.9	3.9
Fe-Wadsleyite	-98.1	3.9	-100.9	3.9	-93.8	3.9
Mg-Spinel	-92.4	2.8	-98.0	3.6	-88.6	4.1
Fe-Spinel	-94.1	2.8	-97.6	3.6	-90.9	4.1
Mg-Enstatite	-91.0	0	-93.9	0	-89.2	0
Mg-Majorite	-205.4	0	-217.0	0	-198.1	0
Mg-Ilmenite	-35.9	0	-39.0	0	-33.7	0
Mg-Perovskite	3.4	0	-0.4	0	6.6	0
Fe-Perovskite	40.5	0	45.5	0	35.2	0
Periclase	0	13.1	0	13.5	0	12.6
Wüstite	0	13.1	0	13.5	0	12.6
Periclase II	139.7	13.1	139.7	13.1	139.7	13.1
Wüstite II	56.8	13.1	56.8	13.1	56.8	13.1
Stishovite	0	0	0	0	0	0
Mg_2SiO_4 Liquid	-422.9	0	-428.6	0	-419.1	0
$MgSiO_3$ Liquid	-345.6	0	-348.5	0	-343.8	0

pressure-temperature conditions of interest here. The cold part remains well constrained by experimental measurements for most of the relevant phases, while the thermal part closely approaches its high temperature limiting behavior ($T/\theta > 4$). In this limit, results are insensitive to the detailed form of the vibrational density of states which is approximately represented in the Debye model. The adequacy of the regular solution model and the assumed volume dependence of γ is more difficult to evaluate. We ensure that our results are insensitive to wide variations in W and q, the second logarithmic volume derivative of θ.

RESULTS

Predicted pseudo-unary Mg_2SiO_4 and $MgSiO_3$ phase diagrams agree well with the experimentally observed topology and the positions and slopes of the various reactions (Figures 4, 5). The predicted dT/dP slope of the Wa=Sp and Sp=Pv+Pe boundaries are somewhat smaller than observed. The calculated majorite field is smaller than observed. This may be caused by the assumed Debye temperature, which is based on elasticity, or because of the assumption of ideal mixing on the octahedral site. A lower Debye temperature, due perhaps to anomalous vibrational properties of the octahedral site or a favorable interaction between Mg and Si on that site may be responsible for the wider observed stability field. A wider majorite field would also reduce disagreement with the observed melting curve near 16 GPa, which may also be caused by a change in thermodynamic properties of enstatite across the ortho- to clino-enstatite transition which have been ignored here. Because experimental pressure and temperature calibration is

difficult under these conditions, differences between calculated and experimental melting curves between 15 and 30 GPa may not be significant. The calculations reproduce both lower and higher pressure measurements [which determine temperature directly, *Knittle and Jeanloz*, 1989a] to within experimental precision.

The predicted topology and temperature dependence of binary Ol-Wa-Sp equilibria also agree well with observation and the position of the univariant lines agree to within .1 GPa (Figures 6, 7). Complete agreement is essentially impossible because the data appear to be mutually inconsistent. At 1473 K observations of coexisting Ol and Wa appear directly above observations of pure Wa near 13.5 GPa, and the observation of Sp at 14.5 GPa is difficult to reconcile with observations of the Wa=Sp end-member reaction. The fact that such inconsistencies are much less frequent in the higher temperature measurements suggests that the lower temperature experiments may not have been sufficiently equilibrated.

Calculated pseudo-binary orthosilicate phase equilibria are compared with the low and high temperature experimental results of *Ito and Takahashi* [1989] in Figures 6 and 7. By

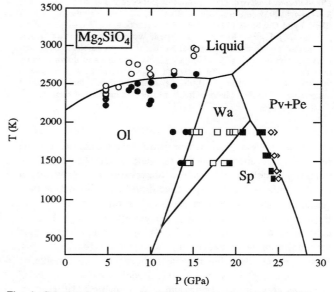

Fig. 4. Calculated pseudo-unary Mg_2SiO_4 phase diagram compared with the experimental data of *Ohtani and Kumazawa*[1981] (Ol=Liquid), *Katsura and Ito* [1989] (Ol=Wa and Wa=Sp) and *Ito and Takahashi* [1989] (Sp=Pv+Pe). For the purposes of illustration we have assumed that the assemblage Pv+Pe melts congruently. Eutectic compositions are expected to lie between orthosilicate and metasilicate stoichiometries at high pressure [*Presnall and Gasparik*, 1989]. Accurate prediction of the eutectic composition requires a multi-component model for the liquid phase, which we do not consider here. Calculated equilibria are identical for the three parameter sets, except for the perovskite forming reactions which are shifted upward and downward by less than 0.5 GPa for the high and low temperature parameters, respectively. Variation of all other parameters within their estimated uncertainties (Table 1) have a smaller effect on calculated equilibria.

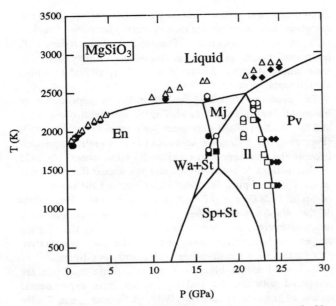

Fig. 5. Calculated pseudo-unary MgSiO₃ phase diagram compared with the experimental data of *Boyd et al.* [1964] (melting below 5 GPa), *Presnall and Gasparik* [1990] (melting between 5 and 20 GPa), *Ito and Katsura* [1991] (melting above 20 GPa) *Ito and Takahashi* [1989] (Il=Pv) and *Gasparik* [1990] (subsolidus reactions). Two distinct triple points (Mj=Il=Pv, Mj=Pv=L) appear coincident on this scale. Calculated equilibria are identical for the three parameter sets except for reactions involving Il, which are shifted upward and downward by less than 0.7 GPa for the low and high temperature parameters, respectively. Variation of all other parameters within their estimated uncertainties (Table 1) have a smaller effect on calculated equilibria.

inverting low and high temperature data separately for the reference free energies of the relevant species (Table 3), we obtained excellent fits to the observations. However, it proved impossible to reproduce both experiments with a single set of reference free energies. Parameters which give the best global fit reproduce the general topology but the calculated coexistence loops shift towards more iron rich compositions with decreasing temperature (dashed lines, Figures 6, 7) while the reverse is observed experimentally. The more complex thermodynamic model of *Fei et al.* [1991a] produces very similar results and is also unable to fit the data. The reason for this discrepancy is unclear. Barring severe deficiencies in two very different thermodynamic formulations, insufficient equilibration or an unquenchable phase transition in one or more of the relevant species may be responsible. Despite different predictions for the Pv=Mw+St reaction, all parameter sets produce a very narrow Sp+Pv+Mw coexistence field. This supports the conclusion of *Ito and Takahashi* [1989] that the spinel breakdown reaction can account for seismic reflections at 670 km depth even in a mantle of uniform composition.

Considering the discrepancies apparent in Figures 6 and 7 to be an additional source of uncertainty, the resulting errors

in our calculated phase diagrams at D'' pressure -temperature conditions are small and will not affect our basic conclusions. Differences between the predictions of high temperature and global parameter sets are reduced by increasing pressure (Figure 8). The stability field of perovskite is very wide at pressures greater than 60 GPa (1500 km depth), regardless of the parameter set used in the calculations; the low temperature parameters produce an even wider stability field of perovskite.

The predicted high pressure phase diagrams satisfy experimental constraints, including the transition of wüstite to a metallic form, observed near 70 GPa in diamond cell [$T \approx 4000$ K; *Knittle and Jeanloz*, 1986, 1991b] and shock wave experiments [$T \approx 1000$ K; *Jeanloz and Ahrens*, 1980] and perovskite syntheses up to 127 GPa and approximately 2500 K [*Knittle and Jeanloz*, 1987]. Our results closely resemble the predictions of *Jeanloz and Thompson* [1983] below 75 GPa. At higher pressures, our calculations show two pseudo-binary coexistence loops, as required by the phase rule, in place of their four-phase field.

The ternary MgO-FeO-SiO₂ diagram summarizes the

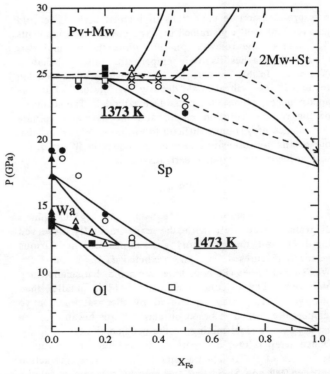

Fig. 6. Low temperature calculated pseudo-binary orthosilicate phase diagram compared with the experimental data of *Katsura and Ito* [1989] and *Ito and Takahashi* [1989]. Closed and open symbols and crosses represent observations of pseudo-unary, -binary, and -ternary coexistence. Solid lines are calculated with the low temperature parameter set, dashed lines with the global parameter set. The Ol-Wa-Sp equilibria differ by less than 0.02 X_{Fe} units for the three parameter sets.

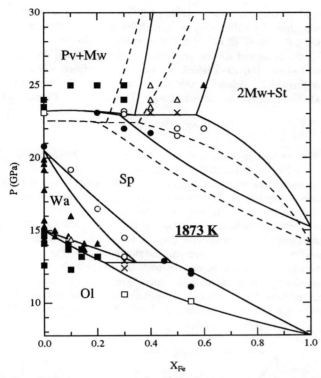

Fig. 7. High temperature calculated pseudo-binary orthosilicate phase diagram compared with the experimental data of *Katsura and Ito* [1989] and *Ito and Takahashi* [1989]. Closed and open symbols and crosses represent observations of pseudo-unary, -binary, and -ternary coexistence. Solid lines are calculated with the high temperature parameter set, dashed lines with the global parameter set. The Ol-Wa-Sp equilibria differ by less than 0.02 X_{Fe} units for the three parameter sets. Variations in all other parameters within their estimated uncertainties (Table 1) have a smaller affect on calculated equilibria.

the entire D" layer are uncertain by approximately 1 %; G. Masters, personal communication, 1992). Uncertainties in the calculations are also much too small to substantially alter the phase diagram. Variations in experimentally constrained parameters within their uncertainties (Table 1), and in W's from 0 to three times their nominal values move the phase boundaries by less than 0.4 X_{Fe} units. For all parameters except q of perovskite, the changes are less than 0.2 X_{Fe} units. Extreme variations in the F_o's, obtained by extrapolating their apparent temperature dependence (difference between low-T and high-T parameter sets) to 5000 K also fails to exclude the Earth from the perovskite stability field. A parametric phase diagram (Figure 10) shows that $MgSiO_3$ perovskite is stable for a wide range of thermal parameters. The stability limit of perovskite in this diagram lies close to the values of θ_o and γ_o adopted here, but outside their 1-σ uncertainty limits. The diagram illustrates an extreme lower bound on the stability of perovskite since temperatures at the top of D" and the average temperature over this layer are almost certainly well below 5000 K, possibly by as much as 2000 K. A temperature of 3000 K produces perovskite stability fields in Figures 8-10 which are 25-60 % wider. Alternative values of K_o and K_o' for perovskite [246 GPa, 4.5; *Yeganeh-Haeri et al.*, 1989] or the transformation of stishovite to its recently discovered high pressure form [*Tsuchida and Yagi*, 1989] have a much smaller effect than

important results (Figure 9). To obtain a conservative estimate of the stability of perovskite with respect to its constituent oxides in D", we have chosen an extreme temperature [5000 K, higher than the melting point of perovskite at this pressure; *Stixrude and Bukowinski*, 1990] since lower values widen the perovskite stability field. Superimposed on the phase diagram are contours of density and bulk modulus calculated with the same thermodynamic potential formalism and the same parameters used to determine the phase diagram. The intersection of the contours corresponding to the observed density and bulk modulus at the base of the mantle [*Dziewonski and Anderson*, 1981], which represent the location of the Earth at this pressure, lies well within the stability field of perovskite.

The uncertainties in seismic properties near the core mantle boundary are larger than those in the rest of the mantle but correspond to an area of the phase diagram much smaller than that of the perovskite stability field (a Backus-Gilbert resolution calculation using an improved free-oscillation data set indicates that the density and bulk modulus averaged over

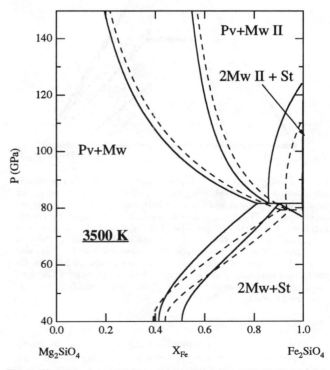

Fig. 8. High pressure orthosilicate phase diagram at 4000 K. Solid and dashed curves are calculated with the high temperature and global parameter sets respectively.

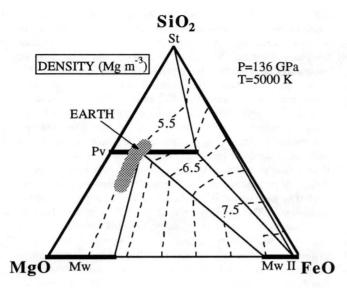

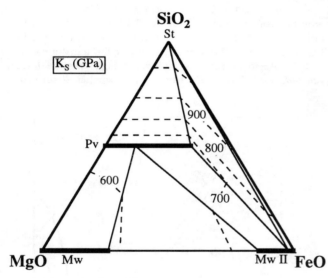

Fig. 9. Ternary MgO-FeO-SiO₂ phase diagram at 136 GPa and 5000 K calculated with the high temperature parameter set. Contours of density (upper figure) and adiabatic bulk modulus (lower figure) are superimposed. The shading, shown only in the upper figure for clarity, indicates the field of possible intersections of the contours corresponding to the observed density and adiabatic bulk modulus of the Earth allowed by uncertainties in calculated and seismically observed densities and bulk moduli. Adopting a value of 205 GPa for the room temperature phase transition in periclase eliminates Mw from the diagram and shifts the Pv+Mw II+St coexistence field towards lower iron contents (X_{Fe}(Pv)=0.58, X_{Fe}(Mw II)=0.83).

variations in thermal parameters on the location of the Earth and the position of phase boundaries on the diagram. Perhaps the largest uncertainty lies in the pressure of the assumed phase transition in periclase. The first principles calculations of *Bukowinski* [1985] provide a probable lower

bound on the transition pressure of 205 GPa at 300 K, much lower than the value adopted here from more recent results [500 GPa, *Mehl et al.*, 1988; this value is also consistent with the results of *Agnon and Bukowinski*, 1990; and *Zhang and Bukowinski*, 1991 which are based on modified electron gas theories]. Adopting the lower value expands the Mw II+St field somewhat but not enough to encompass the Earth.

CONCLUSIONS

Our results indicate that the breakdown of (Mg,Fe)SiO₃ perovskite to its constituent oxides is unlikely in the Earth. If core-mantle boundary temperatures are much lower than those derived from shock wave experiments [*Williams et al.*, 1987; *Jeanloz*, 1990], several alternative explanations for the seismic reflectors at the base of the mantle are possible. Small amounts of stishovite may be responsible, from silica enrichment relative to metasilicate stoichiometry (Figure 9). The phase transition in magnesiowüstite to its high pressure form may produce anomalous velocity gradients near the core-mantle boundary, although it appears much too diffuse to produce a seismic discontinuity (Figure 8). Still unobserved

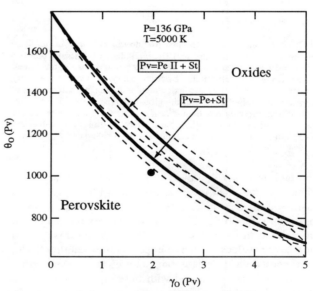

Fig. 10. Parametric phase diagram showing extreme lower bounds on the size of the stability field of MgSiO₃ perovskite relative to its constituent oxides at the pressure-temperature conditions of D". The bold curves indicate the stability limit of perovskite for q(Pv)=2.5, while the lower and upper surrounding dashed lines show the limit for q(Pv)=1.5 and 3.5 respectively. The circle represents the best fitting values adopted here (Figure 3; and Table 1). The size of the symbol is comparable to the 1-σ uncertainties. Adopting different forms for the thermodynamic potentials (e.g. an Einstein model for the thermal energy) may lead to best fit values which fall outside our uncertainties. However, the position of the best fit values relative to the breakdown reactions is unlikely to change by more than a few percent in θ_o or 10 % in γ_o. Lowering the temperature to 3000 K increases the stability of perovskite and shifts the curves towards higher values of γ_o by 0.5.

phase transitions in calcic, diopsidic or $(Mg,Fe)SiO_3$ perovskite, including the predicted orthorhombic to cubic transition in the latter [*Wolf and Bukowinski*, 1985; *Bukowinski and Wolf*, 1988; see also *Hemley and Cohen*, 1992], may produce velocity discontinuities at the appropriate pressures. Partial reaction of Mg-rich perovskite with iron alloys, similar to that observed at lower pressures [*Knittle and Jeanloz*, 1989a, 1991a], may also produce seismic reflectors.

All of these explanations rely on perovskite dominated assemblages and thus cannot reconcile geotherms based on shock wave experiments with the lack of partial melt in D". The melting point of iron under shock loading, and in static compression experiments at lower pressures is currently a matter of active debate [*Ross et al.*, 1990; *Boehler et al.*, 1990]. However, if the shock wave data and the predicted perovskite melting curve are approximately correct, they require either the absence of Mg-rich perovskite from D" or a substantial change in its properties, probably much greater than that associated with the predicted orthorhombic to cubic phase transformation. An exotic bulk composition in D", perhaps dominated by Ca and Al oxide components [*Ruff and Anderson*, 1980], may be required. Alternatively, a breakdown or transformation of $(Mg,Fe)SiO_3$ perovskite to a more diverse assemblage than those considered here remains a possible explanation of the seismic reflectors and the absence of melt in D". Consideration of other chemical components may lower the free energy of the non-perovskite assemblage, inducing a breakdown to a mixture of oxides and metallic alloys (Fe, FeSi). Though speculative, a metallic alloy-oxide layer may be heterogeneous on many length scales because of the disparate physical properties of the two components [*Knittle and Jeanloz*, 1991a]. This provides a possible explanation for observed heterogeneity in D" on length scales ranging from the smallest observable with seismic probes (100 km, roughly one seismic wavelength) to nearly the diameter of the core [*Lavely et al.*, 1986; *Creager and Jordan*, 1986; *Morelli and Dziewonski*, 1987; *Young and Lay*, 1987; *Weber and Körnig*, 1990;].

Acknowledgments. We thank C. Lithgow-Bertelloni and two anonymous reviewers for helpful comments on the manuscript. This work supported by NSF grant EAR-8816819.

REFERENCES

Agnon, A., and M.S.T. Bukowinski, Thermodynamic and elastic properties of a many-body model for simple oxides, *Phys. Rev. B, 41*, 7755-7766, 1990.

Akimoto, S., High-pressure research in geophysics: Past, present and future in *High Pressure Research in Mineral Physics*, edited by M.H. Manghnani and Y. Syono, Terra Scientific Publishing Company, Tokyo, American Geophysical Union, Washington, 1987.

Ashcroft, N.W., and N.D. Mermin, *Solid State Physics*, Saunders College, Philadelphia, 1976.

Ashida, T., S. Kume, and E. Ito, Thermodynamic aspects of phase boundary among α ,β, and γ Mg2SiO4 in *High Pressure Research in Mineral Physics* , edited by M. H. Manghnani and Y. Syono, pp. 427-438, TERRAPUB/AGU, Tokyo/Washington, 1987.

Ashida, T., S. Kume, E. Ito, and A. Navrotsky, MgSiO3 ilmenite: Heat capacity, thermal expansivity, and enthalpy of transformation, *Phys. Chem. Min., 16*, 239-245, 1988.

Bass, J.D., R.C. Liebermann, D.J. Weidner, and S.J. Finch, Elastic properties from acoustic and volume compression experiments, *Phys. Earth Planet. Int., 25*, 140-158, 1981.

Bass, J.D., and M. Kanzaki, Elasticity of a majorite-pyrope solid solution, *Geophys. Res. Lett., 17*, 1989-1992, 1990.

Bloxham, J., and A. Jackson, Fluid flow near the surface of Earth's outer core, *Revs. Geophys., 29*, 97-120.

Boehler, R., N.V. von Bargen, A. Chopelas, Melting, thermal expansion, and phase transition of iron at high pressures, *J. Geophys. Res., 95*, 21731-21736, 1990.

Boyd, R.F., J.R. England, and B.T. Davis, Effects of pressure on the melting and polymorphism of enstatite, MgSiO3, *J. Geophys. Res., 69*, 2101-2109, 1964.

Brown, M.J., and R.G. McQueen, The equation of state for iron and the Earth's core, in *High Pressure Research in Geophysics*, edited by S. Akimoto and M.H. Manghnani, Center for Academic Publications, Tokyo, 611-623, 1982.

Bukowinski, M.S.T., First principles equations of state of MgO and CaO, *Geophys. Res. Lett., 12*, 536-539, 1985.

Bukowinski, M.S.T., and G.H. Wolf, Equation of state and possible critical phase transitions in MgSiO3 perovskite at lower-mantle conditions, in *Structural and Magnetic Phase Transitions in Minerals*, edited by S. Ghose, J.M.D. Coey, and E. Salje, pp. 91-112, Springer-Verlag, New York, 1988.

Bukowinski, M.S.T., and G.H. Wolf, Thermodynamically consistent decompression: Implications for lower mantle composition, *J. Geophys. Res., 95*, 12583-12593, 1990.

Callen, H.B., *Thermodynamics and an Introduction to Thermostatistics*, 2nd ed., 493 pp., John Wiley, New York, 1985.

Courtillot, V., and J. Besse, Magnetic field reversals, polar wander, and core-mantle coupling, *Science, 237*, 1140-1147, 1987.

Creager, K.C., and T.J. Jordan, Aspherical structure of the core-mantle boundary from PKP travel times, *Geophys. Res. Lett., 13*, 1497-1500, 1986.

Duffy, T. S., and D. L. Anderson, Seismic velocities in mantle minerals and the mineralogy of the upper mantle, *J. Geophys. Res., 94*, 1895-1912, 1989.

Duncan, R.A., and M.A. Richards, Hotspots, mantle plumes, flood basalts, and true polar wander. *Revs. Geophys., 29*, 31-50, 1991.

Dziewonski, A.M., and D.L. Anderson, Preliminary reference Earth model, *Phys. Earth Planet. Int., 25*, 297-356, 1981.

Fei, Y., H. Mao, and B.O. Mysen, Experimental determination of element partitioning and calculation of phase relations in the MgO-FeO-SiO2 system at high pressure and high temperature, *J. Geophys. Res., 96*, 2157-2170, 1991a.

Fei, Y., H.K. Mao, J. Shu, G. Parthasarathy, and W.A. Bassett, Simultaneous high P-T x-ray diffraction study of β-(Mg,Fe)2SiO4 to 26 GPa and 900 °K, *J. Geophys. Res., 97*, 4489-4495, 1992.

Gasparik, T., Phase relations in the transition zone, *J. Geophys. Res., 95*, 15751-15769, 1990.

Graham, E., and G. Barsch, Elastic constants of single-crystal forsterite as a function of temperature and pressure, *J. Geophys. Res., 74*, 5949-5960, 1969.

Graham, E., J. Schwab, S. Sopkin, and H. Takei, The pressure and temperature dependence of the elastic properties of single-crystal fayalite Fe2SiO4, *Phys. Chem. Minerals, 16*, 186-198, 1988.

Guggenheim, E.A., *Mixtures*, 270 pp., Clarendon, Oxford, 1952.

Hazen, R.M., J. Zhang, and J. Ko, Effects of Fe/Mg on the compressibility of synthetic wadsleyite: β-(MgxFe1-x)2SiO4 (x<0.25), *Phys. Chem. Minerals, 17*, 416-419, 1990.

Hemley, R.J., and R.E. Cohen, Silicate perovskite, *Annu. Revs. Earth Planet. Sci., 20*, 553-600, 1992.

Hemley, R.J., L. Stixrude, Y. Fei, H.K. Mao, Constraints on lower mantle composition from P-V-T measurements of (Fe,Mg)SiO$_3$ perovskite and (Fe,Mg)O magnesiowüstite, in *High Pressure Research: Application to Earth and Planetary Sciences* edited by Y. Syono and M.H. Manghnani, TERRAPUB, Tokyo, in press, 1992.

Isaak, D.G., O. Anderson, and T. Goto, Elasticity of single-crystal forsterite measured to 1700° K, *J. Geophys. Res., 94*, 5895-5906, 1989.

Ita, J.J., and L. Stixrude, Petrology, elasticity and composition of the transition zone, *J. Geophys. Res., 97,*.6849-6866, 1992.

Ito, E., and E. Takahashi, Postspinel transformations in the system Mg$_2$SiO$_4$-Fe$_2$SiO$_4$ and some geophysical implications, *J. Geophys. Res., 94*, 10637-10646, 1989.

Ito, E., and T. Katsura, Melting of iron-silicate systems, *Union Program and Abstracts, XX General Assembly, International Union of Geodesy and Geophysics*, pg. 119, 1991.

Jackson, I., S.K. Khana, A. Revcolevschi, J. Berthon, Elasticity and polymorphism of wüstite Fe$_{1-x}$O, *J. Geophys. Res., 95*, 21671-21685, 1990.

Jackson, I., and H. Niesler, The elasticity of periclase to 3 GPa, and some geophysical implications, in *High Pressure Research in Geophysics*, edited by S. Akimoto and M.H. Manghnani, pp. 93-113, Center for Academic Publications, Tokyo, 1982.

Jeanloz, R., Universal equation of state, *Phys. Rev. B, 38*, 805-807, 1988.

Jeanloz, R., Shock wave equation of state and finite strain theory, *J. Geophys. Res., 94*, 5873-5886, 1989.

Jeanloz, R., The nature of the Earth's core, *Annu. Rev. Earth Planet. Sci., 18*, 357-386, 1990.

Jeanloz, R., and T.J. Ahrens, Equations of state of FeO and CaO, *Geophys. J.R. Astron. Soc.*, 62, 505-528, 1980.

Jeanloz, R., and A.B. Thompson, Phase transitions and mantle discontinuities, *Revs. Geophys. Space Phys., 21*, 51-74, 1983.

Jeanloz, R., and E. Knittle, Density and composition of the lower mantle, *Philos. Trans. R. Soc. London, Ser. A, 328*, 377-389, 1989.

Katsura, T., and E. Ito, The system Mg$_2$SiO$_4$-Fe$_2$SiO$_4$ at high pressures and temperatures: Precise determination of stabilities of olivine, modified spinel, and spinel, *J. Geophys. Res., 94*, 15663-15670, 1989.

Knittle, E., and R. Jeanloz, High pressure metallization of FeO and implications for the Earth's core, *Geophys. Res. Lett.., 13*, 1541-1544, 1986.

Knittle, E., and R. Jeanloz, Synthesis and equation of state of (Mg,Fe)SiO$_3$ perovskite to over 100 gigapascals, *Science, 235*, 668-670, 1987.

Knittle, E., and R. Jeanloz, Melting curve of (Mg,Fe)SiO$_3$ perovskite to 96 GPa: Evidence for a structural transition in lower mantle melts, *Geophys. Res. Lett., 16*, 421-424, 1989a.

Knittle, E., and R. Jeanloz, Simulating the core-mantle boundary: An experimental study of high-pressure reactions between silicates and liquid iron, *Geophys. Res. Lett., 16*, 609-612, 1989b.

Knittle, E., and R. Jeanloz, Earth's core-mantle boundary: Results of experiments at high pressures and temperatures, *Science, 251*, 1438-1443, 1991a.

Knittle, E., and R. Jeanloz, The high-pressure phase diagram of Fe$_{0.94}$O: A possible constituent of the Earth's core, *J. Geophys. Res., 96*, 16169-16180, 1991b.

Knittle, E., R. Jeanloz, and G.L. Smith, Thermal expansion of silicate perovskite and stratification of the Earth's mantle, *Nature, 319*, 214-216, 1986.

Knopoff, L., and J.N. Shapiro, Comments on the interrelationships between Grünesien's parameter and shock and isothermal equations of state, *J. Geophys. Res., 74*, 1439-1450, 1969.

Krupka, K. M., B. S. Hemingway, R. A. Robie, and D. M. Kerrick, High temperature heat capacities and derived thermodynamic properties of anthophyllite, diopside, dolomite, enstatite, bronzite, talc, tremolite, and wollastonite, *Am. Mineral., 70*, 261- 271, 1985.

Krupka, K. M., R. A. Robie, and B. S. Hemingway, High temperature heat capacities of corundum, periclase, anorthite, CaAl$_2$Si$_2$O$_8$

glass, muscovite, pyrophyllite, KAlSi$_3$O$_8$ glass, grossular, and NaAlSi$_3$O$_8$ glass, *Am. Mineral., 64*, 86-101, 1979.

Lange, R., and I.S.E. Carmichael, Densities of Na$_2$O-K$_2$O-CaO-MgO-FeO-Fe$_2$O$_3$-Al$_2$O$_3$-TiO$_2$-SiO$_2$ liquids: New measurements and derived partial molar properties, *Geochim Cosmochim. Acta, 51*, 2931-3946, 1987.

Lavely, E.M., D.W. Forsyth, and P. Friedemann, Scales of heterogeneity at the core-mantle boundary, *Geophys. Res. Lett., 13*, 1505-1508, 1986.

Liebermann, R.C., Elasticity of olivine (α), beta (β), and spinel (γ) polymorphs of germanates and silicates, *Geophys. J.R. astr. Soc., 42*, 899-929, 1975.

Mao, H.K., R.J. Hemley, Y. Fei, J.F. Shu, L.C. Chen, A.P. Jephcoat, Y. Wu, and W.A. Bassett, Effect of pressure, temperature and composition on lattice parameters and density of (Fe,Mg)SiO$_3$ perovskites to 30 GPa, *J. Geophys. Res., 96*, 8069-8080, 1991.

McQueen, R.G., S.P. Marsh, J.W. Taylor, J.N. Fritz, and W.J. Carter, The equation of state of solids from shock wave studies, in *High Velocity Impact Phenomena*, edited by R. Kinslow, pp. 294-419, Academic, San Diego, 1970.

Mehl, M.J., R.E. Cohen, and H. Krakauer, Linearized augmented plane wave electronic structure calculations for MgO and CaO, *J. Geophys. Res., 93*, 8009-8022, 1988.

Morelli, A., and A.M. Dziewonski, Topography of the core-mantle boundary and lateral homogeneity of the liquid core, *Nature, 325*, 678-683, 1987.

Ohtani, E., and M. Kumazawa, Melting of forsterite Mg$_2$SiO$_4$ up to 15 GPa, *Phys. Earth Planet. Int., 27*, 32-38, 1981.

Olinger, B., Compression studies of forsterite (Mg$_2$SiO$_4$) and enstatite (MgSiO$_3$), in *High Pressure Research: Applications in Geophysics*, edited by M.H. Manghnani and S. Akimoto, pp. 255-266, Academic, San Diego, 1977.

Presnall, D.C., and T. Gasparik, Melting of enstatite (MgSiO$_3$) from 10 to 16.5 GPa and forsterite (Mg$_2$SiO$_4$)-Majorite (MgSiO$_3$) eutectic at 16.5 GPa: Implications for the origin of the mantle, *J. Geophys.. Res., 95*, 15771-15777, 1990.

Ross, M., D.A. Young, R. Grover, Theory of the iron phase diagram at Earth core conditions, *J. Geophys. Res., 95*, 21713-21716, 1990.

Ruff, L., and D.L. Anderson, Core formation, evolution, and convection: A geophysical model, *Phys. Earth Planet. Int., 21*, 181-201, 1980.

Sawamoto, H., D. Weidner, S. Sasaki, and M. Kumazawa, Single-crystal elastic properties of the modified spinel (beta) phase of Mg$_2$SiO$_4$, *Science, 224*, 749-751, 1984.

Sawamoto, H., M. Kozaki, A. Jujimura, and T. Akamatsu, Precise measurement of compressibility of γ-Mg$_2$SiO$_4$ using synchrotron radiation, paper presented at 27th High Pressure Conference Sapporo, Japan, 1986.

Shapiro, J.N., and L. Knopoff, Reduction of shock-wave equations of state to isothermal equations of state, *J. Geophys. Res., 74*, 1435-1438, 1969.

Stebbins, J.F., I.S.E. Carmichael, and L.K. Moret, Heat capacities and entropies of silicate liquids and glasses, *Contrib. Mineral. Petrol., 86*, 131-148, 1984.

Stixrude, L., and M.S.T. Bukowinski, Fundamental thermodynamic relations and silicate melting with implications for the constitution of D", *J. Geophys. Res., 95*, 19311-19325. 1990.

Suzuki, I., Thermal expansion of olivine and periclase and their anharmonic properties, *J. Phys. Earth , 25*, 145-159, 1975a.

Suzuki, I., Cell parameters and linear thermal expansion coefficients of orthopyroxenes, *J. Seismol. Soc. Jpn., 28*, 1-9, 1975b.

Suzuki, I., Thermal expansion of γ- Mg$_2$SiO$_4$, *J. Phys. Earth, 27*, 53-61, 1979.

Suzuki, I., and O. L. Anderson, Elasticity and thermal expansion of a natural garnet up to 1000 K, *J. Phys. Earth., 31*, 125-138, 1983.

Suzuki, I., E. Ohtani, and M. Kumazawa, Thermal expansion of modified spinel, β–Mg$_2$SiO$_4$, *J. Phys. Earth, 28*, 273-280, 1980.

Suzuki, I., K. Seya, H. Takei, and Y. Sumino, Thermal expansion of fayalite, *Phys. Chem. Minerals, 7*, 60-63, 1981.

Tarduno, J.A., W.V. Sliter, L. Kroenke, M. Leckie, H. Mayer, J.J. Mahoney, R. Musgrave, M. Storey, and E.L. Winterer, Rapid formation of the Ontong Java plateau by Aptian mantle plume volcanism, *Science, 254*, 399-403, 1991.

Touloukian, Y. S., R. S. Kirby, R. E. Taylor, and T. Y. R. Lee, in *Thermal Expansion, Nonmetallic Solids*, 1658 pp., TPRC 13, IFI/Plenum, New York, 1977.

Tsuchida, Y., and T. Yagi, A new post-stishovite high pressure polymorph of silica, *Nature, 340*, 217-220, 1989.

Wallace, D.C., *Thermodynamics of Crystals*, John Wiley, New York, 1972.

Watanabe, H., Thermochemical properties of synthetic high-pressure compounds relevant to the Earth's mantle, in *High Pressure Research in Geophysics*, edited by S. Akimoto and M. H. Manghnani, pp. 93-113, Center for Academic Publications, Tokyo, 1982.

Weber, M., and M. Körnig, Lower mantle inhomogeneities inferred from PcP precursors, *Geophys. Res. Lett., 17*, 1993-1996, 1990.

Weidner, D.J., H. Wang, and J. Ito, Elasticity of orthoenstatite, *Phys. Earth Planet. Int., 17*, P7-P13, 1978.

Weidner, D.J., J.D. Bass, A.E. Ringwood, and W. Sinclair, The single-crystal elastic moduli of stishovite, *J. Geophys. Res., 87*, 4740-4746, 1982.

Weidner, D.J., H. Sawamoto, and S. Sasaki, Single-crystal elastic properties of the spinel phase of Mg_2SiO_4, *J. Geophys. Res., 89*, 7852-7859, 1984.

Weidner, D.J., and E. Ito, Elasticity of $MgSiO_3$ in the ilmenite phase, *Phys. Earth Planet. Int., 40*, 65-70, 1985.

Williams, Q., R. Jeanloz, J. Bass, B. Svendsen, T.J. Ahrens, The melting curve of iron to 250 Gigapascals: A constraint on the temperature at Earth's center, *Science, 236*, 181-182, 1987.

Williams, Q., E. Knittle, R. Reichlin, S. Martin, and R. Jeanloz, Structural and electronic properties of Fe_2SiO_4-fayalite at ultrahigh pressures: amorphization and gap closure, *J. Geophys. Res., 95*, 21549-21564, 1990.

Williams, Q., and R. Jeanloz, Melting relations in the iron-sulfur system at ultra-high pressures: Implications for the thermal state of the Earth, *J. Geophys. Res., 95*, 19299-19310, 1990.

Wolf, G.H., and M.S.T. Bukowinski, Ab initio structural and thermoelastic properties of orthorhombic $MgSiO_3$ perovskite, *Geophys. Res. Lett., 12*, 809-812, 1985.

Wood, B.J., and D.G. Fraser, *Elementary Thermodynamics for Geologists*, 303 pp., Oxford University Press, 1977.

Yeganeh-Haeri, A., D.J. Weidner, and E. Ito, Single crystal elastic moduli of magnesium metasilicate perovskite, in *Perovskite: A Structure of Great Interest to Geophysics and Materials Science*, edited by A. Navrotsky and D.J. Weidner, pp. 13-26, Terra Scientific Publishing Company, Tokyo, American Geophysical Union, Washington, 1989.

Young, C.J., and T. Lay, The core-mantle boundary, *Annu. Rev. Earth Planet. Sci., 15*, 25-46, 1987.

Zhang, H., and M.S.T. Bukowinski, Modified potential-induced-breathing model of the potential between closed-shell ions, *Phys. Rev. B, 44*, 2495-2503, 1991.

L. Stixrude, Geophysical Laboratory, Carnegie Institution of Washington, 5251 Broad Branch Rd. NW, Washington, DC 20015.

M.S.T. Bukowinski, Dept. of Geology and Geophysics, University of California, Berkeley, CA 94720.

The History of Global Weathering and the Chemical Evolution of the Ocean-Atmosphere System

LOUIS M. FRANÇOIS

Institut d'Astrophysique, Université de Liège, Liège, Belgium

JAMES C. G. WALKER[1] AND BRADLEY N. OPDYKE

Department of Geological Sciences, The University of Michigan, Ann Arbor, Michigan

A numerical model describing the coupled evolution of the biogeochemical cycles of carbon, sulfur, calcium, magnesium, phosphorus and strontium is developed to describe the changes in chemical weathering, atmospheric CO_2, climate, and oceanic chemistry over Phanerozoic time. Several geochemical data are available that presumably indicate changes in the weathering rates of continents and ocean chemistry: the reconstructed rate of carbonate deposition, the calcite compensation depth (CCD) in the ocean (available for the last 120 my), and the $^{87}Sr/^{86}Sr$ ratio of seawater. A consistent interpretation of these various signals is not obvious. Model results suggest that the strontium isotopic ratio of seawater through time is a valuable indicator of the rate of continental chemical weathering. Indeed, a simple model of the strontium cycle between the ocean, the crustal carbonate pool, the crustal igneous rock reservoir and the mantle can be used to derive the weatherability of the continents through time. If this weatherability history is used in a coupled model of the geochemical cycles and climate, the inferred history of the global mean surface temperature is consistent with the Phanerozoic record of glaciation. Further, the observed Cretaceous-Tertiary changes of the global mean oceanic CCD can be reproduced with the model. The periods of enhanced chemical weatherability of the continents are associated with tectonic events, such as mountain building episodes, rather than with elevated atmospheric CO_2 concentrations or warm climatic conditions.

INTRODUCTION

In the last few years, there have been several attempts to reconstruct the past evolution of the carbon dioxide pressure in the atmosphere and the climate on Cenozoic or Phanerozoic timescales, based on numerical models of the carbon and related geochemical cycles coupled to simple climate models [Walker et al., 1981; Berner et al., 1983; Lasaga et al., 1985; Budyko et al., 1987; Marshall et al., 1988; Kuhn et al., 1989; Berner, 1990; François and Walker, 1992]. The problem is difficult because of our poor quantitative knowledge of the processes involved in the geochemical system to be modelled. Further, the few available geological indicators on which the reconstruction can be based appear inconsistent [Delaney and Boyle, 1988].

[1] Also at: Department of Atmospheric, Oceanic and Space Sciences, The University of Michigan, Ann Arbor, Michigan.

Evolution of the Earth and Planets
Geophysical Monograph 74, IUGG Volume 14

This modelling work is very dependent on our understanding of the changes in chemical weathering over era interval timescales. Unfortunately, the major factors that control the chemical weatherability of the continents over several millions or hundreds of million years have not yet been fully identified. Walker et al. [1981] have suggested that, on a global scale, chemical weathering increases with runoff, surface temperature and atmospheric CO_2 level. They proposed that the Earth's surface temperature is stabilized by a negative feedback loop linking atmospheric CO_2, climate, runoff and the rate of silicate weathering. Thus, in their opinion, climate, runoff and atmospheric CO_2 are the most important factors controlling the rate of chemical weathering. Subsequent modellers of carbon cycle and climate on long-term timescales have usually based their studies on this assumption. Volk [1987] has refined the parametrization of the dependence on the atmospheric CO_2 level, by considering the role of terrestrial productivity in enhancing soil CO_2 pressures. In the same line of thought, the effect on runoff and weathering of the changing distribution of the continents through time has also been addressed [Marshall et al., 1989; Tardy et al., 1989; Berner, 1990], as well as the role of the biological evolution of land plants [Robinson,

1990; Berner, 1990]. However, Raymo et al.[1988] and Raymo [1991] have recently argued that mountain building is the primary factor determining the rate of chemical weathering. According to these authors, the signature of chemical weathering would be left in the history of the strontium isotopic ratio. Molnar and England [1990] also recognize that uplift and the development of mountain glaciers should enhance chemical weathering. In this paper, we develop a numerical model coupling the major geochemical cycles to the climate system, in order to confront these two theories of chemical weathering and to tentatively reconstruct the Phanerozoic history of weathering, atmospheric CO_2 level and surface climate.

Carbon Cycle and Chemical Weathering

The evolution of the oceanic chemical composition on a timescale of a few million to several hundred million years is governed by long-term geochemical processes of which chemical weathering of rocks is one of the most important. Among the various species dissolved in seawater, inorganic carbon (which includes CO_2, H_2CO_3, HCO_3^- and CO_3^{2-}) plays an important role, because it tends to chemically equilibrate with atmospheric CO_2, a greenhouse gas that can substantially affect the Earth's radiation budget and thus change its climatic state. Chemical weathering of continental rocks releases various ions in soil water. These ions are transported into the world ocean through river or groundwater runoff. When added to seawater, they can alter (in the long-term) its carbon content and alkalinity. Of special importance are calcium and magnesium ions which are easily liberated by weathering and can combine with CO_3^{2-} ions to take part in the carbonate cycle. Rather small (relative) variations of calcium and magnesium concentrations in seawater are needed to generate major changes of the alkalinity. Changes of the alkalinity of seawater are potentially important for atmospheric carbon dioxide: at a given concentration of oceanic inorganic carbon, the higher the alkalinity, the larger the proportion of this inorganic carbon in the dissociated form (HCO_3^- and CO_3^{2-} ions) and the lower the carbon dioxide pressure in the water and in the atmosphere. Consequently, the atmospheric carbon dioxide and the alkalinity of seawater are closely linked together. However it is not clear whether on a long-term timescale the alkalinity of seawater controls the atmospheric CO_2 pressure or the reverse.

Walker et al.[1981] have proposed that the rate of chemical weathering increases with surface temperature and/or atmospheric CO_2 concentration. A direct dependence on atmospheric CO_2 is assumed, because abundant atmospheric CO_2 acidifies rainwater through formation of carbonic acid which enhances chemical weathering. The dependence on surface temperature allows both for the kinetics of the chemical reactions involved in weathering and for presumed higher runoff rates under warm climates. These authors also argue that, under these conditions, carbon dioxide should accumulate in the atmosphere (and the climate gradually warm in response to enhanced CO_2 greenhouse) until the rate of silicate weathering on the continents equals the input of volcanic CO_2 to the ocean-atmosphere system. This argument can be understood by considering the simplified cycles of carbon and alkalinity between the crust, the ocean and the atmosphere depicted

in Figure 1. In this simplified diagram, volcanism is assumed to be a source of carbon for the atmosphere. For the sake of simplicity, sub-marine volcanism (e.g. at mid-ocean ridges) is not explicitely shown in the diagram of Figure 1, but its inclusion would not affect the conclusion that will be reached. The magnitudes of the various fluxes are our estimates and will be used throughout this paper. The weathering of carbonate rocks is a source of both carbon and alkalinity (i.e an input of calcium or magnesium in Figure 1) for the ocean. Half of the carbon added to the ocean originates from the rock and the other half is in fact atmospheric carbon dioxide dissolved in rain water percolating the soil or assimilated by land plants during photosynthesis and released to the soil upon decomposition of organic matter. This atmospheric carbon dioxide is used up by weathering as carbonic acid H_2CO_3 (actually the reacting species is more usually an organic acid, but the global budget is the same, as discussed by Berner and Berner [1987]) according to the overall chemical reaction:

$$CaCO_3(rock) + H_2CO_3(atm.)$$
$$\rightarrow Ca^{2+}(river) + 2\ HCO_3^-(river) \tag{1}$$

where "Ca" can be substituted by "Mg" in the case of magnesium carbonate weathering. When one mole of calcium carbonate is weathered, one mole of calcium (or magnesium) ions is released in the river system and transported to the ocean, corresponding to an input of two equivalents of alkalinity to seawater. The other oceanic source of alkalinity in Figure 1 is provided by silicate weathering of which the chemical reaction can be schematically written as (in the case of Ca-silicates):

$$CaSiO_3(rock) + 2\ H_2CO_3(atm.) + H_2O$$
$$\rightarrow Ca^{2+}(river) + 2\ HCO_3^-(river) + H_4SiO_4(river). \tag{2}$$

The weathering of one mole of $CaSiO_3$ releases one mole of calcium ions which is transported to the ocean. During this process, ocean water is thus enriched by two equivalents of alkalinity. The reaction also involves two moles of carbon which are transferred from the atmosphere to the river system and ultimately to the ocean in the form of HCO_3^- ions. For the atmospheric reservoir to be in balance, an air-sea exchange of CO_2 from the water to the atmosphere is necessary. Finally, the only sink of carbon and alkalinity out of the ocean is assumed to be the deposition of carbonates in marine sediments followed by their incorporation in the crustal pool of carbonate rocks. On a timescale of several million years, oceanic carbon and alkalinity must remain very close to their steady state values, that is their cycles must be simultaneously balanced. In the simplified system of Figure 1, this simultaneous balance of both cycles is only possible if the rate of silicate weathering [i.e. the global rate of reaction (2), or the rate of release to the ocean of Ca^{2+} and Mg^{2+} ions from silicate weathering] is equal to the rate of volcanic release of CO_2.

If, as proposed by Walker et al.[1981], the rate of silicate weathering is an increasing function $F_{sw}(T_s, P_{CO_2})$ of the surface temperature T_s and the atmospheric CO_2 pressure P_{CO_2}, the value of P_{CO_2} can then be extracted from the equation:

$$F_{sw}[T_s(P_{CO_2}), P_{CO_2}] = F_{volc} \tag{3}$$

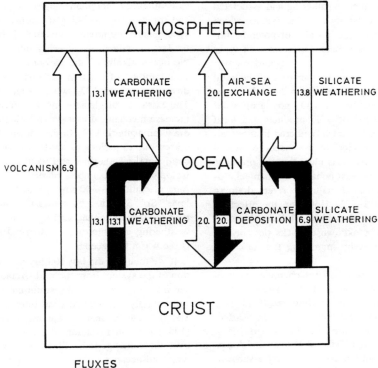

Fig. 1. Simplified representation of the carbon cycle between the crust, the ocean and the atmosphere. White arrows are fluxes of carbon, while black arrows are fluxes of calcium and magnesium which govern the alkalinity of seawater.

where F_{volc} is the rate of CO_2 release from volcanoes and where the dependence of T_s on P_{CO_2} is calculated from a climate model. Consequently, the value of the atmospheric CO_2 pressure is determined as soon as F_{volc}, the form of the weathering function F_{sw} and the $T_s(P_{CO_2})$ relationship are known. It follows that P_{CO_2} does not depend on the ocean chemistry at the time considered. Rather, it is the ocean chemistry which adjusts itself to the value of P_{CO_2}. For instance, the oceanic alkalinity must respond to the value of P_{CO_2}. Indeed, during times of high atmospheric CO_2 pressures, the weathering rates are higher and result in an enhanced delivery of calcium and magnesium ions to the oceans, thus leading to increased seawater alkalinity. Conversely, during times of low atmospheric CO_2, the weathering rates are reduced and the ocean alkalinity is lower. In summary, in this system, the partial pressure of CO_2 in the atmosphere is fixed by purely geochemical and climatic requirements, but the alkalinity is dictated by the intensity of weathering, that is by atmospheric CO_2 level. The total amount of dissolved inorganic carbon and the pH in the ocean water are then such that the charge balance is satisfied and that the concentration of CO_2 in surface water $[CO_2]_s$ is consistent with the atmospheric CO_2 pressure [i.e. only a small difference should exist between $[CO_2]_s$ and the equilibrium concentration (through Henry's law) to create a small flux of CO_2 from the ocean to the atmosphere (see Figure 1) compensating for the atmospheric CO_2 consumption in weathering].

The dependence of the rate of chemical weathering on T_s and P_{CO_2} postulated by Walker et al. [1981] is highly speculative. For example, part of the temperature dependence relies on a presumed increase of continental runoff with global mean surface temperature. However, continental runoff is the difference between precipitation and evaporation on land. Both of these quantities should increase with global mean surface temperature, but the fact that their difference would increase is less obvious: our quantitative knowledge of the global hydrological cycle remains surprisingly poor [Chahine, 1992], and a reliable decription of the variation of global hydrological processes associated with climate changes remains a challenge even for most sophisticated GCMs. Actually, continental runoff must balance the atmospheric flux of water from ocean to land, which depends on both the wind speed and the moisture content of the air. Runoff rates are thus directly related to the atmospheric circulation whose response to climate warming is not easy to predict and which depend on many factors, such as the distribution of land masses over the globe. Further, atmospheric CO_2, surface temperature and runoff are not the only factors controlling the rate of chemical weathering and associated denudation of the continents. A variety of other factors which have been changing over geologic time are also of potential importance: the geology of the surface rocks (different rock types have different susceptibilities to weathering), the vegetation cover (vegetation can fix the soil and prevent the transport of the solid weathering products) and the biological activity in the soil (microbial and root respiration enhances the soil CO_2 pressure), relief (slopes can facilitate the evacuation of solid products and local precipitation rates and runoff can be strongly altered in mountainous areas) and the presence of glaciers or ice sheets (enhanced breakdown of rocks), etc. Of course, it is clear that the presence of glaciers in mountainous regions must enhance mechanical erosion, which transfers more suspended matter to rivers draining these areas [Berner and Berner, 1987]. However, the effect on chemical weathering (and thus the dissolved solids in rivers) is less clear, although mechanical breakdown of rocks possibly alters the chemical weathering rates by increasing the rock area exposed to weathering.

The importance of relief and glaciation for chemical weathering has been put forward by Raymo et al. [1988]. These authors suggest that the late Cenozoic mountain building events not only have enhanced mechanical erosion, but have also strongly affected the rate of chemical weathering and thus atmospheric CO_2. They focus on uplifts of the Himalayas, Tibetan Plateau and Andes mountains, because the rivers from these regions carry a substantial proportion of the total dissolved and particulate load tranported to the oceans today. They mention that other areas, such as the southwest United States, the Swiss Alps, and the Southern Alps of New Zealand, also underwent recent increases in uplift rates since the late Neogene. They argue that chemical weathering is increased in mountainous areas, due to rapid mechanical breakdown of rocks and enhanced exposure of primary minerals to chemical weathering, to the presence of abundant easily eroded sedimentary rocks in areas of active tectonism, and to higher runoff rates resulting from orographic effects on precipitation. Consequently, the dissolved load of rivers and the rate of ion input to the ocean should have increased substantially during recent uplift episodes. One indicator of that increase would be provided by the trend toward larger values exhibited by the evolution of the strontium isotopic ratio ($^{87}Sr/^{86}Sr$) of seawater over the last 5 million years (my), as recorded in marine carbonates [see also Capo and DePaolo, 1990]. Clearly, the source of radiogenic strontium to the oceans is the river input from rock weathering, while that of nonradiogenic strontium is principally hydrothermal (see Figure 2). If the mean isotopic ratio of the weathered material and the hydrothermal source have not changed markedly during the last 5 my, then the increase of the $^{87}Sr/^{86}Sr$ ratio over the same period would indeed record an increase of the global rate of chemical weathering.

Raymo [1991] has generalized these concepts by applying them to the Phanerozoic (i.e. the last 570 my) revitalizing Chamberlin's [1899] theory of glaciations. The strontium isotopic ratio of seawater is known for the whole Phanerozoic (Figure 2) and the highs of that curve seem to approximately correlate with known periods of glaciation. According to Raymo's view, uplift and erosion of major mountain ranges might control atmospheric CO_2 and the onset of global glaciations. Periods of uplift would be accompanied by intense erosion and chemical weathering. In particular, the silicate weathering rate would be higher, increasing the ocean alkalinity and carbonate ion concentration (which favors removal of calcium carbonate from the ocean) and driving down atmospheric CO_2, which results in a global climate cooling. The existence of glaciers and widespread icesheets would further increase mechanical erosion and global chemical weathering and draw atmospheric CO_2 further down. Glacial periods would then correspond to periods of uplift, low atmospheric CO_2 and intense continental weathering, and would be marked by high $^{87}Sr/^{86}Sr$ isotopic ratios. Examples of such periods are the late Carboniferous glaciations ~300 my ago and the present time. Periods of low $^{87}Sr/^{86}Sr$ ratios, by contrast, would be those when atmopheric CO_2 was abundant, and the climate was warm, with reduced weathering and intense hydrothermal activity. The best example is the warm Cretaceous.

If we focus on the last 100 my of Earth's history, on the evolution of the system from the mid-Cretaceous to the present, we can see from Figure 2 that the strontium isotopic ratio has increased substantially during that time interval, whereas Figure 3 shows that the rate of seafloor spreading (as reconstructed by Gaffin [1987] through inversion of the sealevel history) has decreased over the same period. Since most volcanic CO_2 is associated with mid-ocean ridge or plate margin volcanism, it is reasonable to assume that the rate of release of volcanic CO_2 to the ocean and atmosphere is more or less proportional to the seafloor spreading rate. Consequently, on one hand, Raymo's interpretation of the strontium isotopic record, implies that the weathering rate of continental silicate rocks has increased from the Cretaceous to the present, while, on the other hand, the history of seafloor spreading suggests a decrease of the volcanic release rate of CO_2 during the same period. This conclusion contradicts the conventional assumption of Walker et al. [1981] that these two rates must have remained equal to each other at any time in the past and points out the major implications of Raymo's theory for our understanding of the global geochemical evolution. In the conventional view, high weathering rates are predicted under warm climates and the strontium isotopic record is interpreted as purely lithologic in origin, implying drastic temporal changes in the ratio of the weathered products coming from old shield rocks versus young volcanics [Brass, 1976; François and Walker, 1992; Berner and Rye, 1992]. On the contrary, under Raymo's interpretation of the strontium signal, periods of intense weathering correspond to glacial conditions.

If, as implied by Raymo's arguments, the rate of silicate weathering has differed in the past from the input rate of CO_2 from volcanoes, then the carbon and alkalinity cycles must have been more complex than in the system presented in Figure 1. Indeed the difference between the two rates must be compensated by an additional flux of carbon (not carrying alkalinity), if the two cycles are assumed to be at or near steady state (which must be true on several million year timescales, i.e. much longer time intervals than the carbon residence time in the ocean). François and Walker [1992] have attempted to reconstruct the Phanerozoic history

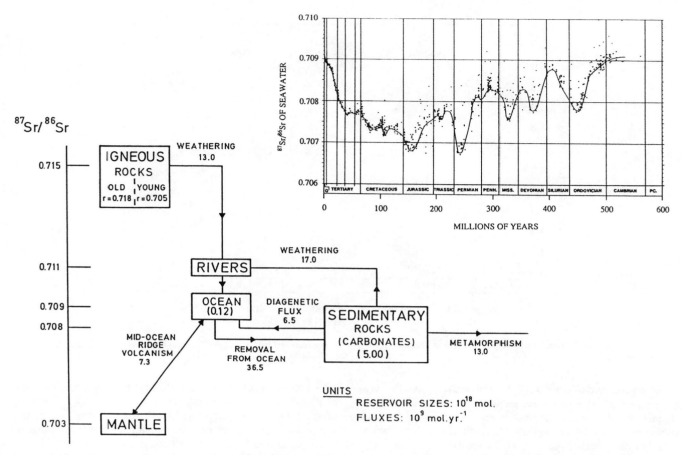

^{87}Sr/^{86}Sr

Fig. 2. The cycle of oceanic strontium. The present-day state of the cycle is shown on the bottom left [from François and Walker, 1992]. The vertical axis shows the approximate ^{87}Sr/^{86}Sr ratios of the various reservoirs. The graphic [from Burke et al., 1982] on the upper right shows the Phanerozoic evolution of the strontium isotopic ratio in carbonates deposited from seawater as a function of age. Since there is no isotopic fractionation during carbonate formation, this ratio reflects that of seawater at the same age, provided the ratio has not been altered by diagenesis.

of the atmospheric CO_2 pressure from Raymo's assumption that the strontium isotopic curve is an index of the intensity of past weathering. They assumed that the carbon flux compensating for possible past differences between silicate weathering and release of volcanic CO_2 was provided by weathering of basaltic seafloor. Deep oceanic water is slightly corrosive for the basaltic rocks of the ocean floor. During the reaction, calcium ions from the rock are released to seawater. These ions then react with carbonate ions of the water to precipitate calcium carbonates [Staudigel and Hart, 1983; Staudigel et al., 1981; 1989]. If, as assumed for simplicity in the model of François and Walker [1992], all calcium ions released from the basalt are precipitated as carbonates, the budget of the overall process is just a sink of oceanic carbon. The detailed mechanism is not well established and the magnitude of the flux is poorly known. Nevertheless, the important point is that such a flux of carbon which has been varying in the past is required to accomodate the data, regardless of the exact nature of that flux. For example, we have noticed that, during the Cretaceous, the input rate of volcanic CO_2 was higher than today, while the

silicate weathering rate was lower. This means that the seafloor weathering sink of carbon had to be larger at that time, so that the excess volcanic CO_2 not consumed by silicate weathering (followed by carbonate deposition) was removed from the system via seafloor weathering.

The role played by seafloor weathering could equivalently be accomplished by imbalances in the sub-cycle of organic carbon (weathering of kerogen carbon from shales followed by deposition of organic matter on land or on the seafloor). Under that hypothesis, the rate of organic carbon deposition should have exceeded that of kerogen carbon weathering at Cretaceous time. The organic carbon sub-cycle was included in the model of François and Walker [1992], but did not play a major role in compensating the difference between volcanic CO_2 release and silicate weathering, because this organic sub-cycle did not undergo major imbalances in the model simulations performed. Indeed imbalances of at most $1-2 \times 10^{12}$ mol yr^{-1} are observed in these simulations during Late Carboniferous and Cretaceous, when the global deposition rate of organic matter was highest. The reason for such small imbalances

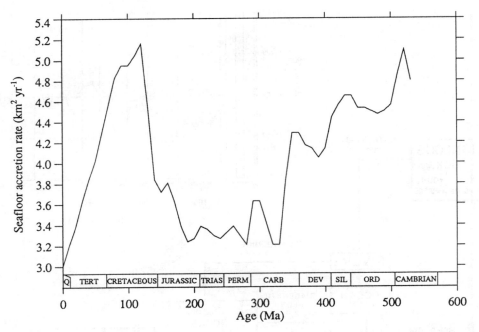

Fig. 3. The Phanerozoic history of the rate of seafloor spreading, as reconstructed by Gaffin [1987] through inversion of the history of sealevel.

of the organic cycle is that François and Walker make the weathering rate of kerogen carbon proportional to the atmospheric O_2 pressure. This model assumption creates a negative feedback stabilizing the model atmospheric O_2 level: if the global rate of organic carbon burial (the long-term source of atmospheric O_2) happens to increase at some time in the past, the atmospheric O_2 pressure becomes higher and the rate of kerogen weathering is increased, thus limiting strongly any imbalance in the organic sub-cycle, i.e. any global source or sink of atmospheric O_2. In the model, the small imbalances in the organic carbon sub-cycle are partially compensated for by imbalances of opposite sign in the cycle of reduced sulfur, tending to further limit the excursions of atmospheric O_2. Larger imbalances of the organic sub-cycle would of course be possible if kerogen weathering was made independent of atmospheric O_2. However, we feel that imbalances large enough to compensate for the difference between silicate weathering and volcanic CO_2 release are unlikely, because they would imply too large excursions of atmospheric O_2. This will be shown later in the paper. It is the reason why we prefer the seafloor weathering sink as the required flux of carbon out of the system.

MAIN CHARACTERISTICS OF THE MODEL

In this paper, we want to confront the conventional theory of chemical weathering (in which the global rate of weathering is assumed to increase with surface temperature, runoff and atmospheric CO_2) with the new theory emphasizing the importance of mountain uplift and proposed by Raymo et al. [1988] and Raymo [1991]. Model simulations based on both theories will be performed. In the case of the conventional theory, the weathering

rates will be calculated from standard relationships, as a function of surface temperature, runoff and atmospheric CO_2. The role of soil biosphere in enhancing chemical weathering will be taken into account. In this case, the strontium isotopic evolution is not used, as it is implicitly assumed that these isotopic fluctuations are mainly lithologic, reflecting a varying amount of radiogenic material from old shield rocks in the weathered products. To simulate the new uplift theory, we assume that the relative contribution of the old shield versus young volcanics in the weathered products coming from silicate rocks is constant through time, i.e. the mean strontium isotopic ratio of the silicate rocks weathered is held constant. After substraction of the contribution of the hydrothermal exchange of strontium with the mantle (see Figure 2), the observed fluctuations of the seawater strontium isotopic ratio then reflect past changes in the intensity of continental weathering. Consequently, in the framework this hypothesis, the intensity of past continental weathering can be estimated from the strontium isotopic record. Periods of high (radiogenic) seawater $^{87}Sr/^{86}Sr$ ratio are periods intense weathering, periods of low seawater $^{87}Sr/^{86}Sr$ are periods of reduced weathering.

These two methods of estimating the past weathering history of the continents will be compared and their implications for the evolution of the carbon cycle and atmospheric CO_2 will be discussed. A climatic evolution will be reconstructed from the history of atmospheric CO_2 obtained for each method and a comparison with the succession of cold and warm periods in the Phanerozoic geological record will be undertaken. Also, deposition rates of carbonates on the shelf and pelagic environments will be compared to observations when they are available, that is for the last 140 Ma. For that most recent period, data on past oceanic calcite compensation depths (see below) exist and will be employed for validation of the model results. The consideration of such geological proxies in carbon cycle models is essentially new and, we feel, is an very important matter that needs to be carefully addressed.

The model used is an updated version of that developed by François and Walker [1992]. It calculates the evolution over geologic time of the cycles of carbon, calcium, magnesium, sulfur, phosphorus (to estimate biological productivity in the ocean), and strontium. This geochemical model is coupled to an energy-balance climate model determining at each time in the past the yearly averaged zonal mean surface temperature as a function of latitude. This climate model takes into account the changing latitude distribution of continents over time and uses the atmospheric CO_2 concentration calculated in the geochemical model. The solar luminosity is assumed to increase linearly with time from a value at the beginning of the Cambrian period 3.12% lower than at present to allow for the 25% increase over the age of the Earth. The calculation of the weathering rates in the geochemical model also uses the climatic output when temperature-dependent weathering rates are assumed. In this case, runoff and weathering rates are calculated at each latitude. The model uses the past distribution of land and shelf areas versus time and latitude for the whole Phanerozoic, as tabulated by Parrish [1985]. The variation of global land and shelf areas with time is illustrated in Figure 4. All volcanic and hydrothermal fluxes considered in the model are made proportional to the seafloor accretion rate of Figure 3. The burial rate of organic carbon on land is estimated from the percentage of coal basin sediments relative to total clastics as tabulated by Berner and Canfield [1989]. The burial rates of organic carbon and pyrite on the shelf are obtained from a two-layer sediment model.

The calculation of carbonate deposition rates on the seafloor will be described later in the text.

The main differences of the model with respect to the earlier version are summarized below. Figure 5 shows the present-day state of the carbon and alkalinity cycles on which the model is based. The processes are the same as those considered by François and Walker [1992], but the present-day values of some fluxes are slightly modified and carbonate deposition has been split into three (shelf, continental slope, pelagic) rather than two components. The distribution of the deposition fluxes between these three environments is close to recent estimates by Wilkinson and Algeo [1989]. The system contains the cycles of both inorganic and organic carbon. Hydrothermal (ridge) volcanism and seafloor weathering are taken into account.

The present-day rates of carbonate and silicate weathering are slightly increased with respect to the values derived by François and Walker [1992] on the basis of data from Meybeck and Helmer [1989]. The reason is that estimates of the present-day deposition rate of carbonates on the seafloor suggest a value of 20×10^{12} mol yr^{-1} or even larger [e.g. Wilkinson and Algeo, 1989] for total (shelf+slope+pelagic) deposition. On long-term timescales, carbonate deposition must balance weathering, but the measure of carbonate sedimentation is more likely a long-term average than the estimate of weathering from the dissolved load of rivers. The rates used by François and Walker [1992] for carbonate and silicate weathering have thus been multiplied by a constant factor, such that the total input to the ocean of calcium and magnesium from these processes is 20×10^{12} mol yr^{-1}. As in this previous work, the sink for magnesium originating from carbonate and silicate weathering is hydrothermal exchange of seawater Mg for basaltic Ca at mid-ocean ridges (note that this exchange does not

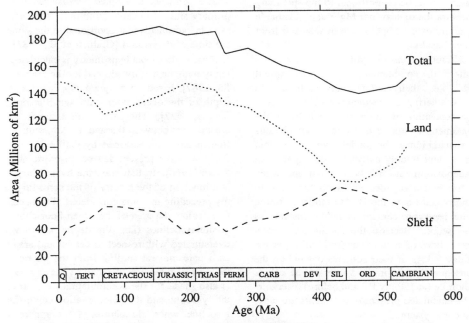

Fig. 4. The Phanerozoic history of shelf, land and total cratonic area from Parrish [1985].

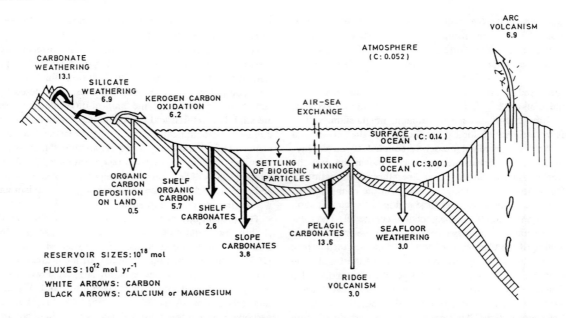

Fig. 5. The present-day fluxes in the cycles of carbon and alkalinity used in the model. White arrows are fluxes of carbon, while black arrows are fluxes of calcium and magnesium.

alter the budget of ocean alkalinity), but no magnesian calcite or dolomite is deposited in the model. Indeed, the calcium ions that are released to seawater from the hydrothermal exchange are removed from the ocean in the form of carbonates. The parameterization of that exchange is here slightly modified with respect to our previous work [François and Walker, 1992] or the BLAG model [Berner et al., 1983; Lasaga et al., 1985]: the exchange rate is still assumed to be proportional to the spreading rate, but not to the concentration of seawater Mg, rather it is made proportional to the departure of the Mg/Ca ratio in seawater from the same ratio in average basaltic rocks.

Another important difference from our previous calculation is that calcite and dolomite in the continental crust are now explicit reservoirs of the model, i.e. their contents vary with time in response to deposition, weathering and metamorphism-volcanism. Note, however, that no deposition of dolomite is included in the model, because the parameterization of that rate is not obvious and would add much speculation to the model. For that reason, the dolomite reservoir content is only calculated during the last $\sim$150 my, when the deposition rate has been small and when the reservoir content has essentially decayed with time due to weathering and metamorphism (as in the BLAG model). Before that time, the dolomite reservoir content is held constant, as it was in our previous work. Because the continental crustal reservoirs have residence times of several hundred million years, the calculated present-day values of their contents depend on the initial conditions. Consequently, the initial values of the calcite (at 600 my BP) and dolomite (at 150 my BP and before) reservoirs are chosen such that the calculated present-day contents are close to estimates for the modern system.

Additional minor improvements include the calculation of the

saturation depth (or lysocline) not only for calcite, but also for aragonite, allowance for borate ion chemistry in the carbonate speciation scheme, and a few refinements in the parameterization of carbonate deposition on the shelf (mainly through coral reef buildup). Also, the flux of strontium removal out of the ocean is now proportional to the total deposition rate of carbonates, rather than only proportional to the concentration of oceanic strontium. That dependence is adopted because removal in carbonates is the primary sink of oceanic strontium. It is assumed that the Sr/Ca ratio in carbonates is proportional to the same ratio in seawater at the time of deposition [Graham et al., 1982].

Finally, the ocean hypsometry is calculated in the model consistently with the history adopted for the seafloor spreading rate [Gaffin, 1987], based on a square root law for the variation of mean depth of the ocean floor with age [Parsons, 1982; Parsons and Sclater, 1977]. The calculation assumes a distribution of ridge crest depths between 0.3 and 3.3 km, with a maximum at 2.5 km (the mean value assumed by Gaffin [1987]) and can reproduce fairly well the present-day ocean hypsometry from Menard and Smith [1966]. In that way, the hypsometry varies with time and is a function of the history of the spreading rate during the $\sim$200 my preceding the time considered. The ocean hypsometry is used to calculate the area of the ocean floor above the calcite and aragonite lysoclines (i.e. the depths above which ocean water is oversturated with respect to calcite and aragonite). Each calcium carbonate mineral settling from the photic zone accumulates on the area of the ocean floor above its own lysocline. A correction is also made for the accumulation in the transition zone between the lysocline and the compensation depth (i.e. a depth below the lysocline, where dissolution of the mineral exactly balances the rain rate from above).

TABLE 1. Model runs and their characteristics

	Seafloor Weathering	Silicate Weathering	Carbonate Weathering
Run #1	No	depends on A_{land}, T_s and P_{CO_2}	depends on A_{land}, T_s and P_{CO_2}
Run #2	Yes	depends on A_{land}, T_s and P_{CO_2}	depends on A_{land}, T_s and P_{CO_2}
Run #3	Yes	calc. from $^{87}Sr/^{86}Sr$	calc. from $^{87}Sr/^{86}Sr$
Run #4	Yes	calc. from $^{87}Sr/^{86}Sr$	proportional to A_{land}

A_{land} = land area
T_s = surface temperature
P_{CO_2} = atmospheric CO_2 pressure

CONTINENTAL WEATHERABILITY OVER PHANEROZOIC TIME

In this section, we first discuss the model results for the last 530 my, a time interval when strontium isotopic data are available from Burke et al.[1982]. This period covers almost the whole Phanerozoic and thus extends far back into the Paleozoic when global glaciations have been recorded. It is a long time period during which we can expect the weathering rates to have varied widely; correspondingly, atmospheric CO_2 has almost certainly fluctuated also. Several time forcing functions are common to all runs: the seafloor accretion rate (Figure 3) driving all volcanic and hydrothermal fluxes, the percentage of coal basin sediments relative to total clastics used to estimate organic carbon deposition on land and the areas of land and shelf at each latitude.

Four runs of the model have been performed. They are summarized in Table 1. Run #1 is the "classical" way of modelling long-term atmospheric CO_2 and climate evolution [e.g. Walker et al., 1981; Berner et al., 1983; Marshall et al., 1988; Kuhn et al., 1989], with no consideration of the seafloor weathering process in the carbon cycle, and with all continental weathering rates proportional to land area and increasing with surface temperature T_s (directly and through increased runoff) and atmospheric CO_2 pressure P_{CO_2}. These weathering rates are calculated in each band of latitude. The dependence on T_s and P_{CO_2} is that adopted by François and Walker [1992]. Silicate weathering is proportional to runoff and to the factor $\exp[(T_s - 285/21.7)]$. Carbonate weathering is just proportional to the square root of runoff. Runoff in each latitude band is parameterized as a function of continental fraction, surface temperature and latitude. Both silicate and carbonate weathering are proportional to power 0.3 of the soil CO_2 pressure, calculated according to the model proposed by Volk [1987]. To simulate the origin of land plants near 400 Ma, the maximum soil productivity (i.e. the limiting productivity at high CO_2, see Volk's paper) is increased linearly from zero before 440 Ma to the present value at and after 360 Ma. The strontium isotopic data are not used. Run #2 is the same as run #1, except that seafloor weathering is now taken into account. Run #3 is a simulation of Raymo's [1991] hypothesis that the $^{87}Sr/^{86}Sr$ ratio

of seawater through time indicates the intensity of weathering. Of course, in this simulation, seafloor weathering has to be included to compensate for the difference between silicate weathering and volcanic CO_2 release. The strontium cycle is modelled according to the diagram of Figure 2. Relative changes through time of igneous (silicate) and sedimentary (carbonate) rock weathering on the continents are assumed to be the same, i.e. the weatherabilities (see below) of both rocks are equal. These temporal changes of continental weathering are determined by the strontium model from the condition that, at each time, the model $^{87}Sr/^{86}Sr$ ratio of seawater is exactly equal to the observed value. Run #4 is the same as run #3, except that only silicate weathering is calculated from the strontium isotopic ratio of seawater. Carbonate weathering (as well as weathering of any other sedimentary material such as kerogen carbon weathering) is here assumed simply proportional to the global land area of Figure 4.

For the rest of this paper, we will define the (chemical) weatherability ξ_w^{sil} of continental silicates as the rate of silicate weathering per unit area of continent normalized to present-day, i.e.

$$\xi_w^{sil} = \frac{A_{land}(t=0)}{A_{land}} \frac{F_w^{sil}}{F_w^{sil}(t=0)} \qquad (4)$$

where A_{land} and F_w^{sil} are respectively the global land area (from Figure 4) and the global rate of (chemical) weathering of silicate rocks, and where "t=0" refers to present-day. Defined in that way, ξ_w^{sil} measures the rate of chemical denudation of the continents relative to silicates. The history of ξ_w^{sil}, the chemical weatherability of silicate rocks on the continents, calculated in the four model runs of Table 1 is presented in Figure 6. The corresponding Phanerozoic evolutions obtained for the atmospheric CO_2 level and the global mean surface temperature are shown in Figures 7a and 7b.

In the case of the "classical" run #1, because the organic carbon sub-cycle of the model remains almost exactly balanced throughout the Phanerozoic (except in the Carboniferous, when deposition of organic carbon on land is important), the silicate weathering rate is always very close to the release flux of volcanic CO_2

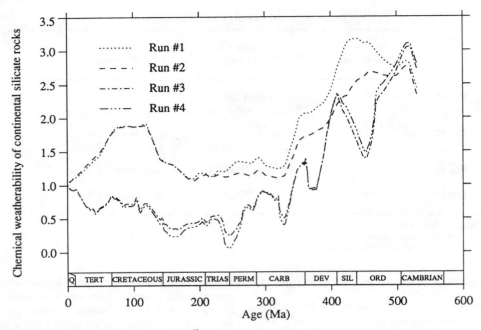

Fig. 6. The evolution of the silicate weatherability ξ_w^{sil} of the continents during the last 530 my, calculated in the four model runs summarized in Table 1.

according to equation (3). As a result, the value of ξ_w^{sil} obtained in this run is just a combination of the volcanic forcing of Figure 3 with the land area variation of Figure 4 [according to equation (4) where F_{volc} is substituted to F_w^{sil}]. Since the present-day period corresponds to maximum land area and minimum volcanic outgassing, ξ_w^{sil} is today smaller than at any other time in the Phanerozoic. The weatherability may have been much higher in the past. For example, between 500 and 400 Ma (Ordovician-Silurian), the silicate weatherability is enhanced by a factor of 2.5 to 3, due to a combination of small land area and relatively high volcanic activity. To support such an intense level of weatherability at that time, run #1 requires rather elevated pressures (up to 10 times the present) of atmospheric CO_2, with a relatively warm climate. Actually, it is during this time period that model temperatures reach their Phanerozoic maximum in run #1, in contradiction with the record on several continents of glaciations during Late Ordovician and Early Silurian [Caputo and Crowell, 1985]. Further, Figure 7b suggests the existence of an extensive cold period between ~330 and ~170 Ma, when surface temperatures were close to present (i.e. mean Quaternary). The first half of this period is more or less coincident with Late Carboniferous glaciations, but no important glacial event is known near 200 Ma [Crowell, 1983; Caputo and Crowell, 1985]. Consequently, run #1 presents significant discrepancies with the geological record of glaciation.

The comparison between runs #1 and #2 shows the effect of the inclusion of the seafloor weathering process in the carbon cycle system. Although the present-day rate for this process is small compared with silicate or carbonate weathering on the continents, the model value of this rate is increased substantially

in the past. More importantly, the difference between the carbon fluxes of seafloor weathering and mid-oceanic ridge volcanism is small during Mesozoic and Cenozoic ages, but is slightly larger during Paleozoic time when the silicate weatherabilities of the two runs differ most substantially. The effect on atmospheric CO_2 and surface temperature is most important during the early Paleozoic. At that time, the global temperature drops by ~2 K from run #1 to run #2. As shown in Figure 7b, the duration of the Carboniferous-Permian glaciation is again too long in run #2, since it extends into the Triassic and Jurassic periods.

As mentioned above, the weatherability history of continental silicates in runs #3 and #4 is calculated from the condition that the strontium isotopic ratio of model seawater remains always equal to its observed value on the curve from Burke et al. [1982] displayed in Figure 2. The only difference between these two runs resides in the calculation of the carbonate (and other sedimentary rock) weatherability. In run #3, carbonate weatherability is the same as for silicates, whereas run #4 assumes it is constant through time (i.e. carbonate weathering is only proportional to land area). The results of these two runs do not differ much from one and other. The two weatherability histories are very close, because the strontium isotopic ratio of seawater is not very sensitive to the intensity of carbonate weathering. Indeed, the mean isotopic ratio of the sedimentary carbonate reservoir (≃0.708 today) is usually close to that of seawater (0.709 today), because sedimentary strontium has been deposited out of earlier seawater. As a result, weathering of average carbonate rock and addition of its Sr content to seawater does not alter substantially the oceanic isotopic composition. It is only when the strontium isotopic ratio of seawater varies rapidly that the isotopic compositions of the oceanic and sedimentary re-

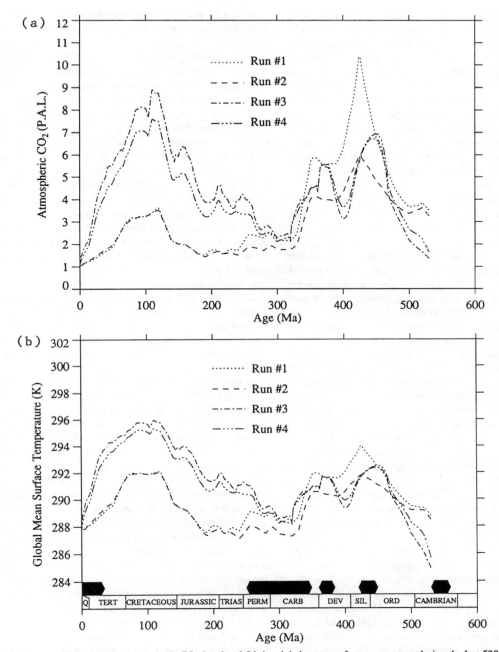

Fig. 7. The evolution of (a) the atmospheric CO_2 level and (b) the global mean surface temperature during the last 530 my, calculated in the four model runs summarized in Table 1. Black bars at the bottom indicate times when glaciations have been reported [Crowell, 1983].

servoirs can be expected to differ markedly, such as during Late Permian and Early Triassic for instance, when the weatherabilities of runs #3 and #4 are subtantially different. It also follows that the histories of atmospheric CO_2 and surface temperature are not much different in the two runs, since atmospheric CO_2 is essentially controlled by silicate weathering (on the continents or on the seafloor), not by carbonate weathering.

However, the continental silicate weatherability histories in runs #3 and #4 are much different from those calculated in the two previous runs. The present-day continental weatherability is rather high by comparison with Mesozoic and Early Cenozoic times. Weatherability appears to have been higher than today only in the Paleozoic era. In terms of Raymo's ideas, the high weatherability at that time would be due to the succession of orogenic episodes during the Paleozoic (Caledonian and Hercynian cycles). Nevertheless, the weatherability obtained for the early Paleozoic is higher than its value today (a time of rather intense orogenic activity) by a factor of 2 to 3. It is thus possible that other factors than mountain building have influenced continental weathering at that time, or maybe the $^{87}Sr/^{86}Sr$ isotopic ratio is not the perfect indicator of weathering for that remote period, because changes in continental lithology (or more specifically the outcrop distribution over the drainage basins of old and young igneous rocks which have different strontium isotopic ratios) are possibly not negligible over the 400 my that separate that time period from the present. It might be indeed that the observed seawater strontium isotopic signal is a mixture of both a change in the intensity of weathering and in the mean isotopic composition of the igneous source. This problem will not examined more fully here because information regarding the lithology of the drainage basins of major world rivers in the past is missing.

The Late Carboniferous was a time of continental collision between Gondwanaland in the South and the Laurentian continent in the North [e.g. Stanley, 1986]. This collision was accompanied by uplift of mountain chains in Europe (Hercynides), Northwest Africa (Mauritanides) and North America (Appalachians). This suturing of land masses was the first step in the formation of the Pangaean supercontinent. At the same time, climate was colder with widespread glaciations extending from Late Carboniferous into Early Permian (Figure 7b). During that period, in runs #3 and #4, the model weatherability of continental silicates is comparable to present-day, although slightly smaller, with values peaking at ~0.9 near 300 Ma. The model atmospheric carbon dioxide pressure reaches minimum values of ~2 times the present-day and the global mean surface temperature is close to present. The model thus correctly predicts glaciations during Late Carboniferous. During Permian time, the model surface temperature first remained low to rise later near mid-Permian.

During Late Permian, the model global mean surface temperature had risen by 3 K with respect to the Late Carboniferous value, and gradually continues to rise throughout Triassic and Jurassic periods into the Early Cretaceous. This behaviour is fully consistent with the disappearance of glacial records during Late Permian and their absence during Mesozoic and early Cenozoic, in contrast with the history obtained in runs #1 and #2. The Phanerozoic minimum of continental silicate weatherability in runs #3 and #4 is reached during Late Permian, Triassic and Jurassic periods, when model values are less than half those for the present-day. At that time, the Late Carboniferous mountains were probably already substantially eroded. Also, the Pangaean supercontinent had formed and its breakup already began. In the presence of diverging continental plates, mountain building can be expected to be reduced, resulting in a simultaneous decrease of the weathering rates. Further, as speculated by François and Walker [1992], dry conditions must have prevailed in the center of Pangaea and consequently have reduced global weathering. Also, as noted by the same authors, intracontinental seas are likely on a supercontinent and can collect the products of chemical weathering and prevent them from reaching the world ocean (of which the strontium isotopic ratio is used in the model).

In runs #3 and #4, the silicate weatherability starts rising during Early Cretaceous, when the Pangaean supercontinent began to be broken apart. At the same time, however, a sharp increase of the seafloor spreading rate and volcanic activity occurred (Figure 3). For that reason, model P_{CO_2} does not go down, but rather carbon dioxide accumulates in the atmosphere, to reach values of ~8 times the present near mid-Cretaceous, approximately when volcanic CO_2 outgassing is maximum. The global mean surface temperature at that time is about 7 K higher than today. This is the warmest global climatic conditions throughout the Phanerozoic in runs #3 and #4. Subsequently, near 85 Ma, the atmospheric CO_2 pressure and the surface temperature begin to be reduced in response to a decreasing volcanic activity. A similar decrease of P_{CO_2} and T_s is obtained in runs #1 and #2. However, in runs #3 and #4, the cooling is much less rapid during Late Cretaceous and Early Cenozoic and the major temperature drop occurs after the Eocene. As a result, in runs #3 and #4, the model Eocene temperature is only 1–2 K lower than at the Cretaceous peak, compared with a 3 K difference in runs #1 and #2. Thus, runs #3 and #4 are also more consistent with the observed warm Eocene period. Also, in the same runs, the temperature drops quickly approximately after the Eocene-Oligocene transition, when ice is thought to have begun to form over Antarctica [Miller et al., 1987]. The silicate weatherability over the continents also increases strongly from that time to the present, consistent with the progressively higher rates of uplift in various parts of the world.

In conclusion, the comparison between Phanerozoic geochemical models differing in their calculation of continental weatherability is compatible with the idea proposed by Raymo et al. [1988] and Raymo [1991] that the strontium isotopic ratio of seawater is an index of the intensity of chemical weathering. The general climatic evolution calculated from atmospheric CO_2 variations based on this assumption is consistent with the climatic record from today up to at least Carboniferous time. The derived weatherability of the continents can then be interpreted in term of tectonic and mountain building events, with periods of continental collision and uplift corresponding with times of intense chemical weathering on land, and epochs of diverging land masses being more likely times of reduced chemical denudation of the continents. Note that the results of this section a posteriori justifies our choice of the seafloor weathering sink of carbon as the compensator for the departure between volcanic CO_2 release and silicate weathering, with respect to a possible imbalance in the sub-cycle of organic carbon. Indeed, the spreading rate history of Figure 3 indicates that, near mid-Cretaceous, the volcanic activity was higher than at present by a factor of ~1.6, while, at the same epoch, Figure 6 suggests a weatherability of about 0.6 the present. In terms of fluxes, using the present-day values of Figure 5, these numbers mean that, in runs #3 and #4, the volcanic release of non-mantle carbon was ~ 11×10^{12} mol yr^{-1} at mid-Cretaceous time, while the rate of weathering of continental silicates was ~ 4×10^{12} mol yr^{-1}. To balance the difference between the two fluxes, a seafloor weathering sink of carbon higher than at present by ~ 7×10^{12} mol yr^{-1} is required. On a long-term view, the magnitude of this carbon flux is rather important. It must be emphasized that it is unlikely that such a high flux can be created by a desequilibrium in the sub-cycle of organic carbon. Indeed, a molecular oxygen source of the same amount would result (fluxes of reduced sulfur are too small for an imbalance of its cycle to compensate for such a large desequilibrium) and atmospheric O_2 would double in only ~5 my!

THE CARBONATE CYCLE OVER THE LAST 140 MY

Abundant data on the history of carbonate deposition exist for Cretaceous and Cenozoic times. In particular, the history of the calcite compensation depth (CCD) is known in the major oceanic basins for approximately the last 120 my [Van Andel, 1975]. These data can be used to test the capacity of our model to calculate the CCD. Further, Opdyke and Wilkinson [1988] have reconstructed from observed data the past deposition rate of carbonates in pelagic, slope and shelf environments, for the last 140 my. Total calcium-magnesium carbonate deposition rates represent an excellent independent proxy for global weathering rates, which we can compare with what would be predicted by the model. In the reconstruction, pelagic carbonate accumulation rates were calculated from the high and low accumulation values multiplied by the appropriate seafloor surface area that would have existed above the given saturation depth for that time interval (CCD) [Whitman and Davies, 1979]. Slope deposition rates were taken from Hay [1985], and the shallow water data were obtained from the work of Ronov and coworkers (see Opdyke and Wilkinson for specific Ronov references). Ronov's platform data of carbonate volumes is within 20% of volumes calculated independently for North America using data from the Cook and Bally atlas [1975]. The largest uncertainty contained is associated with the slope carbonate deposition. The percentage of slope carbonates increases throughout the Tertiary, making the Neogene one of the most difficult intervals to confidently quantify in terms of the total carbonate mass deposition. Global average CCD puts overall constraints on the carbonate saturation of the deep ocean and allows us to calculate when pelagic carbonates became an important component of the total mass flux. Mass of pelagic carbonate did not reach 25% of the total flux until the Miocene [Opdyke and Wilkinson, 1988].

In the model, following these authors, we use the Mesozoic-Cenozoic history of the area A_{acc}^{shelf} of platform carbonate accumulation from Ronov [1980]. Today, only about 16% of the tropical shelf (i.e. in the latitude band between 28°S and 28°N, in which reef formation is observed at present) are areas of carbonate accumulation. In the past, reefal carbonate accumulation was higher by a presumably big factor. First, the total shelf area was larger (Figure 4). Second, the northern and southern latitude limits within which carbonate accumulation occured were moved poleward. Third, as implied by Ronov's data, the percentage of the tropical shelf being sites of carbonate accumulation has decreased markedly throughout the Cenozoic. The cumulative effect of these factors, is that the area of platform carbonate accumulation A_{acc}^{shelf} in Ronov's data varies by a factor of ~20 over the last 140 Ma. From an environmental point of view, this change is not easy to explain, especially in the presence of presumably higher atmospheric CO_2 pressures in the past. The larger area of platform carbonate accumulation in the past is possibly associated with a warmer global climate and different oceanic circulation patterns, as well as with possible changes in platform hypsometry.

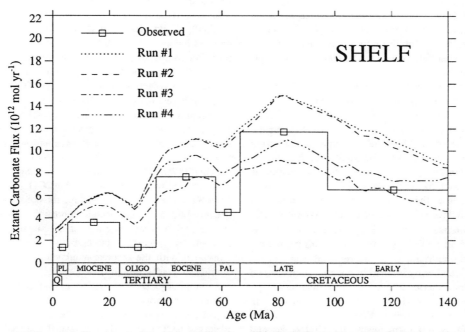

Fig. 8. The extant flux of shelf carbonate deposition as a function of Mesozoic-Cenozoic age, for the four model runs summarized in Table 1. Model results are compared with estimates from data of Ronov and colleagues (see Wilkinson and Algeo [1989] and references therein) for the various geological epochs.

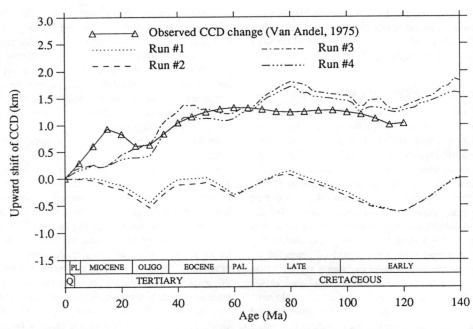

Fig. 9. The evolution of the calcite compensation depth (CCD) over the last 140 my calculated in the four model runs summarized in Table 1 and comparison to observations. The observed signal is a global average of the data of Van Andel [1975] for the different ocean basins.

A detailed modelling of these factors is not yet possible in the framework of a simple model. Thus, here, we assume that the rate of carbonate deposition on the shelf is proportional to the values of A_{acc}^{shelf} from Ronov [1980]. Further, to allow for the effect of a higher atmospheric CO_2 pressure and a lower surface water pH in the past, this rate is also made proportional to $(\Omega_S^{ara} - 1)^{2.5}$, where Ω_S^{ara} is the aragonite solubility ratio of the model surface water. The power 2.5 used here is a fit to the laboratory precipitation rates reported by Morse [1983].

To compare model results with the present remaining mass of platform carbonates published by Ronov and his colleagues, we have calculated in the model the extant deposition flux of shelf carbonates as a function of age, i.e. the deposition flux on the shelf at the age considered corrected for subsequent weathering or metamorphism of carbonate rocks (consistent with the adopted weathering and volcanic laws). The result of this calculation is shown in Figure 8 for each of the four model runs of Table 1 and compared with estimates of the mean extant flux for the various geological epochs from data of Ronov's group (see Wilkinson and Algeo [1989] and references therein). The best fit to the data is obtained in the case of runs #3 and #4. Indeed, in runs #1 and #2, too many calcium and magnesium ions are released to the ocean from the very intense weathering implied by warmer climates and higher pressures of atmospheric CO_2 in the geological past.

For the same reason, the model CCD is by far too deep in runs #1 and #2. Indeed, as shown in Figure 9, these two runs predict a deeper CCD by ~0.5 km during the Cretaceous and Early Cenozoic with respect to present, whereas the observed data show clearly an opposite trend. By contrast, the observed CCD change is fairly well reproduced in runs #3 and #4, with a shallower CCD by 1.0–1.5 km in the past. Note also that the deepening of the CCD at the Eocene-Oligocene transition is well reproduced by the model, although the more recent changes of the observed signal in Miocene time are missing in model results.

Finally, Figure 10 shows the total (shelf+slope+pelagic) rate of carbonate deposition on the seafloor calculated in the four model runs and compared to the same flux reconstructed from observed data (this reconstructed flux is slightly revised with respect to that of Opdyke and Wilkinson [1988]). The curves displayed are original fluxes, not extant. Given the errors inherent in the global compilation of calcium carbonate deposition, the trends indicated by the data are clear: there has been no significant change in the total rate of calcium carbonate deposition during Cenozoic and Cretaceous times. Rather, just a transfer of depositional sites from the shelf to the pelagic realm occured during the Cenozoic and was accompanied by a deepening of the CCD globally (see Figure 9). Runs #1 and #2 miss to reproduce the constancy of the total flux, because both shelf and pelagic depositions are overestimated, due to too high weathering rates of continental rocks. In runs #3 and #4, by contrast, total carbonate deposition is much more consistent with the reconstructed values, although run #3 show a small increase through time.

In summary, the calculation of the continental silicate weathering rates from the strontium isotopic ratio of seawater also appears consistent with past carbonate deposition rates and the observed CCD signal. By contrast, the rate of ion delivery to the ocean obtained from the assumption that global chemical weathering is enhanced in the presence of warmer climate or higher

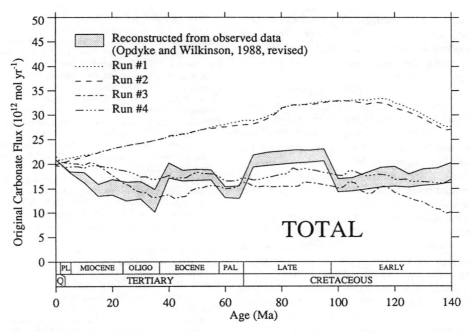

Fig. 10. The original flux of total (shelf+slope+pelagic) carbonate deposition on the seafloor as a function of Mesozoic-Cenozoic age, for the four model runs summarized in Table 1. Model results are compared with original fluxes reconstructed from observed data.

atmospheric CO_2 levels looks too high and implies a deeper CCD in the past with respect to the present.

CONCLUSIONS

In this paper, we have presented a numerical model calculating the evolution of the major geochemical cycles on timescales of several million years. We have used this model to discuss the history of chemical weathering during the last 530 my. More specifically, the idea proposed by Walker et al. [1981] that the global chemical weathering rate of continental silicates should increase with the surface temperature and the atmospheric CO_2 level has been tested and compared to the more recent argument developed by Raymo et al. [1988] and Raymo [1991] that the strontium isotopic ratio of seawater can be used as an index of the intensity of chemical weathering (the uplift hypothesis). Following these latter authors, chemical weathering would be more influenced by mountain building events or other tectonic factors than by atmospheric CO_2 or warm climatic conditions.

The model runs performed in this study clearly favor the uplift theory of chemical weathering proposed by Raymo and colleagues. Indeed, in the framework of this theory, the evolution of atmospheric CO_2 and global mean surface temperature calculated by the model appears more consistent with the geological record of glaciations. The Phanerozoic evolution of the continental weatherability derived from these model runs can be interpreted in connection with tectonic events. Further, the history of the calcite compensation depth and the observed age distribution of continental carbonates are successfully reproduced for the last 140 my, the period when data are available. By contrast, a model assuming that global weathering increases with surface temperature

and atmospheric CO_2 would predict a slightly too warm (compared with the rest of the Phanerozoic) period during early Paleozoic when some glaciations have been reported, non-observed glacial events during Triassic and early Jurassic, and a deeper calcite compensation depth (for the global ocean) in Cretaceous and Tertiary periods with respect to present (in contrast with the shallower observed CCD).

A consequence of the uplift theory is, contrary to the conventional view, that the silicate weathering rate cannot have always been equal in the past to the volcanic release of CO_2. We suggest that the difference between the two rates is compensated for by weathering of the seafloor basaltic crust. In our simulation of the uplift theory, this pH-dependent sink of carbon is the principal mechanism that controls atmospheric CO_2, since continental weathering is assumed independent of T_s and P_{CO_2}. The negative feedback provided is efficient in reducing excursions of atmospheric CO_2, but is of no use in limiting non-CO_2-induced climatic fluctuations. In particular, seafloor weathering does not provide a negative feedback to prevent the Earth's surface from freezing under reduced solar luminosity conditions and the feedback stabilizing the Earth's surface temperature must be searched for elsewhere in the climate system. In any case, even the conventional negative feedback associated with continental weathering of silicate rocks [Walker et al., 1981] was not very powerful during the earliest stages of the planet's evolution, when the solar luminosity was lowest (according to standard solar evolutionary models assuming no mass loss throughout the history of the Sun), but when continental masses were not yet formed.

The model used in this study, despite its relative success, must be considered as being a first stage in the reconstruction of the history of global weathering, atmospheric CO_2 level and climate on several million to hundred million year timescales. Too many processes (such as seafloor weathering) are poorly known and need to be better parameterized or modelled. The same is true for some evolutionary forcings (such as the history of seafloor spreading) used in the calculation, which need to be refined, especially for the more remote times. Moreover, in our modelling of the strontium cycle, we have assumed that the mean isotopic composition of the igneous source has remained constant throughout the Phanerozoic. This assertion might not be true and the consequences for the model results are possibly important. In the future, we plan to address this problem of the composition of the source rock. Also, the model must be confronted with additional geological data, such as the carbon and sulfur isotopic composition of ancient seawater which would better constrain the organic sub-cycle of the model. Previous workers [Garrels and Lerman, 1981; Berner and Raiswell, 1983; Lasaga et al., 1985; François and Gérard, 1986] have run coupled calculations of the carbon and sulfur cycles incorporating these latter isotopic data. However, their models did not consider neither the strontium isotopic history of seawater, nor the carbonate deposition data discussed here, and we think that the only way to get better confidence into model results is to take all the available data simultaneously into consideration in the same calculation, because of the large number of feedback processes presumably involved in the Earth's geochemical system.

Acknowledgments. The first author (L. F.) was supported by the Belgian National Foundation for Scientific Research (F.N.R.S.). The same author acknowledges travel support from F.N.R.S. and N.A.T.O. for visits to The University of Michigan. He is also grateful to The Department of Geological Sciences of that university for its hospitality. This research has been supported in part by the National Aeronautics and Space Administration under Grant NAGW–176.

REFERENCES

Berner, E. K., and R. A. Berner, *The global water cycle: geochemistry and environment,* Prentice Hall, Englewood Cliffs, N. J., 1987.

Berner, R. A., Atmospheric carbon dioxide levels over Phanerozoic time, *Science, 249,* 1382–1386, 1990.

Berner, R. A., and D. E. Canfield, A new model for atmospheric oxygen over Phanerozoic time, *Am. J. Sci., 289,* 333–361, 1989.

Berner, R. A., A. C. Lasaga, and R. M. Garrels, The carbonate-silicate geochemical cycle and its effect on atmospheric carbon dioxide over the past 100 million years, *Am. J. Sci., 283,* 641–683, 1983.

Berner, R. A., and R. Raiswell, Burial of organic carbon and pyrite sulfur in sediments over Phanerozoic time: a new theory, *Geochim. Cosmochim. Acta, 47,* 855–862, 1983.

Berner, R. A., and D. M. Rye, Calculation of the Phanerozoic strontium isotope record of the oceans from a carbon cycle model, *Am. J. Sci., 292,* 136–148, 1992.

Brass, G. W., The variation of the marine $^{87}Sr/^{86}Sr$ ratio during Phanerozoic time: interpretation using a flux model, *Geochim. Cosmochim. Acta, 40,* 721–730, 1976.

Budyko, M. I., A. B. Ronov, and A. L. Yanshin, *History of the Earth's Atmosphere,* Springer-Verlag, Berlin, 1987.

Burke, W. H., R. E. Denison, E. A. Hetherington, R. B. Koepnick, H. F. Nelson, and J. B. Otto, Variation of seawater $^{87}Sr/^{86}Sr$ throughout Phanerozoic time, *Geology, 10,* 516–519, 1982.

Capo, R. C., and D. J. DePaolo, Seawater strontium isotopic variations from 2.5 million years ago to the present, *Science, 249,* 51–55, 1990.

Caputo, M. V., and J. C. Crowell, Migration of glacial centers across Gondwana during Paleozoic era, *Geol. Soc. Am. Bull., 96,* 1020–1036, 1985.

Chahine, M. T., GEWEX: the global energy and water cycle experiment, *EOS, Transactions of the American Geophysical Union, 73,* p. 9, 1992.

Chamberlin, T. C., An attempt to frame a working hypothesis of the cause of glacial periods on an atmospheric basis, *J. Geol., 7,* pp. 545–584, 667–685, 751–787, 1899.

Cook, T. D., and A. W. Bally, *Stratigraphic atlas of North and Central America,* Princeton University Press, Princeton, N. J., 1975.

Crowell, J. C., Ice ages recorded on Gondwanan continents, *Geol. Soc. South Africa Trans., 86,* 237–262, 1983.

Delaney, M. L., and E. A. Boyle, Tertiary paleoceanic chemical variability: unintended consequences of simple geochemical models, *Paleoceanography, 3,* 137–156, 1988.

François, L. M., and J.-C. Gérard, A numerical model of the evolution of ocean sulfate and sedimentary sulfur during the last 800 millions years, *Geochim. Cosmochim. Acta, 50,* 2289–2302, 1986.

François, L. M., and J. C. G. Walker, Modelling the Phanerozoic carbon cycle and climate: constraints from the $^{87}Sr/^{86}Sr$ isotopic ratio of seawater, *Am. J. Sci., 292,* 81–135, 1992.

Gaffin, S., Ridge volume dependence on seafloor generation rate and inversion using long term sealevel change, *Am. J. Sci., 287,* 596–611, 1987.

Garrels, R. M., and A. Lerman, Phanerozoic cycles of sedimentary carbon and sulfur, *Proc. Natl. Acad. Sci. USA , 78,* 4652–4656, 1981.

Graham, D. W., M. L. Bender, D. F. Williams, and L. D. Keigwin Jr., Strontium-calcium ratios in Cenozoic planktonic foraminifera, *Geochim. Cosmochim. Acta, 46,* 1281–1292, 1982.

Hay, W. W., Potential errors in estimates of carbonate rock accumulating through geologic time, in *The Carbon Cycle and Atmospheric CO_2: Natural Variations Archean to Present,* edited by E.T. Sundquist and W.S. Broecker, pp. 573–583, Geophys. Monogr. 32, American Geophysical Union, Washington, D. C., 1985.

Kuhn, W. R., J. C. G. Walker, and H. G. Marshall, The effect on Earth's surface temperature from variations in rotation rate, continent formation, solar luminosity and carbon dioxide, *J. Geophys. Res., 94,* 11129–11136, 1989.

Lasaga, A. C., R. A. Berner, and R. M. Garrels, An improved geochemical model of atmospheric CO_2 fluctuations over the past 100 million years, in *The Carbon Cycle and Atmospheric CO_2: Natural Variations Archean to Present,* edited by E.T. Sundquist and W.S. Broecker, pp. 397–411, Geophys. Monogr. 32, American Geophysical Union, Washington, D. C., 1985.

Marshall, H. G., J. C. G. Walker, and W. R. Kuhn, Long-term climate change and the geochemical cycle of carbon, *J. Geophys. Res., 93,* 791–801, 1988.

Menard, H. W., and S. M. Smith, Hypsometry of ocean basin provinces, *J. Geophys. Res., 71,* 4305–4325, 1966.

Meybeck, M., and R. Helmer, The quality of rivers: from pristine stage to global pollution, *Palaeogeogr. Palaeoclimatol. Palaeoecol. (Global Planet. Change Sect.), 75,* 283–309, 1989.

Miller, K. G., R. G. Fairbanks, and G. S. Mountain, Tertiary oxygen isotope synthesis, sealevel history, and continental margin erosion, *Paleoceanography, 2,* 1–19, 1987.

Molnar, P., and P. England, Late Cenozoic uplift of mountain ranges and global climate: chicken or egg? *Nature, 346,* 29–34, 1990.

Morse, J. W., The kinetics of calcium carbonate dissolution and precipitation, in *Carbonates: Mineralogy and Chemistry,* edited by R.J. Reeder, pp. 227–264, Review in Mineralogy 11, Mineral Society of America, 1983.

Opdyke, B. N., and B. H. Wilkinson, Surface area control of shallow cratonic to deep marine carbonate accumulation, *Paleoceanography, 3,* 685–703, 1988.

Parrish, J. T., Latitudinal distribution of land and shelf and absorbed solar radiation during the Phanerozoic, *Open-File Report 85-31,* United States Department of the Interior, Geological Survey, 1985.

Parsons, B., Causes and consequences of the relation between area and age of the ocean floor, *J. Geophys. Res., 87,* 289–302, 1982.

Parsons, B., and J. G. Sclater, An analysis of the variation of ocean floor bathymetry and heat flow with age, *J. Geophys. Res., 82,* 803–827, 1977.

Raymo, M. E., Geochemical evidence supporting T.C. Chamberlin's theory of glaciation, *Geology, 19,* 344–347, 1991.

Raymo, M. E., W. F. Ruddiman, and P. N. Froelich, Influence of late Cenozoic mountain building on ocean geochemical cycles, *Geology, 16,* 649–653, 1988.

Robinson, J.M., Lignin, land plants, and fungi: Biological evolution affecting Phanerozoic oxygen balance, *Geology, 18,* 607–610, 1990.

Ronov, A. B., The Earth's sedimentary shell — quantitative patterns of its structure, composition and evolution, 20th V.I. Vernadskiy Lecture, in *The Earth's sedimentary shell* (in Russian), edited by A.A. Yaroshevskiy, pp. 1–80, Nauka, Moscow, 1980 (Reprint Serv. 5, pp. 1–73, *Am. Geol. Inst.,* Alexandria, Va., 1983).

Stanley, S. M, *Earth and life through time,* Freeman and Company, New York, 1986.

Staudigel, H., and S. R. Hart, Alteration of basaltic glass: Mechanism and significance for the oceanic crust-seawater budget, *Geochim. Cosmochim. Acta, 47,* 337–350, 1983.

Staudigel, H., S. R. Hart, and S. Richardson, Alteration of the oceanic crust: Processes and timing, *Earth Planet. Sci. Lett., 52,* 311–327, 1981.

Staudigel, H., S. R. Hart, H. U. Schmincke, and B. M. Smith, Cretaceous ocean crust at DSDP site 417 and 418: Carbon uptake from weathering versus loss by magmatic outgassing, *Geochim. Cosmochim. Acta, 53,* 3091–3094, 1989.

Tardy, Y., R. N'Kounkou, and J.-L. Probst, The global water cycle and continental erosion during Phanerozoic time (570 my), *Am. J. Sci., 289,* 455–485, 1989.

Van Andel, T. J., Mesozoic/Cenozoic calcite compensation depth and the global distribution of calcareous sediments, *Earth Planet. Sci. Lett., 26,* 187–194, 1975.

Volk, T., Feedbacks between weathering and atmospheric CO_2 over the last 100 million years, *Am. J. Sci., 287,* 763–779, 1987.

Walker, J. C. G., P. B. Hays, and J. F. Kasting, A negative feedback mechanism for the long-term stabilization of Earth's surface temperature, *J. Geophys. Res., 86,* 9776–9782, 1981.

Whitman, J. M., and T. A. Davies, Cenozoic oceanic sedimentation rates — how good are the data? *Mar. Geol., 30,* 269–284, 1979.

Wilkinson, B. H., and T. J. Algeo, Sedimentary carbonate record of calcium-magnesium cycling, *Am. J. Sci., 289,* 1158–1194, 1989.

L. M. François, Institut d'Astrophysique, Université de Liège, 5, Avenue de Cointe, B-4000, Liège, Belgium.

J. C. G. Walker and B. N. Opdyke, Department of Geological Sciences, The University of Michigan, C. C. Little Science Building, Ann Arbor, MI 48109.